Green Technologies for Sustainable Agriculture

Green Technologies for Sustainable Agriculture

Editors
Professor (Dr.) Arvind Kumar
Dr. Pushpalata Dubey

2024
Daya Publishing House®
A Division of
Astral International Pvt. Ltd.
New Delhi – 110 002

ISBN : 9789359191355 (HB)

Published by : **Daya Publishing House®**
A Division of
Astral International Pvt. Ltd.
– ISO 9001:2015 Certified Company –
4736/23, Ansari Road, Darya Ganj
New Delhi-110 002
Ph. 011-43549197, 23278134
E-mail: info@astralint.com
Website: www.astralint.com

Preface

As against the recommendation of Green Restoration Policy (1988) to keep 33 per cent geographical area under forest, India has only 19 per cent forest area (Usha Rai, 2000). High demographic pressure is further deteriorating the situation due to increasing pressure on forests for firewood, fodder, and timber. Area wise India stands as 7th largest country of the world (2.3 per cent land area) but accommodates 1027 and 4708 million of human and livestock population, which ranks 2$^{nd.}$ (16.6 per cent of world population) and 1$^{st.}$ in the world, respectively (Anonymous, 2001). Although the country stands 1st in rice, pulses and oilseeds production but projected food grain, fodder and fire wood requirement of country is quite huge i.e. 295 million tons (by 2020 AD), 676 million tons and 46 M M^3 (by 2025), respectively. Per capita land has decreased from 0.33 ha in 1951 to 0.15 in 2002. Although rain-fed/dry farming areas cover 70 per cent of crop area, but contribute hardly to 40 per cent of national food grain production. Shrinking crop area, fast degenerating natural resources, depleting soil fertility, declining use efficiencies of agricultural inputs, dwindling output input ratio have rendered crop production less and less remunerative (Singh *et al.*, 2002). Nearly 60 per cent population of the country is dependent on agriculture in which rain-fed agriculture supports 40 per cent of our population and 75 per cent of poor in the country (Pant, 2001). Therefore, maintenance of health of production base stands as prime goal to be achieved for sustainable production, prosperity and posterity. Green technology is necessitated highly for amelioration of climate, production of firewood, fodder, timber, fibre, fruit, gum etc. Green technology is a viable option particularly in Indian tropics and sub tropics.

But the temporal and spatial scale of human-induced changes in natural systems and the magnitude of the disturbances have increased exponentially over time. Increases in human population densities and the increasing effluence of human population contribute to the scale of disturbance. The consequence is that cumulative anthropogenic change has overwhelmed many repair mechanisms of natural systems. While previous generations could rely on natural systems to bounce back without human intervention, current generations cannot. So, green technology with ecological restoration is needed to facilitate the sustainable agriculture. So, the aim of this book is to provide a wide – ranging and authoritative coverage of topics of Green technologies for sustainable agriculture. The all research articles are written by authors who are internationally acknowledged experts in the different fields of agriculture science with the intention to provide a treasure in the form of book to the students, researchers and academicians as well.

My special thanks and appreciation go to the scientists whose contributions have enriched this volume. I wish to express my sincere gratitude to Dr. P. C. Hembram, Hon'ble Vice Chancellor, S. K. M. University, Dumka who has been a source of constant inspiration. I am especially thankful to Professor (Dr.) B. N. Jha, Hon'ble Pro Vice-Chancellor, S. K. M. University, Dumka for his encouragement. I owe my special thanks to Professor M. C. Dash, Hon'ble Vice Chancellor of Sambalpur University, Professor N. C. Datta of Calcutta University, Professor S. K. Konar of Kalyani University, Professor P. S. Murthy of Bangalore University, Professor P. Natarajan of Trivandrum University, Professor S. P. Roy of Bhagalpur University, Professor A. K. Quereshi of Bhopal University, Professor Tanmay Bhattacharya of Midnapore University, Professor P. C. Mishra of Sambalpur University, Professor B. D. Joshi of Hardwar University, Professor G. C. Pandey of Faizabad University, Professor K. C. Sharma of Ajmer University, Professor M. Raziuddin of Hazaribag University, Professor U. S. Bagde of Mumbai University, Dr. M. P. Sinha of Ranchi University, Professor Gurdeep Singh of I. S. M., Dhanbad, Dr. P. K. Goel of Karad, Dr. M. C. Varma and Shri T. Poddar of Bhagalpur University. I also acknowledge the incentives provided by one of my research scholars, Dr. Chandan Bohra, Head of Zoology, B. S. K. College, Barharwa. for helping me actively in bringing out this book.

I also express my deep sense of gratitude to my parents whose blessings have always prompted me to pursue academic activities deeply. I am also thankful to my sweet wife, Professor Kumari Bimla and my two lovely sons, Kumar Pallav Shivshankaran and Kumar Prasun Ramakrishnan whose natural smiles extended to me relief all through this tiresome endeavour.

Last but not the least, I am also thankful to Mr. Anil Mittal, Proprietor, Daya Publishing House, Delhi for taking keen interest in bringing out of this book. Finally, I will always remain a debtor to all my well-wishers for their blessings, without which this book would not have come into existence.

Dumka **Professor Arvind Kumar**

Contents

Chapter 1

Green Technology in Relation to Sustainable Agriculture

Arvind Kumar and Chandan Bohra

Environmental Science Research Unit, Post Graduate Department of Zoology, S.K.M. University, Dumka – 814 101 (Jharkhand)

Introduction

Indiscriminate use of chemicals in everyday life has threatened the bio-world by presenting a deteriorating image of chemicals and especially of synthetic chemicals causing serious health and environmental hazards. Apart from the pharmaceutical usage, the synthetic chemicals have been accommodated in the life of human in such a way that it seems rather impossible to get aside from them. Frequent reports are published about violating the statutory norms for the usage availability of hazardous chemical limit in eatables, mineral waters and beverages especially soft-drinks. For example, recently, a Joint Parliamentary Committee (JPC) set up by Government of India probed the pesticide limits in some popular soft drinks following the complaints of exceeding the limit. The JPC furnished the report on 4th February, 2004 and found the allegations true. Another example may be quoted of the use of food colours in eatables. In Indian rural markets especially, various colours are encountered in soft drinks, sweets, sausages, pickles, bakery and confectionary products etc. A general rural and illiterate consumer is not aware that what to eat and how much. He is attracted towards the sparkling colour which may even be the potent carcinogen. Moreover in vegetables, green colour, malachite green, a potent carcinogen is frequently used to polish the tinda, karela, bhindi, parval etc. to show them green and fresh. A survey in early seventies (Khanna *et al.*, 1973) conducted by IRTC, Lucknow, revealed that more than 70 per cent of colours used in eatables were non-permitted by Prevention of Food Adulteration Rules (PFAR, 1955) in India.

Various chemicals, pesticides, insecticides and colours are drained out from the agricultural lands, which ultimately contribute to water, air and soil pollutions. Every human born on this earth is now subjected to the contact of dangerous chemicals from the moment of conception to death (Carson, 1962). The increased use of chemical fertilizers in agriculture helped the country in achieving self-sufficiency in food grain production. However, it has also polluted the environment and caused slow deterioration of soil health. The chemical residues in the food product are also causing injury to human beings and cattle population.

To combat these problems and in the light of sustainable agriculture, green technology is now being greatly used (Sharma and Sharma, 2004).

Biofertilizers and Green Manures: An Eco-friendly Alternative to Chemical Fertilizers

Biofertilizers are the source of microbial inoculants, which have brought hopes for many countries both economically and environmentally. The term "biofertilizer" denotes all the nutrient inputs of biological origin of plant growth. Therefore, in developing countries like India, biofertilizers can solve problems of high cost of chemical fertilizers and thus can save the economy of the country. Biofertilizers primarily aims at cultivating the land and raising crops in such a way so as to keep the soil alive and in good health. Biofertilizers helps in maintaining environmental health by reducing the level of residues in the products. It helps in keeping agricultural production at a higher level and reduces its cost and makes it sustainable. It not only saves energy of both animal and machine but also reduces the risk of crop failure. Green manures also play a vital role in increasing the soil fertility and productivity. The objectives of green manuring are to increase the organic matter content of soils, to maintain and improve soil structure, to reduce loss of nutrient particularly nitrogen, to provide a source of N and P for the succeeding crop and to reduce soil erosion (Greenland, 1979).

Biofertilizer and green manures acts as reserve source for all nutrients. It adds organic matter to the soil and thus increases soil fertility. The importance of green manuring had been recognized as early as 500 BC in India (Kadke, 1965). Among the 100 leguminous green manures surveyed, only *Crotolaria juncea* and *Sesbania aculeate* are acceptable to farmers of India (Vanchani and Murthy, 1964). *S. aculeate* is widely grown as green manure because of its high nitrogen fixing capacity and its ability to withstand a wide range of soil conditions like salinity (Dargan *et al.*, 1982)., drought (Brewbabaer and Glover, 1987), alkalinity and water logging (Ghai *et al.*, 1988).

The benefits of biofertilizers for crop production and seedling establishment are also immense. Biofertilizers inoculated plants exhibit an increased plant growth, high nutrient status including that of phosphorus besides offering resistance to pathogenic and disease causing microbes (Gupta *et al.*, 2003).

Thus we are now well acquainted about the immense role played by various biofertilizers in varying fields. Intensive agriculture needs massive application of fertilizers to boost the yields, sustainable agriculture is being continuously modified and refined to reflect current agriculturally related environmental concerns and economic pressure of farmers. The term defined by various authors by and large, points out that sustainable agriculture renders around the use of low cost inputs less dependent on chemicals maintenance of soil fertility and ecological harmony. Further these practices should presence and enhance the biological productivity.

So, there is an urgent need to translate our understanding of biofertilizer and green manure viable agro-technology that will help the economic improvement of low-income groups and it is also essential for sustainable agriculture.

Generally, the beneficial affects of microbial system, particularly those of free living systems are not dramatically visual, which means that the farmers are to be educated, convinced about the benefits of the technology, which are low cost, economically and ecologically profitable. This requires a close involvement of the research and extension personnel at different levels. Biofertilizer inoculums production in bulk in necessary, but, availability of quality, inoculums is a constraint. Thus, there is an urgent need to translate our understanding of biofertilizers into viable agro-technology that will help the economic improvement of low-income groups as the potential of productivity of biofertilizers has sustainable role in global economy. The commercial, cultivation of important medicinal plants aromatic crop, pulses etc. needs special attention to improve agronomic practices for qualitative and quantitative improvement.

Biofertilizers (especially mycorrizal fungi) have become important in stabilizing forests. At field level, agronomists, foresters, plant breeders, geneticists, ecologists and pedologists are aware that mychorrizal fungi play an important role in plant nutrition and growth, species diversity and soil structure in the system on which they work.

Biofertilizer needs strengthening and appears necessary attention prior to manipulation of bacteria, actinomycetes or fungi in order to obtain genetically improved strains for forestry practice, horticultural and environmental crop production and agricultural crop productivity.

The need for the use of biofertilizers has arisen, primarily, for the two reasons. First, because increase in the use of fertilizer leads to increased crop productivity. Second, because increased usage of chemical fertilizer leads to damage in soil texture and raises another environmental problems. Therefore, the use of biofertilizers is both economical and eco-friendly. Realizing the importance of biofertilizers is supplementing the use of chemical fertilizers, the Government of India has launched the "National Project on Development and Use of Biofertilizers" during the 6th five year plan. Under the project national center, 6 regional centers (Table 1.1) and 40 BGA projection centers have been established. These centers will produce 800 tonnes of *Rhizobia* and 600 tonnes of BGA annually.

Table 1.1: Location of Nutrition and Regional Biofertilizers production and Development Centres

1.	National Biofertilizers Development Centre, Ghaziabad, U. P.).
2.	Regional Biofertilizers Development Center (North) Hisar, Haryana.
3.	Regional Biofertilizers Development Centre (Central) TNKW, Jabalpur, M. P.
4.	Regional Biofertilizers Development Centre (West), MPKV, Pune, Maharashtra.
5.	Regional Biofertilizers Development Centre (South), UAS, Bangalore, Karnataka.
6.	Regional Biofertilizers Development Centre (East) QUAT, Bhubaneswar, Orissa.
7.	Regional Biofertilizers Development Centre (North-East), ICAR, Shillong, Meghalaya.

Bacterization

Bacterization is a technique where seed dressing is done by applying various bacteria like *Azotobacter, Azospirillum, Rhizobium, Pseudomonsa* etc. These bacteria secrete bacterial fertilizers which have been used for the preparation of biofertilizers like Azotobacterin by *Azotobacter chrococcum*, phosphobacterin by *Bacillus megaterium*. In addition to this *B. firnus, B. cirulans, B. subtilis, pseudomonas striata* are considered very suitable for inclusions in Indian condition. In field trails Berseem maize, wheat and paddy responded to the inoculation with phosphobacterin and showed a significant increase in the yield. In addition both C3 and C4 plants have also been benefited from bacterial inoculation and the ectosymbiotic association appears to be worth exploiting for improvement of the growth and yield of these plants. These bacteria invades cortical and vascular tissues of the host plant and enhance number of lateral root hairs, increase mineral uptake, nitrogen fixation and growth stimulating substances, improve the growth of host plants and structural status of the soil.

Marked improvement in the growth of host plant are also seen as the nitrogenase activity evoke growth promoting responses, produce vitamins, auxins, cytokinins gibberellic acid. If such vitamins and hormonal substances are produced within the plant, they would certainly improve extension growth, chlorophyll content, dry-weight of the plant, enhanced leaching of electrolytes, phosphosyntheteses and higher CO_2 fixation. These bacteria can also hydrolyse leaf polymers and lipids (Pal, 1984; Samanta, 1984; Maiti, 1984) and also the leaf wax (Samanta and Sen, 1986) for growth.

Subha Rao (1982) has summarized the beneficial affects of inoculation of seeds/roots of crop plants with *Azotobacter*, particularly *Azotobater chroccum* and *Azospirillum brasilens* recorded by Indian workers. In case of *Azotobacter*, the increase over uninoculated controls varied between 2 per cent and 62 per cent depending on the host plants, the method of inoculation, the locality of work and the agroclimatic condition. In case of *Azospirillum brasilens* inoculation, the results obtained were more spectacular. With sorghum carrier based inocula gave increase in yield varying between 17 per cent and 105 per cent in different regions. Significant promotions were also observed with pearl millet, the maximum increase being about 70 per cent in some regions.

The pulse crops have good nutritive value, high protein content but their average yield is low. Thus, the inoculum of *Rhizobium* strain should be used as an associative symbiosis to boost up the yield. In *Arachis hypogea* (major oil seed leguminous crop of the country), productivity is recorded very low (972 kg/ha) but when inoculated with *Rhizobium*, positive enhancement in productivity was found to vary from 11.36-32.2 per cent (Kulkarni *et al.*, 1986). *Rhizobium* strains NC 92 and *Azospirillum lipoferum* significantly increases the grain yield (8-29 per cent) straw yield (2-14 per cent) and additional nitrogen uptake (1.5-6.0 kg/ha).

Rhizobial biofertilizers are able to fix 50-150 kg N/ha annum. *Azotobacter* and *Azospirillum* besides nitrogen supply to soil secrete antibiotics which acts as pesticides. Improvement in growth and yield of crops have also been observed with foliar spray of other strains of *Azotobacter* belonging to *A. chrococcum, A. vinelandii, A. paspali, A. brasilens* etc. (Nandi and Sen, 1981; Samanta and Sen, 1986).

The crops which have markedly benefited from such associations include rice (8 cultivars), wheat (2 cultivars), maize, sugarcane, sugarbeet, jute, soyabean, chickpea, greengram, blackgram, pea, groundnut, mustard, sunflower, radish, pumpkin, potato, tomato, brinjal, cauliflower, cabbage, onion, tobacco, spinach, mulberry etc (Sengupta and Sen, 1981; Sengupta *et al.*, 1981; Nandi and Sen, 1982; Nandi *et al.*, 1982,1983; Mishra, 1982; Somchouary, 1982; Sen *et al.*, 1985).

The beneficial effects of associative symbiosis are certainly encouraging. The variability of the result suggests that more intensive studies should be undertaken in different agro-climatic regions, not only for the confirmation of the claims but also for development of the proper conditions for optimization of beneficial effects.

The introduction of the infective capacity of *Rhizobium*, *Agrobacterium* or *Frankia* into non-infective diazotrophs have also proved to be useful in associative symbiosis. Extensive survey work for isolation of more active strains is an urgent need. Some combinations yield better results than others and it is important that the right host and the right conditions for a particular diazotrophs to give optimal effects are to be identified. This technique has found to increase the yield and productivity about 10-20 per cent in Russia and many other countries.

Such free living and symbiotic bacterias are commonly known as PGPR is now-a-day being promoted to reduce the dependence on expensive less ineffective fertilizers, pesticides, insecticides and can save the national economy and protect the environment for sustainable agriculture.

Algal Biofertilizer

Algalization is a mode of applying blue-green algae cultures in the field as biofertilizers. Extensive work on algal biofertilizers is being promoted and practiced in China, USSR, Philippines, India etc.

Various blue-green algae such as *Anabaena, Nostoc, Cyanobacteria, Azolla, Cylindrospermum* etc. have promontory qualitative as well as quantitative role and they have been widely seen in the paddy fields (about 25-30 kg/N/ha/cropping). They have minimized the harmful effects and shown tolerance against heavy metals and chemical fertilizers. It has been found after 3-4 consecutive years that the algal effects become consisted and there is no need of using this practice, i.e. the parental inoculum is sufficient for growth and multiplication.

In India, Singh (1961) was the first to observe the characteristic succession of blue-green algae in the rice fields of U.P. The role of BGA in nitrogen economy of rice fields and on the yield of rice has been well demonstrated and widely documented (Singh, 1961 and Venkatraman, 1981).

In addition to accretion of atmospheric nitrogen and release of growth promoting substances, BGA can influence the soil property and nutrition of rice plants in many ways. The organic compounds present in BGA have chelating properties which can form chelates with several micronutrient elements like Fe, Mn, Cu, Zn etc. and thus help them to remain in an available form for the plants by protecting them from reactions in their ionic form with various soil constituent.

Blue-green algae also liberate large amount of oxygen during photosynthesis by which standing water in rice field gets highly oxygenated, use of *Azotobacter* with blue-green algae makes use of the excess as well as the easily oxidisable organic matter liberated by blue-green algae for their respiration and as a source of energy material. This may minimize the loss of nitrogen through denitrification and also augment atmospheric nitrogen addition to the soil. Besides enrichment of soil, blue-green algae add a considerable amount of biomass to the soil which influence the availability of nutrients. Blue-green algae influence crop growth, tillering, panicle initiating, etc. The succession of blue-green algae improves the soil aggregation status, fixes nitrogen (Roy Choudhary *et al.*, 1979), improve N content (Watanabe, 1965; Chopra and Dube, 1971) and lesser extent the organic matter content (Shankran, 1971).

A significant reduction in reduced substances like oxidisable matter, sulphides and ferrous ion due to algal, inoculation has also been reported (Aiyer *et al.*, 1972). Similarly, *Azolla* is also an active system which is very useful as a biofertilizer for rice production (Watanabe *et al.*, 1981 and Kannaiyan *et al.*, 1982). It has been well established that *Azolla* can supply around 25-30 kg N/ha/crop (INSFFER, 1982 and Kannaiyan, 1982). The nitrogen fixed by *Azolla, Anabaena* symbiosis become the source of nitrogen for the rice crop and appears to have considerable potential for increasing rice yields (Talley *et al.*, 1977). Besides nitrogen fixation to rice crop, *Azolla* is known to suppress the weed population in wet land rice (Moody, 1981 and Kannaiyan *et al.*, 1982).

Mycorrhizal as Biofertilizer

A mycorrhiza (fungus root) is a symbiotic and mutualistic association between a non-pathogenic or weakly pathogenic fungus and living root cells, primary cortical and epidermal cells.

The fungi receive organic matter from the plant and in turn, improve the mineral salt and water absorbing properties of roots. There are seven types of mycorrhizas-Ectomycorrhizae, Endomycorrhizae or Vesicular arbuscular mycorrhizae (VAM), Ectoendomycorrhizae, orchid mycorrhizae, Arbotoid mycorrhizae, Monotropoid mycorrhizae, Ericoid mycorrhizae. Table 1.2 depicts the mycorrhizal type, their characters and distribution within plant kingdom.

Table 1.2: Mycorrhizal Types, their Characters and Distribution within Plant Kingdom (Marks, 1991)

Sl.No.	*Mycorrhizae*	*Host Range*	*Types of Relationship*
1.	Vesicular arbuscular Mycorrhizae	All groups of plant kingdom	Coiled intracellular hyphae, Vesicles and Arbuscales present.
2.	Ectomycorrhizae	Gymnosperms, Angiosperms	Sheath, intercellular hyphae
3.	Arbutoid mycorrhizae	Very restricted Ericales	Sheath, inter and coiled, Intracellular hyphae
4.	Monotropoid Mycorrhizae	Very restricted	Sheath, inter & coiled, Intracellular hyphae
5.	Ericoid mycorrhizae	Very restricted Ericales	No sheath, No intercellular hyphae, long coiled.
6.	Ectoendo-Mycorrhizae	Gymnosperms, Angiosperms	Sheath option, inter & intracellular hyphae
7.	Orchid mycorrhizae	Very restricted	Only coiled, intracellular hyphae

Mycorrhizae increase the absorptive surface of roots resulting in increased uptake of water, nutrients from the soil, increase translocation of P 32 (Melin and Nilson, 1953 & 1955), Zn, Na and other minerals (Browen *et al.*, 1974) to the hosts. Recently Thapar and Pokhriyal (1990) reported nitrogenase activity in 6 ectomycorrhizal fungi out of which *Laccaria laccata* and *Suillus brevipes* were found the most active fungi. Mycorrhizae make the host plants drought and frost resistant, increase tolerance to trees against high soil temperature, soil toxins and heavy metal toxins and provide protection to host roots against pathogenic invasion in different ways (Gupta *et al.*, 2003).

Green Manuring

It is a farming practice where a leguminous plant, which has derived enough benefits from appropriate species (*Rhizobium*) is ploughed under the soil and then a non-legume is grown and is allowed to take the benefits of the already fixed nitrogen. Several leguminous plants can be used as green manures like *Cassia hirsuta, Acacia nilotica, Glycine wightii, Cicer arietinum, Crotalarioa juncea* (Sunhemp), *Vigna catjong* (cowpea), *Sesbania aculeata* (Dhanicha), *Mucuna deeringianum* (velvet bean), *Cyamopsis tetragobnolobe* etc. There green manures in addition to nitrogen, also provide organic matters like NPK, minimize the number of pathogenic microorganism in soil. In India for small and marginal farmers green manuring is being re-emphasized as cheaper and is efficiently capable of reclaiming the **"USER LANDS"**.

Higher response in rice for green manures have been observed than other organics such as FYM and *Azospirillum* (Saravana pandian and Raniperumal, 1994). The physiological efficiency in rice was higher with green manuring alone and it was better than fertilizer and integrated nitrogen sources (Budhar and Palaniappan, 1996). Green manure acts as reserve source of all nutrients. It adds organic matter to the soil and thus increases soil fertility. Better utilization of phosphorus and potassium to an extent of 10-12 per cent was observed in rice due to green manure incorporation (Lehhasrekantha and Palaniappan, 1990). The percentage of recovery of nitrogen from green manure was 48.6 per cent compared to 6.6 per cent from compost (Tanaka, 1977). The contribution of nitrogen from leguminous green manures is the most commonly observed benefit and almost certainly their primary advantage.

Green manures, particularly the leguminous crops, are great value for improving soil fertility. A green manure that decomposes slowly, releasing little of its nitrogen for the first crop, will have a largest effect on organic matter content and some residual effects on N supply to future rops. The beneficial effects of green manures include the build up of organic matter content and available plant nutrients and improvement in the microbiological and physical properties of soil. Besides supplying nitrogen, green manures favourably influence the physical properties of the soil, water relention, reduce the leaching loss of nutrients and improve the availability of other plant nutrients like phosphorus, sulphur and zinc through their impact on chemical and biological properties of soil (Singh *et al.*, 1991). Green manure can increase soil N and P content, maintain and renew soil organic matter, improve soil structure and physical characteristics (Jainobin, 1983). Inclusion of leguminous green manures in the cropping system is known to improve the soil conditions for enhanced and sustained rice production during the wet season (IRRI, 1988).

Thus, it seems that multiple and potential benefits derived from biofertilizers and green manures in sustainable agriculture and global economy renders a saving in terms of time, money and energy for small and marginal farmers. They are eco-friendly tools which have tremendous scope to improve the qualitative and quantitative production of the plants. They are free from pollution hazard and increase soil fertility (Solaimalai *et al.*, 2003).

Phytopesticides: Biological Control of Pest

Miraculous achievement made in the agricultural production through Green Revolution during last three decades, converted India of the fifties, as an importer of the food grain to that of an exporter. The technological advancement in agriculture is brought about through the increased production of new high yielding varieties of crops by means of the application of chemical fertilizers and pesticides. Over the past three decades, increase in crop yields has largely been due to the production and use of enormous quantities of pesticides each year. The tendency to look at only higher production of food crops for the ever-increasing population has no doubt saved mankind from hunger and pestilance but at the same time neglecting the hazardous consequences of ecological disturbance has led to severe environmental degradation.

Pesticides cause different types of hazards and toxicity. Pesticides residue may constitute a significant source of contamination of air, water, soil and food which could become a threat to the continued existence of many plant and animal communities of the ecosystem. Regardless of the method of application, large amounts of pesticides ultimately reach the soil, as a result, world soil are accumulating ever increasing amount of residues of wide variety of pesticides which can move into the bodies of animals, pass into air or water, are absorbed by plants etc. The presence of pesticides in soil must therefore continue to be of major interest to environmental scientists. Less than 0.1 per cent pesticides reach the target pest and remainder negatively affect humans, livestock and natural biota (Pimental, 1992). Indiscriminate and heavy use of pesticides has contaminated the food grains, dairy products, fruits, vegetables, fodders, horticulture land, drinking water and the living environment as a whole.

The continual addition of large amount of persistent pesticides to the environment has caused great anxiety among many ecologists. In the most natural situation, the plants, animals and micro-organisms of the soil are absolutely essential for its fertility. The soil contains micro-organisms that are responsible for the conversion of nitrogen, phosphorus and sulphur to the forms available for plants. Recognizing the fact that most of the complex physical and chemical processes responsible for soil fertility are dependent on soil microorganisms use of DDT has been banned in many countries.

Thus under such an alarming situation, phytopesticides should be used to control agricultural pests for sustainable agriculture (Deshmukh, 2001).

Biological control in its broadest sense existed as a natural phenomenon since the beginning of life. Bio-pesticides have a long history. Even in Neolithic times (about 7000 BC) farmers used bio-pesticides prepared from seeds of resistant plants. Cooke and Baker (1983) pointed out that Chinese farmers practiced crop rotation, which was involved a form of biological control, as early as 5000 years ago. Mechnonoff (1879) and Krassilstschik (1880) in Russia successfully demonstrated bio-control for insect pests. As early as in 1690 the water extract of tobacco leaves were being used to kill sucking insects on garden pest. The active principle in tobacco extract was later shown to be the alkaloid nicotine which is the first naturally accruing insecticide isolated in 1828. Nicotine functions as a non persistent contact insecticide against aphids, capsids, leaf miner, codling moth and thrips on a wide variety of crops (Busbey, 1950-51). At 1850, two important natural insecticides were introduced namely rotenone and dihydro rotenone from the roots of the plant *Derris elliptica* (Fukami and Nakajima, 1971) being used for the control of caterpillars and pyrethrum from flower heads of *Chrysanthemum cinerariaefolium* (Matsui and Yamamoto, 1971) which are still widely used insecticides.

Advantage of Biocontrol

1. Bio or phytopesticides do not persist for long time in the environment. They are highly photodegradable.
2. Most important property of bio-pesticide is that they are most specific, they kill only target pest where as they do not harm other pest in the same fields.
3. The raw materials for the production of the biopesticides are agroproducts which are easily affordable and production cost will be minimum.
4. Biopesticides are non-toxic to other living creatures than the targeted pest. They are normally present in the nature. Thus they do not cause any pollution or do not harm the environment. Enhance they help in sustainable agriculture.
5. Bio-pesticides being living micro-organisms propagate in the nature, thus increase in number.

Biocontrol Agents

Many biological agents can be used to control insects and plant pathogen. These can be listed as below:

1. Plant origin pesticides
2. By using resistant plant varieties
3. Microbial pesticides.

Plant Origin Pesticides

Different products of plant origin like oils, extracts or powders obtained from seeds, leaves, fruits, flowers, barks, stems can control harmful insects. These products are easily available and eco-friendly in nature. Neem today is regarded as natures own answer to effective insect control. Its eco-friendliness easy availability and renewable nature helped to prepare different pesticides from its various parts. Neem extract have been found quite effective against more than 200 pests (Mariappan, 1995). The presence of biologically active principles in seed and other parts of certain plants and their extra ordinary pest management traits have, in recent years, arose considerable interest among the scientists

Table 1.3: Important Commercially Available Neem Based Pesticidal Products in India

Sl.No	Trade Name	Active Ingredients	Manufacture	Biocidal Activities
1.	Goderej Achook	Azadirachtin Azadiradione Nimocinol, Nimbocinol, Epinimbocin	Bahar Agrochem & Feed (P) Ltd. Lote Parshuram, Taluka Khed, Ratnagiri–415 722.	Antifeedant, male sterlity, chitin inhibitor, growth disruptent, ovicidal
2.	Field Marshal	Azadirachtin	Khetiwadi Shiyabave Vadodara–390 001	Antifeedant, Repellant
3.	Jawan Crop	Azadirachtin	MCDA Agro Pvt. Ltd. Bombay–400 001	Repellant, Growth disrupter
4.	Margocide CK 20% Macrocide CK 80%	Azadirachtin (0.30% W/W) (0.15% W/W)	Monofix Agro Prod. Ltd., V.G. Limbikal Bld. Gokul Road, Hubli–580 030	Antifeedant, Repellant, Growth disrupter
5.	Neemark	Azadirachtin (0.30%)	West coast Herbochem Pvt. Ltd. 105/B Rajaji Nagar, IIIrd Crose Road, Tumkur Road, Bangalore–560 022	Antifeedant, nematicidal, Synergist
6.	Neemgold	Azadirachtin (0.15% EC)	Southern Petrochemical Industries Crop. Ltd. Madras–600 001	Antifeedant
7.	Nimbicidine	Azadirachtin (0.03%)	T. Stanes & Co. Ltd. 8/23-24, Race Course Road, Coimbatore–600 018	Antifeedant, Metamorphosis disruptor, Synergist
8.	RD–9 Ropellin	Neemoil 93 EC containing Azadirachtin (300 ppm)	ITC Ltd, ILTD Div. IBD, 6-3-1110, Amrutha Mall, Begumpect, Hyderabad–500 016	Antifeedant, Repellant
9.	Wellgro	Neem Kernel Powder	ITC Ltd, ILTD Div. IBD, 6-3-1110, Amrutha Mall, Begumpect, Hyderabad–500 016	Repellant, Fungus inhibitor, Antiviral, N-loss Prenentor
10.	Sukrina	Azadirachtin (1500 ppm)	Canster Pvt. Ltd. Chennai–600 116.	Antifeedant, Repellant
11.	Neemol	Azadirachtin	Meenal oil & Agro Ind. Ahmedabad–380 009	Antifeedant
12.	Neem guard	Kernel Extract	Akshay Chemicals Mumbai–400 052	Repellant, Metar Morphosis disruptor
13.	Nethrin	Neem oil	Amitul Agrochem Pvt. Ltd, Gorakhpur	Pesticidal
14.	Bioneem	Azadirachtin (0.03% EC)	Ajay Biotech Lab. A/12/5, Kubera Park, Kanchawa, Pune–411 040	Aphid, Jassid, Spotted Bollworm
15.	Nimbasol	Azadirachtin (1500 ppm)	Nimba Food & Chem Pvt. Ltd. 14-A/10 Western Extn, Pusa Road, New Delhi	Whitefly, Bollworm, Leaf roller
16.	Neemazal	Azadirachtin	EID Parry (I) Ltd. Farm Input Div. 234 NSC Bose Road, Chennai–600 001	Aphid, Leaf Hopper, Whitefly
17.	Rakshak	Azodirachtin (1500 ppm)	Murkumbi Manufacturing 1438/2, Kalmath Road, Belgaum–590 002	Thrip, Aphid, Jassid

Table 1.4: Plants and their Products Used in Controlling Different Pests

Plant	*Pesticidal Product*	*Target Crops/Pest & Diseases*	*Dose/Mode of Application*
Abelnoschus esculentus (Lady's finger)	Extract of Leaves	Rice & Wheat/ Brown Rust	25-50 gms Leaves/lit of water/ springs
Aegle marmelos (Bel)	Extract of Leaf & fruit in boiled water	Various crops/various pests	Spray
Allium sativum (Garlic)	Bulb & Leaf dust, Leaf extract, Leaf dust mixed with ash	Pea, Grains, Fruit, Tree, Ornamental plants/ Powdery Mildew, Fungal fruit flies	Dust/Extract spray 10 gm/kg
Allium cepa (Onion)	Bulb dust & Leaf Extract.	Rice, Wheat, Pea, Fruit tree/Leaf Curl, Powdery mildew, Fungal	Dust/Spray
Amaranthus spinosus (Kanta nate)	Leaf dust & Leaf extract	Rice & wheat/Brown Rust	Spray with 25-50 gms leaves/lit of water
Andropogon sorghum (Jowar)	Leaf Extract	Ground nut/Twig rot.	Spray
Annona squamosa (Custard apple)	Aqueous extract of seeds and leaves	Brinjal, Rice, Wheat/ Diamond Black moth Pod borer	Spray
Apios Americana (Ground nut)	Leaf extract	Vegetables/Root rot.	Spray on soil around the root zone
Azadirachta indica (Neem)	Seed Kernel/Cake	Rice, Gingelly, Red gram Greengram/Sheath Blight BPH, leaf folder, Ear-head, bug, Sheath rot, Pod borer, Powdery moldew.	150 kg/ha of soil 5% F.S.
		Cotton, caster/White fly, leaf hopper, Semi hopper	Spray with neem oil 2%, 5% FS 3% FS
		Coconut, Rice/Rhinocerous Beetle BPH, Sheath rot, Sheath blight	Spray with sand 1:2 ratio 3% F.S. 15 lit/ha FS.
		Redgram, Greengram, Coconut/Pod borer, Powdery mildew, Termite	2% F.S. 3% F.S., 5% Spray on the base upto 2 m hight.
		Cotton,Bengal Gram/Whitefly, Pod borer, Shoot webber & Pod Webber	3% F.S., 5% F.S. & 0.01% Combined with endosulfan 0.07%
	Leaf extract & Oil emulsion	Tomato & Chilli, Groundnut, Jute, Brinjal, Onion, Ornamental Crops/Leaf Curl, Root rot, Stem rot.	Spray (5%)
Brassica nigra (Black Mustard)	Leaf extract in boiled water	Jute/Stem rot of Jute	Spray

Contd...

Table 1.4–Contd...

Plant	*Pesticidal Product*	*Target Crops/Pest & Diseases*	*Dose/Mode of Application*
Carica papaya (Papaya)	Leaf dust & extract	Rice, Wheat, Maize, Ornamental Crops/Caterpillar, Aphid, Garden bug, Weevil, Leaf eater	100 gm leaves/lit of water with a little K. oil.
Cassica sophera (Kalkasunda)	Leaf decuction	Rice & Wheat, Vegetable/Root Rot.	Spray 100 gm leaves/L of water
Catharanthus rosens (Periwinkle)	Extracts of leaf flower & root	Cotton/White fly	Spray
Chrysanthemum chrysanthemem	Pyrethrium from dried flowers	Various crops, Ornamental Crops/Flying insects, House hold pests	Spray
Cocos nucifera (Coconut)	Leaf extract	Banana, Groundnut/Top leaf curl, Twig rot	Spray or jiject inside the trunk of banana tree
Corchonus capsularis (Jute)	Stem dust or extract	Rice, Wheat and other crops/Brown rust and various pests	Applied in Soil or spray
Curcumon domestica (Turmeric)	Seed dust	Rice, Wheat & Vegetables/Root rot	Spray
Cymbopogon nardus (Citronella grass)	Oil, Bark and Leaf extracts	Groundnut, Jute, Vegetables, Ornamental Crops/ Stem & Root rot	Spray on soil around the root zone.
Datura metal (Dhatura)	Leaf dust & extract.	Grains, Wheat, Rice/Fungal infection, Brown rust	Spray dust @ 10 gm/kg of grain & extract in 1 lit of water
Derris elliptica (Tropical legumes)	Rotenone derived from rhizomes	Vegetables, ornamentals, Kitchen garden pest	Spray
D. indica (Pungam)	Oil extract	Groundnut, Jute/Root & Stem rot.	Spray 2.5-3% oil emulsion in water
Euphorbia pulcherrima (Latpate)	Leaf extract	Rice/Burn spot	Spray with boiled water
Faericalum vulgare (Fennel)	Leaf extract	Vegetables, Ornamental, Many agricultural pests	Spray

Contd...

Table 1.4–Contd...

Plant	*Pesticidal Product*	*Target Crops/Pest & Diseases*	*Dose/Mode of Application*
Holarrhaena antidysenteria (Kurchi)	Bark extract	Vegetables, ornamentals/Root Knot.	Spray on soil around the root zone.
Hydrocarpus kurzil (Chalmugra)	Oil extract	Jute, Potato, citrus fruits/Stem rot, Rotting cancer of fruits	Spray with 2.5-3% oil in water and emulsifier at 0.1% with some little detergent
Impomoea sp.	Leaf powder/Leaf extract	Cotton, Rice/Whitefly sheath rot and ear-head bug	50 gm leaves/l of water, 5% F.S. 10% F.S.
Jatropha carcus (Ratanjot)	Plant extract	Various crops/pests	Spray
Lantana camara (Lantana)	Leaf extract	Potato, Brinjal, Tomato, Chillies, Crucofers and Onion/Leaf miners	Spray
Lonchocarphus sp. (Tropical legumes)	Rotenone derived from rhizomes	Vegetables/kitchen garden pests	Spray
Madhuca indica (Mahua)	Leaf extract, oil	Groundnut, Jute, Potato/Stem rot, Root rot	Spray with 2.5-3% oil emulsion in water
Mentha sp. (Pudina)	Leaf extract	Rice, Wheat, Ornamental Crops, Brown Rust	Spray with 25-50 Leaves/l of water
Michelia champaca (Champac)	Flower extract	Various crops/Mosquito larval	Spray
Moringa oleifera (Drum stick)	Root & Leaf extract	Vegetables & Fruit Ornamental/Root rot	Spray
Nerium indicum (Karabi)	Leaf extract	Rice & Wheat/Brown Rust.	Spray with 25-50 gms leaves/l of water
Nicotiana tabecum (Tabacco)	Leaf extract	Various crops, ornamentals/So many Insects	Spray
Ocimum basillicum (Tulsi)	Leaf extract	Potato, Brinjal, Tomato, Chilli, Crucifer, Onion, ornamental/Leaf miners, leaf curls	Spray
Opuntia sp.	Plant extract	Various crops & Pests	Spray
Piper nigrum (Black peeper)	Seed extract	Various crops/Various pests	Spray
Ricinus communis (Caster)	Oil extract from seed.	Maize, other Crops/Weevil, Aphids, Cater-piller	Spray with water

all over the world and a fairly good amount of data has been generated on several plant species regarding their pest control potential. As regard to pesticidal action, the chemical constituents of neem act in different ways under different circumstances. They produce their various biocidal action by deterring migrating locust from feeding repelling larval and adults, disrupting mating and sexual communications, deterring females from laying eggs, disrupting/inhibiting the development of eggs, larvae an/or pupae, poisoning larvae and adults, blocking the moulting of larvae or nymphs, inhibiting the formulation of chitin, inhibiting the degradation of pesticide in soil by micro-organism and through phagodeterrency. The multilateral mode of action of neem constituents makes practically impossible for insects to develop immunity to its pesticidal products. Almost every part of neem is bitter but seeds possess maximum deterrency. Presently two types of pesticide formulations are being produced all over the world. The oil based formulations in which the active constituent is azadirachtin. Other formulations are alcohol based in which azadirachtin and other bitters are extracted with alcohol and the concentrated extracts are used for spray on crops. Taking the pesticidal potential into account and keeping an eye on Indian as well as western market, some Indian companies have launched neem based pesticides during past few years (Table 1.3) (Tripathi, 1999).

Pest Control with Resistant Plant Varieties

Development of hybrid variety of plant with increased resistance to attack of insects is the major way of pest control. Some developed resistant varieties are as follows:

1. Pusa Jawala of Chilli is resistant to aphid and jassid.
2. Pusa samrat of brinjal and N-2-41 of Onion are resistant to jassid.
3. Apple variety, merton 778 and fanny are resistant to woody aphid.
4. Okra is resistant to yellow vein mosaic virus.

Microbial Pesticide

These includes bacteria, viruses fungi etc. There have been much achievements in developing many new bio-pesticides by using various bacteria, fungi, viruses, antibiotics and plant extracts. Nearby representative of all types of biological groups can be effectively used in controlling. The hazardous pests of different crops. Such agents would spread through insect population and would wipe out insects in target area. Extensive further research and field trials are essential for wide and satisfactory application of bio-pesticides. Genetic engineering is an effective tool for developing biological agent with more desired properties.

In developing countries like India, there is a more opportunity for mass application of this green technology in agriculture. Thus, bio or phyto-pesticides is relatively recent promising and reliable technology having a wide scope specially in agro-based nations for sustainable agriculture.

References

Aiyer, R., Salahudeen, S. and Venkatraman, G. S. 1972. Use of algae as biofertilizer. *Indian J. Agric. Sci.,* 42: 280–390.

Brewbabaer, J.L. and Glover, N. 1987. *Sesbania aculeate*: a potent green manure. In: *Green Manure in Rice Farming*. IRRI, Los Banos, p. 29–43.

Budhar, M.N. and Palaniappan, S. P. 1996. Green Manuring in Rice. *J. Agron. Crop. Sci.,* 76: 183–187.

Busbey, R.L. 1950–51. Plants that help to kill insects. In: *The Year Book of Agriculture: Crop in Peace & War.* United States Department of Agriculture, Govt. Printing Office, Washington, D.C. pp. 765–771.

Chander, S. and Munshi, A.D. 1995. Phytopesticides in controlling pests. *E. News*, 20: 32.

Chopra, T.S. and Dube, J.N. 1971. Impact of Blue-green algae on nitrogen content of soil. *Plant and Soil*, 35: 453–463.

Dargan, K.S. 1982. *Crop Production in Salt Affected Soil*. Oxford & IBH Publ. Co,. New Delhi.

Deshmukh, A.M. Biological control of pests and plant pathology. In: *Current Topics in Environmental Sciences*, (Eds.) Tripathi, G. and Pandey, G.C. ABD Publ., Jaipur, p. 296–309.

Fukami, H. and Nakejima, M. 1971. Rotenone and the Rotenoids. In: *Naturally Occurring Insecticides*, (Eds.) Jacobson, M. and Crosby, D.G. Marcel Dekker, New York, p. 71–79.

Ghai, S.K. 1988. Green manure in relation to soil condition with special reference to alkalinity and water logging. *Biol. Fert. Soils*, 6: 22–25.

Gopalan, M. 1996. Phytopesticides in controlling pests and diseases of agricultural crops. *Spice India*, 9: 10–20.

Greeland, D. 1979. Green manuring in improving soil fertility. *J. Soil. Sci.*, 26: 294–296.

Gupta, R.K., Kaushik, S., Sharma, P. and Jain, V.K. 2003. Biofertilizers: An eco-friendly alternative to chemical fertilizers. In: *Environmental Challenges of the 21st Century*, (Ed.) A. Kumar. APH Publ. Corp., New Delhi, p. 275–287.

INNSFER. 1982. In report on the INSFER *Azolla* study tour in Vietnam. Interm. Rice Res.

IRRI. 1988. The role of green manure in rice farming system. In: *Proc. International Symposium on Sustainable Agriculture*, May 25–29, IRRI, Philippines.

Jaino, B. 1983. *Green Manures in China*. Publishing House of Agriculture, Beijing, China.

Kadke, J.R. 1965. Green Manuring in Indian farming. *Indian J. Agron.*, 10: 443–446.

Kanniyan, S. 1982. In multiplication and use of *Azolla* biofertilizer for rice productions training. Tamil Nadu Agric. Univ., Coimbatore, Tamil Nadu.

Kanniyan, S., Thangarajan, M. and Oblisami, G. 1982. Use of Azolla biofertilizer in rice crop with special reference to control of weed population. *Nat. Acad. Sci. Newslett.*, 5: 60.

Khanna, S. K., Singh, G. B. and Singh, S. B. 1973. Use of colours in eatables: A survey and its harmful effects. *J. Fd. Sci. Tech.*, 10: 33–37.

Lekha, S. and Palaniappan, S.P. Green Manuring in rice cultivation. *J. Agron. Crop. Sci.*, 165: 14–18.

Maiti, T. 1984. *Ph.D. Thesis*, Kalyani University, Kalyani.

Mariappan, V. 1995. *Neem for Management of Crop Disease*. Associated Publ. Co., New Delhi.

Matsui, M. and Yamamoto, J. 1971. Pyrethrods. In: *Naturally Occurring Insecticides*, (Eds.) Jacobson, M. and Crosby, D.G. Marcel Dekker, New York, p. 3–70.

Mishra, R.K. 1982. *Ph.D. Thesis*, Kalyani Univ., Kalyani.

Moody, K. 1981. Use of *Azolla* biofertilizer in Rice Crop. *Intern. Rice Res. Inst. Res. Paper Series*, 64: 35.

Nandi, A.S. and Sen, S.P. 1981. *Azotobacter*: An eco-friendly biofertilizer. *Arch. Microbial.*, 130: 147–156.

Nandi, A.S. and Sen, S.P. 1982. Crop improvement by the use of biofertilizer. *J. Agri. Sci. Comaridge*, 98: 179–189.

Nandi, A.S., Sengupta, B. and Sen, S.P. 1982. Impact of biofertilizer on agricultural Crops. *J. Agri. Sci.*, 98: 167–175.

Nandi, A.S., Sengupta, B. and Sen, S.P. 1983. Biofertilizer: Its uses and benefits. *J. Hort. Sci.*, 58: 147–550.

Pal, D. 1984. *Ph.D. Thesis*, Kalyani Univ., Kalyani.

Prevention of Food Adulteration Rules. 1955. SRO. 2106. Government of India.

Pimintel, D. 1992. Foreword. In: *Ecological Processes in Agro-ecosystems*, (Eds.) Sniyomi, M., Yano, E., Koizumi, H., Andow, D. A. and Hokyo, N. Yokendo Publ., Tokyo, p. 1–2.

Roychoudhury, P., Kaushik, B.O., Krishnamurthy, G.S.R. and Venkatraman, G.S. 1979. Use of blue-green algae in agroecosystem for improving soil fertility. *Curr. Sci.*, 48: 454–464.

Samanta, R. K. 1984. *Ph. D. Thesis*, Kalyani University, Kalyani.

Samanta, R. K. and Sen, S. P. 1986. Role of *Azotobacter* in agricultural crops. *J. Agri. Sci.*, 107: 673–677.

Sarvanapandian, P. and Raniperumal, A. 1994. Role of Green Manures in rice field. *Oryza*, 31: 123–126.

Sen, S. P. 1982. *Frontiers Research in Agriculture*. Golden Jublee Symp. Indian Statistical Institute, Kolkata, pp. 224.

Sen, S.P., Maiti, T., Pal, D., Samanta, R.K., Sen, M. and Somchoudhury, M. 1985. *Frontiers on Applied Microbiology* (Eds.) Mukherjee, K.G., Pathak, N.C. and Singh, V.P., Vol. I. Print House, Lucknow, pp. 13.

Shankaran, A. 1971. *Role of Blue-green Algae in Improving the Soil Fertility*. ICAR, New Delhi, pp. 28.

Sengupta, B., Nandi, A.S., Samanta, R.K., Pal, D., Sengupta, D.N. and Sen, S.P. 1981. Biofertilizer: Role in sustainable agriculture. *Ann. Bot.*, 48: 705–710.

Sharma, R.K. and Sharma, S.D. 2004. Green Chemistry. A present day need for sustainable future. *Everyman's Science*, 39(2): 93–99.

Singh, R.N. 1961. *Role of Blue-green Algal Biofertilizers: A Low Cost Technology for Rice*. FAO Soils Bull., 46: 102–105.

Solaimalai, A., Ramesh, P.T. and Baskar, M. 2003. Use of Green Manure and green leaf manure in rice agroecosystem. In: *Perspectives in Agroecosystem*, (Ed.) Kumar, A. Daya Pub. House, Delhi, p. 11–35.

Talley, S.N., Talley, B.J. and Rains, D.W. 1977. *Genetic Engineering for Nitrogen Fixation*. Plenum Pub. Co., New York, pp. 250.

Tanaka, A. 1977. *Soil Rice Symposium*. IRRI Los Banos, Philippines.

Tripathi, Y.C. 1999. Potential of Neem constituents in pest management. *Everyman Science*.

Tripathi, Y.C. and Tripathi, G. 1999. Importance of phytopesticides in environmental protection: In: *Modern Trends in Environmental Pullution and Ecoplanning*, (Ed.) Kumar, A. ABD Publ., Jaipur, p. 82–96.

Usher, G. 1984. *A Dictionary of Plants*. CBS Publ. & Dist., Delhi.

Vanchani, M. B. and Murthy, S. K. 1964. *Green Manuring for Rice*. Res. Report Series, 170, ICAR, New Delhi, pp. 50.

Venkatraman, G.S. 1981. Role of blue-green algae on rice field. In: *Proc. National Work Algal Systems Indian Soc.*, Biotechnol. IT., New Delhi.

Watanabe, A. 1965. Use of blue-green algae in rice crop in relation to improvement of nitrogen content. In: *Proc. Nat. Acad. Sci.*, India, 35A: 361.

Watanabe, J., Bai, K., Berja, N.S., Espines, C.R. and Subudih, B.P.R. 1981. *Use of Azolla as Biofertilizer in Rice Crop*. IRRI Res. Paper Ser. 69, Intern Rioce Res. Inst. Philippines, pp. 11.

Chapter 2

Soil and Groundwater Pollution by Agrochemicals: A Review

D.S. Kler, Navneet Kaur and R.S. Uppal

Department of Agronomy and Agrometeorology, Punjab Agricultural University, Ludhiana – 141 004

ABSTRACT

Increased consumption of agrochemicals for attaining self-sufficiency in the developing countries, resulted in soil and groundnut pollution. Agrochemicals *i.e.*, pesticides and herbicides disrupts the activity of soil micro-organisms like bacteria, fungi, algae and nematode which affect the nutritional quality of soils. These agrochemicals affect the important biological processes (Ammonification, Nitrification, Denitrification and Nitrogen fixation) and enzymatic activities of soil. The leaching of nitrogenous fertilizers to the groundwater also has many biological consequences. These agrochemicals affected the health of human beings. Use less persistent pesticides. Run off, volatilization and leaching may be reduced to the maximum extent possible. To achieve the desired goal we will have to take special measures to solve the problems of the future considering future ecology more important than present economy.

Introduction

An increased scale of man's agricultural, industrial and social activities has brought with it an undesirable change in the chemical, physical and biological properties of the environment. According to Miller (1991), "Any undesirable changes in the characteristics of the air, water, soil and food that can adversely affect the health, survival or activities of humans or other living organisms is called pollution". In intensive cropping system, the increased consumption of fertilizers and herbicides has created many environmental problems among which groundwater and soil pollution are main. The pesticides and fertilizers reaching the soil environment may persist and alter the structure and function of these beneficial non-target organisms. Nutrients and pesticides are the primary contaminants of groundwater.

Fertilizer Application in Relation to Environment

Some major concerns of fertilizer application in relation to pollution problems are pollution of drinking water, eutrophication of lakes and rivers, volatilization of NH_3 and nitrogen oxides, imbalanced quality of plant products and agricultural land as sinks for methane. Nitrate affects the quality of drinking water. Nitrate itself is not toxic but nitrite originating from the reduction of nitrate induces methaeno-globinemia in infants by inhibiting oxygen transport in the blood. Nitrate and amines can react to form nitrosamines which are cancinongenic. Acceptable values for NO_3 concentrations in drinking water have been set at 50–100 mg nitrate per litre (WHO), a maximum of 50 mg nitrate per litre (EL) and a maximum of 45 mg nitrate/litre (US Public Health Service). The recovery of fertilizer N is incomplete. Nitrogen balance in terms of per cent of fertilizer N showed that 58 per cent was taken up by the crop while 5 per cent was leached out of the rooting profile and 21 per cent was found as residual soil mineral N and 16 per cent was considered as unaccounted (Volatile) losses (Mengel *et al.*, 2001). For arable crops 20–60 per cent of applied N is taken up while for grass is 40–80 per cent depending soil, season and rates. Nitrate leaching in arable soils can be reduced by catch crops. Most of the excessive N applied is lost either by leaching or by volatilization in the form of NH_3 or after denitrification in the form of N_2 and N_2O. Dinitrogen oxide (N_2O) is involved in the destruction of the ozone layer in the stratosphere. Ammonia released from the intense livestock farms in to the atmosphere is brought back in to the soils by precipitation when it may be oxidized to nitrate and thus also contribute to the nitrate in lakes, rivers and acquifers. Nitrate leaching, N_2O release in to the atmosphere and the loss of phosphate because of too high rates of application are a hazard to the environment. Phosphorus may also become a limiting factor along O_2 due to eutrophication and the eutrophic threshold level below which algal growth is limited is considered to be in the region of 10 microgram per litre of P. Soil phosphates contribute only to about 4 to 5 per cent to the total phosphate level leached in to the surface water and other comes from detergents and urban wastes.

Pesticides in Relation to Environment

Edward (1971) told that pesticides may be directly toxic to animal in soil and affect them genetically to produce population resistant to the pesticides. The soil microorganisms like bacteria, fungi, algae and nematodes play important role in soil nutrient through their role in decay of plant and other organic matter in soil. Most of the fungicides, nematicides and insecticides applied in the soil also suppress the microbial activity in soil. The chemicals that reach the soil in drift from sprays or washed off from foliage may also be harmful to the microorganisms. Earthworms, *Enchytraeid* worms, *Collembola*, some *Acarina* and Dipterous larvae help to disintegrate plant material. Various factors as shown below like adsorption to clay and organic matter, leaching with the downward percolation of water, volatilization to the atmosphere, uptake by soil organisms, microbial degradation, chemical and photolysis affect the fate of pesticide in soil (Mishra *et al.*, 1989).

Herbicides affect the important soil biological processes like ammonification, nitrification, denitrification, nitrogen fixation and soil enzymatic activities (Rao *et al.*, 1993 and Kulshrestha, 1993). Phenoxy acids, Triazines and urea type compounds have positive affect on these microorganisms. 2,4-D has neither an adverse nor a favourable effect on ammonification in soils at field application rates but at 10 times the field rate ammonification was inhibited. Soil application of Simazine @ 6 kg/ha stimulates both *Nitrosomonas* and *Nitrosococcus*. Debona and Andus (1970) established a decreasing order of inhibiting effectiveness for ten herbicides against nitrification as–Propanil > Ioxynil >

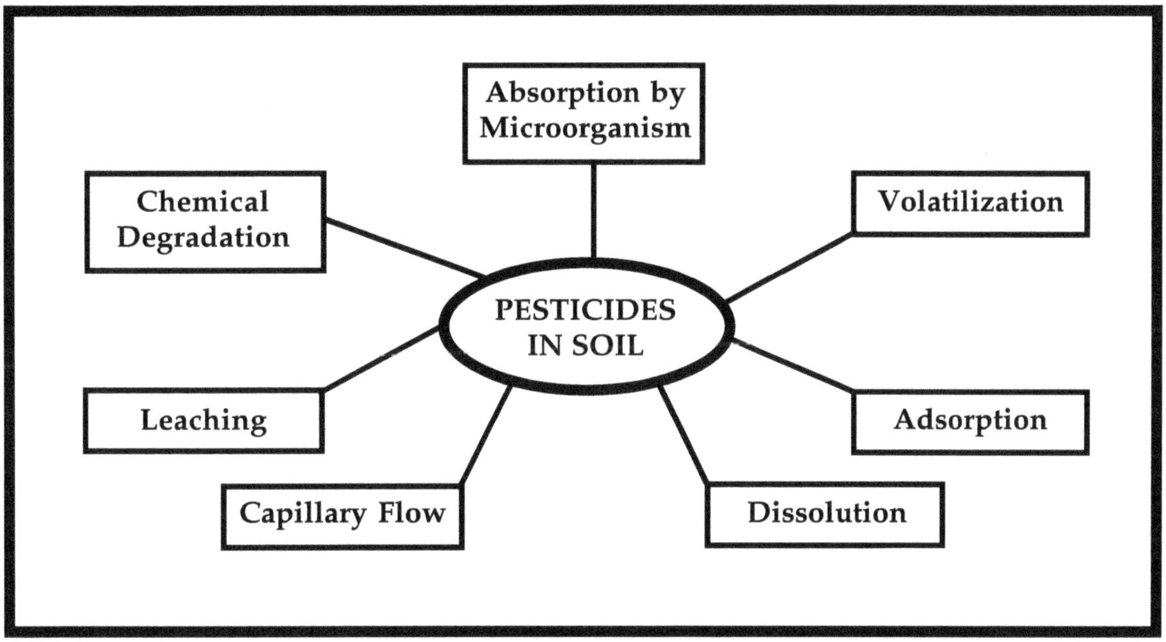

Chlorflurazole > Bromoxynil > Chlorthiamid = 2,3,6-TBA > Picloram > Paraquat > Dchlobenil. Atrazine at concentrations of 3.4 and 2.6 g/kg causes an increase in the NO_3^- reduction to NO_2 compared to N_2. Halogen substitution in phenyl urea and aniline herbicide affected differential inhibiting reaction on the activity of denitrifying bacteria. Propanil and EPTC at low concentrations stimulated algal growth and N_2 fixation. Butachlor and Alachlor have detrimental effect on growth of *Nosto muscorum* and *Anabena doliolum*. Nitrofen and MSMA at excessive rate inhibited growth and activity of *Azotobacter* sp. in cotton but at normal rate they were higher (100 g/ml) than blue green algae (10 or 20 g/ml). DNOC, Dinosoeb and Linuron are most inhibiting herbicides to *Rhizobium* under the influence of Dinoseb, *Rhizobium trifoli* showed uncoupling of oxidative phosphorylation. Similarly Dinitroanilines and Carbamates showed negative effects on *Rhizobia* and nodulation (Chopra and Magu, 1985).

Soil enzymes are often referred as an index of soil fertility and soil health, because of their sensitivity, accuracy and ease of assay. Certain herbicides inhibit the enzymatic activities while some stimulate their activities (Table 2.1). Kuhnelt (1963) was among the first to state that the soil pollution by industrial waste adversely affects the soil fauna. Many of these pollutants can increase in concentration while passing through the food web of ecosystem and therefore, become serious threat to organisms occupying higher trophic levels (Figure 2.1).

Effect of Agrochemicals on Population Structure

Herbicides

The herbicides like Paraquat and Atrazine have toxic effects on the secondary decomposers like microarthropods. The herbicide combination of Praquat, Glyphosate, Alachlor, Linuron, Fluanzifopbutyl, Acifluorfen and Bentazon decreased the decomposition rate of wheat straw residue and lowered the total number of microarthropods. At field conditions, Dichlobeini (7.5 kg/ha), Dalapon (1.5–7.5 kg/ha), Flumeturon (1 kg/ha) inhibited the growth of bacteria (El-Shahawy *et al.*, 1982). Five years of annual application of Diuron in sugarcane was inhibitory to *Actinomycetes* in soil (Heydrich and Fernandez, 1984). Certain pathogenic species such as *Sclerotium erolfsii* and *Pythium* sp. are susceptible to most herbicides, while *Trichoderrna* and *Penicillium* are resistant. Most pathogenic *Fusarium* sp. are susceptible while *Foxysporum* is not (Pope and Hort, 1981). Algae are also affected by various herbicides. Atrazine and Metauron inhibited the photosynthesis. 2, 4-D significantly decreased the heterotophic algae while *Perometrayne* decreased the population of *Anabaena* sp. (Yee *et al.*, 1985).

Table 2.1: Effect of Herbicide on Soil Enzymatic Activity (Hicks *et al.*, 1990)

Compound Class	*Enzyme*							
	Dehydro-genase	*Urease*	*Catalase*	*Phospha-tase*	*Protease*	*Inver-tase*	*Aspara-ginase*	*Protein*
Subsituted urea	I	I	I	S	S	I	I, S	U
Triazines	I	I	I, S	S	I	S	S	I
Nitrophenols	I, S, N	I, S	I	U	U	N	U	U
Carbamates	N	I	U	S	I	I	U	I
Hetero Cyclic Compounds	I	I	U	S	I	I	U	I
Benzoicacid phenyl acetates	I	I	S	S	U	S	U	U

I: Inhibition; S: Stimulation; U: Effects unknown; N: No effect.

Fertilizers

The fertilizers change the physicochemical properties of the soil affecting soil arthropods both qualitatively and quantitatively and tend to increase the population of *Collembola* (Miller, 1991).

Nitrogen

Application of NPK increased the number of *Collembola* two-fold while N decreased *Aptergola* insects and other soil arthropods. Such changes are attributable to the changes in pH and toxicity caused by high concentration of ammonia and other salts. The abundance of microarthropods increased considerably due to the application of NPK incorporations under maize and wheat, as well as under the monoculture of wheat, paddy and jute cultivation. Such increase in total population was mainly due to the increase in *Tarsonemina, Acaridal, Oribataidae* after 4 and 8 years of fertilizer application. However, *Collembola* and *Proshgmata* and *Endoshgmata* changed significantly in abundance compared to the control plots (Miller, 1991).

Lime and Trace Elements

Lime application decreased the abundance of cryphostigmated mites and increased that of *Gamasina, Uropodina* and *Trembidoidea* in pure stand which was probably due to changes in soil pH. The oribated mite *Tectocepheus clatus* was significantly reduced due to the effect of lime. Lime treatment increased the population of *Diptera* larvae more than two-fold while *Chilopda* was slightly reduced in the fertilized plots. Cadmium, Arsenic, Lead etc. and other non-essential poisoning elements being impurities in the ore may exist in the materials that were used to make the fertilizers (Table 2.2). The Cd is readily taken up by plants (a high bioavailability for plant uptake) and food chain transfer is the primary route exposure (Table 2.3). The Cd toxicity causes renal dysfunction and *itai-itai* disease. The disease is characterized by Cd induced bone loss that produces localised, severe pain in the joints of victims. Pb concentrations are generally much higher than Cd concentrations. Despite the higher Pb concentrations in soil, plant uptake of Pb is very limited (a low bioavailability for plant uptake). Lead

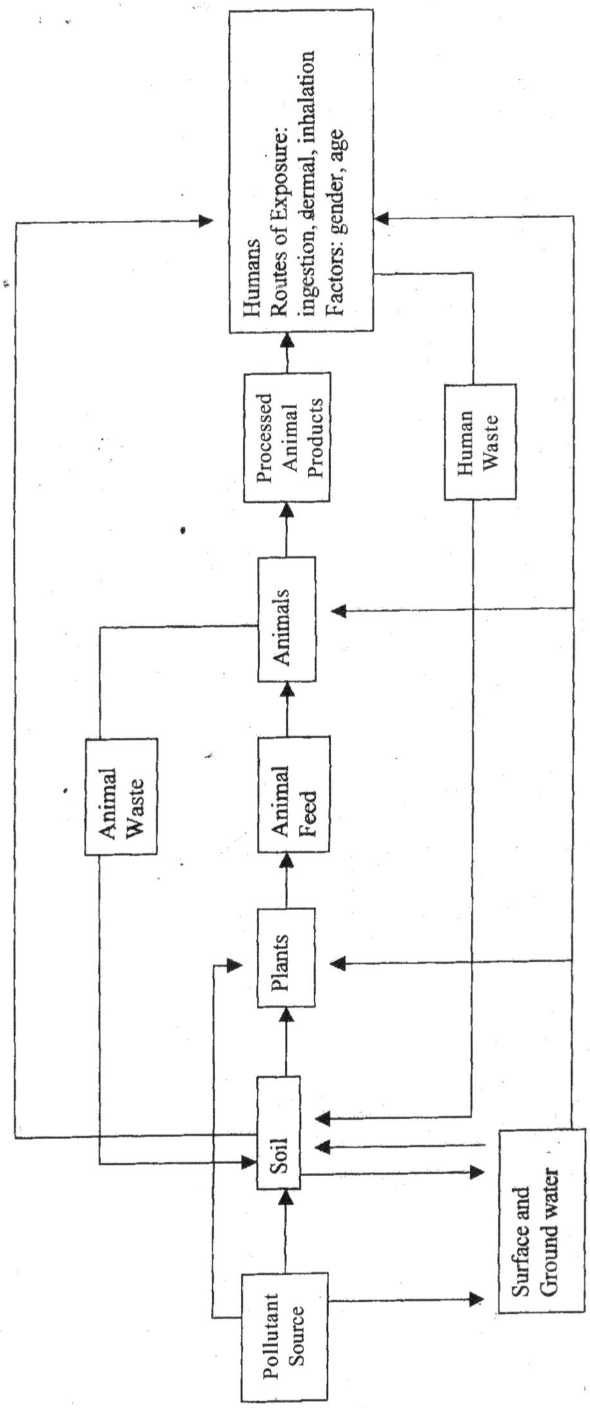

Figure 2.1: Pathways for Human Exposure to Pollutants Applied to Soils (Adapted from Brams *et al.*, 1989)

has a number of physiological effects on the human body and infant children are at much greater risk than adults as is impairment of mental development. The Hg acids caused *Minamata* disease manifested itself in mental retardation, cerebral palsy and fetus mortality in the children.

Table 2.2: Average and Range of Arsenic, Cadmuim and Lead Concentrations (mg/kg) in Selected Inorganic Commercial Fertilizers and Soil Amendments, Liming Materials Sold in California (Department of Food and Agriculture, 1998)

Material	*Grade*	*As*	*Cd*	*Pb*
Limestone	–	2.2 (0.3–5.1)	6.2 (4.0–8.1)	33.9 (3.2–52.0)
Gypsum	–	1.3 (0.2–3.4)	0.9 (0–2.5)	4.4 (0.8–9.2)
Triple super phosphate	0–20–0	4.8 (0.5–14.0)	110 (0–163)	2.0 (0–4.0)
Monoammonium phosphate	11–23–0	8.1 (1.5–16.0)	74.1 (0–166)	2.6 (0–9.0)
Diammonium phosphate	18–20–0	4.9 (0.3–8.5)	38.1 (5.0–125)	4.0 (0–15.6)
Soft/rock phosphate	–	13.0	130	13.0
Muriate of potash	0–0–50	0.4 (0.1–1.6)	0.8 (0–1.6)	3.3 (2.5–4.0)
Urea	49–0–0	0	0	0
Blended	11–5–0	11.2	95.0	3600
Zinc sulfate	–	2.6 (0–29.5)	32.3 (0–233)	203 (0–1430)

Table 2.3: Estimated Magnitude of the Extent of Trace Element Poisonings (Nriagu, 1988)

Elements	*Global Emissions (1000 mg/year)*			*People Affected*	*Comments*
	Air	*Water*	*Soil*		
Pb	332	138	796	> 1 billion	Blood Pb > 20 g/dL
Cd	7.6	9.4	22	500,000	Producing renal dysfunction
Hg	3.6	4.6	8.3	80,000	Certified Hg poisonings
As	18.8	41	82	> 100,000	Skin disorder and H_2O As > 2 g/L

Similarly human health is also affected from As. Exposure most often occurs from drinking water contaminated with arsenical pesticides, such as orchards and on consumption become primary route of exposure. In plants trace elements cause phytotoxicity *i.e.* reduced yields or death of plants by substances in soil. Symptoms for trace elements induced phytotoxicities can include stunting, chlorosis, necrosis and death of plants. Trace elements most often associated with phytotoxicity are B, Cu, Ni and Zn leading to reduced both the quantity and quality of the food produced from the soil (Nriagu, 1988).

Groundwater Pollution

Groundwater seeps from upland to lowland areas, eventually discharging in streams, lakes, ponds and wetlands. The rates of movement of groundwater vary depending upon size of cracks in rocks and size of the pores between soil and rock particles. Most precipitation which recharges groundwater moves only a few miles from the point of recharge to the point of discharge. That means agricultural activities could contaminate groundwater. Untreated industrial waste effluents when discharged in unlined drains or low-lying area can percolate down thereby affecting the quality of groundwater. Although a considerable portion of metallic cations (Cr, Cu, Pb, Cd, Ni etc) are likely to be retained in the top soil, however, if the soil is light textured and rate of down movement of water is high, then migration of these pollutants to the saturated zone may occur. A study of groundwater of industrial area of Ludhiana revealed pH was as low as 2.0 (characteristic of strong acids). Totawat (1987) reported in Rajasthan, the well water near a Zn smelter plant had very high concentration of flouride and zinc.

Depending on the rate and method of application of fertilizer nitrogen, the kind of crop, soil and climatic variables, 35–60 per cent of the applied N is usually recovered by crop plants, 10–20 per cent may be converted to N_2 gas, nitrous oxide or ammonia and volatilize into the atmosphere and if there is excess water percolating through the soil, part of the remaining N may move below the root for leaching to groundwater. Certain circumstances like rate, source, timing etc. could affect nitrogen leaching to the. groundwater. *Nitrogen rate*–at one pound of nitrogen per 1000 square feet, no leaching occurred, *Nitrogen source*–slow release fertilizers can reduce the leaching, *Application timing*–in late fall, plants take up less nitrogen for a greater chance for leaching to occur, *Irrigation practices*–the more irrigation, the greater the chances for nitrate leaching, *Soil texture*–the sandier the soil the more chances for nitrate leaching and *Age of site*–younger sites with less organic matter are highly fertilized for more leaching. However, the United States Environmental Protection Agency established the current drinking water standards (Table 2.4) as Health State Advisory level of 10 mg/l nitrate nitrogen (equivalent to 10 ppm nitrate-nitrogen or 45 ppm nitrate) based on the human health risks due to nitrate consumption (Kross *et al.*, 1993). *Blue Baby Syndrome:* Methemoglobinemia is the condition in the blood, which causes infant cyanosis or blue baby syndrome. Methenoglobin is probably formed in the intestinal tract of an infant when bacteria converts the nitrate ion to nitrite ion. One nitrite molecule then reacts with two molecules of hemoglobin to form methemoglobin. In acid mediums, such as the stomach, the reaction occurs quite rapidly (Comly, 1987). This altered form of blood protein prevents the blood cells from absorbing oxygen which leads to slow suffocation of the infant which may lead to death (Gustafson, 1993). Because of oxygen deprivation, the infant will often take on a blue purple tinge in the lips and extremities, such as vomiting and diarrhoea, relative absence of dietress when severely cyanetic but irritable when mildly cyanotic and chocolate-brown coloured blood (Comly, 1987). Animals can also be susceptible to methemoglobinemia, although the health advisory level for livestock is much higher, ~ 40 mg NO_3–N/l (180 mg NO_3^-/l). *Gastric cancer and other health affects*: Nitrosamines and nitrosamides are carcinogenic and carcinogenic nitros compounds are involved in the etiology of some types of cancers in humans. The high nitrate intake is increasing incidence of goiter due to reduced assimilation of iodine by human body. Once the pesticide has been applied to a field, its movement is controlled by several processes like retention, transformation and plant uptake. The more the pesticide is adsorbed by the soil, the less will be transported. Microbes break down the pesticide into H_2O and CO_2 molecules. The microbial transformation of pesticides include biodegradation, cometabolism, polymerization or conjugation, accumulation and secondary effects. Photochemical transformation is an important degradation pathway for many pesticides, especially the surface applied. The most common transportation is the mass flow and diffusion. Mass flow is when a pesticide molecule is transported by a flow of water very quickly over large distances. Diffusion occurs when a pesticide travels from an area of high concentration to flow concentration by way of random molecular movement. The accumulation of pesticide through plant uptake can have severe consequences through the food chain if the pesticide is translocated into the section of the plant that will be subsequently harvested. Most herbicide concentrations are observed in the range of 0.1–100 µg/l. Atrazine was the most commonly detected herbicide in groundwater (Chopra and Magu, 1985 and Brahms *et al.*, 1989).

Table 2.4: Maximum Permissible Limits of Nitrate in Potable Water Prescribed by Different Countries and Organizations (Lunkad, 1994)

Country/Organization	*Concentration as NO_3–N (mg/l)*	*Concentration as NO_3 (mg/l)*
WHO	10	45
US Environmental Protection Agency	10	45
ICMR (India)	10	45
Canada	10	45
Poland	10	45
EEC	11.30	50
Bulgaria	6.7	30
Belgium	11.30	50
Denmark	11.30	50
Finland	6.8	30
Hungary	9.0	40
UK	11.3	50
	22.6	100
USA	10	45

Management Practices to Control Soil and Groundwater Pollution

Nutrient Management

Split applications of N, fertigation, slow release fertilizers, injections and deep placement of volatile N fertilizers could improve synchrony between N availability and crop N uptake pattern. Use of a nitrification and adding nitrapyrin (N-serve) to applied nitrogen reduced nitrate leaching on an average by 10–15 per cent.

Integrated Pest Management

Integrated pest management like use of crop rotation than continuous corn is advised to reduce the need for corn rootworm insecticides and use of crop rotation with tolerant varieties could help to control plant diseases.

Conservation Tillage

Reducing tillage and retaining crop residues on the soil surface limit the runoff and overland flows that carry pesticides and nutrients out of the field. Conservation tillage methods, ridge tillage system placing chemicals on and in the ridges and timing of application as per crop's growth cycle significantly reduced the amount that moves to groundwater. The practices that reduce surface losses of pesticides by increasing infiltration may actually increase the amount of pesticide that reach groundwater supplies.

Cover Crops

The effectiveness of cover crops in controlling erosion is well documented. Small grain cover crops have shown some efficiency at retrieving fertilized corn or vegetable crops. Cover crops may recover significant quantity of soil profile NO_3 to prevent N transport to groundwater.

Chemical Properties and Selection of Pesticide

Use less persistent chemicals. There is little rise of groundwater contamination by those chemicals that absorb strongly to soil particles. A pesticide influences leaching potential with solubility in water, ability to bind with the soil (adsorption) and the rate at which it breaks down. High solubility (dissolves readily), low binding ability and slow breakdown help increase a pesticide ability to move to the groundwater. Among the frequently used herbicides that have a greater potential to leach and are labelled with Groundwater Advisories are those that contain acetochlor, alachlor, atrazine, sulfetrazone, disulfuron, acifluorfen, dimethenamid, chloransulam, carbofuran, flumetsulam, simazine, metribuzin, metolachlor and clopyralid.

Precautions for Irrigation and Fertigation

Chemigation is a management tool for groundwater protection. Backflow prevention devices are mandatory on fertigation irrigation systems that inject fertilizers and pesticides.

References

Brams, E., Anthony, W. and Witherspoon, L. 1989. Biological monitoring of an agricultural food chain: soil cadmium and lead in ruminant tissues. *J. Environ. Qual.*, 18: 317.

California Department of Food and Agriculture 1998. Development of risk based concentrations for arsenic, cadmium, and lead in inorganic commercial fertilizers, prepared for the California Department of Food and Agriculture by Foster Wheeler Environmental Corporation, Scramento, CA.

Chopra, P. and Magu, S.P. 1985. Effect of selected herbicides and city compost on the rhizospheric microflora of wheat and maize. *Indian J. Agron.*, 30(1): 5–9.

Comly, H.H. 1987. Cyanosis in infants caused by nitrates in well water. *J. American Med. Assoc.*, 257: 2788–2792.

Debona, A.C. and Andus, L.J. 1970. Studies on the effects of herbicides on soil nitrification. *Weed Res.*, 10: 250–263.

Edward, C.A. 1971. *Pesticides and the Soil Fauna*. Report Rothamated Experiment Station, 1970: 194.

El-Shahawy, R.M., Mostafa, M.T., Retad, M.S. and Saleh, S.A. 1982. Response of cotton plant and soil microflora to cotoran Multi. *Ann. Agric. Sci., Mashtohar*, 18: 352.

Gustafson, D.I. 1993. *Pesticides in Drinking Water*. Van Nostrand Reinhold, New York, pp. 241.

Heydrich, M. and Fernandez, C. 1984. Charges in the composition of the actinomycete community in a soil treated with diuron. In. *Soil Biology and Conservation of Biosphere*, (Ed.) J. Szegi. Akademiai-Kiado, Budapest, Hungary, p. 209–215.

Hicks, R.J., Stotzky, G. and Van Voris, P. 1990. Review and evaluation of the effects of xenobiotic chemicals on microorganisms in soil. *Adv. Appl. Microbiol.*, 35: 195–253.

Kross, B.C., Hallberg, G.R., Bruner, R., Cherryholmes, K. and Johnson, K.J. 1993. The nitrate contamination of private well water in Iowa. *American J. Pub. Hlth.*, 83: 270–272.

Kuhnelt, W. 1963. Soil inhating arthropoda. *Ann. Rev. Entomol.*, 8: 115–136.

Kulshrestha, G. 1993. Ecological implications of herbicide use in Agriculture. In: *Pesticides: Their Ecological Impact in Developing Countries*, (Eds.) G.S. Dhaliwal and Balwinder Singh. Commonwealth Publishers, p. 190–216.

Lunkad, S.K. 1994. Rising nitrate levels in groundwater and increasing N-fertilizer consumption. *Bhu Jal News*, p., 4–10.

Mengel, K., Kirkby, E.A., Kosegarten, H. and Appel, T. 2001. *Principles of Plant Nutrition*. Kluwer Academic Publishers, London, p. 383–390.

Miller, G.T. 1991. *Environmental Science: Sustaining the Earth*. Wadsworth Publishing, Belmont, CA.

Mishra, S.D., Prasad, D. and Dwivedi, B.K. 1989. Pesticide residue in soil and soil organisms. In: *Soil Pollution and Soil Organisms*. p. 17–29.

Nriagu, J.O. 1988. A silent epidemic of environmental metal poisoning. *Environ. Pollut.*, 50: 139–161.

Pope, P.E. and Holt, H.A. 1981. Paraquat influences development and efficacy of the mycorrhizal fungus. If Glomus fasciculatus. *Can. J. Bot.*, 59(4): 518–521.

Rao, V.R., Adhya, T.K. and Sethunathan, N. 1993. Effect of pesticides on soil health. In: *Pesticides: Their Ecological Impact in Developing Countries*, (Eds.) G.S. Dhaliwal and Balwinder Singh. Commonwealth Publishers, p. 112–130.

Totawat, K.L. 1987. Effect of treated effuent discharged from zinc smelter on field crops and well waters adjoining to effluent stream. In: *Tech. Report*, Sukhadia University Udaipur, p. 4–12.

Yee, D., Weinberger, P., Johnson, D.A and Dechain, C. 1985. *In vitro* effects of s-triazine herbicide prometryne, on growth of terrestrial and aquatic microflora. *Arch. Envir. Contam. Toxicol.*, 14 (1): 25–31.

Chapter 3

Resource Productivity and Allocation Efficiency in the Production of Sunflower and Groundnut in Andhra Pradesh

Y. Sudhakar Reddy and G.P. Reddy***

**Agricultural Officer (FCO) Lab, Government of Andhra Pradesh, Rajendranagar, Hyderabad – 560 030*
***Senior Scientist (Agricultural-Economics), National Academy of Agricultural Research Management, Rajendranagar, Hyderabad – 560 030*

ABSTRACT

India has the largest area (21 million hectares in 2002–2003) in the world under oil seeds and yet it has emerged as the largest importer of edible oils. It is well known that only about 20 years ago India was a net exporter of vegetable oils. But, the increased internal demand, due to improved standards of living and rapid population growth have led to a ban of exports of edible oil seeds and oils. The tremendous increase in the demand for oil seeds and vegetable oils, coupled with the drain on the exchequer of the country by way of imports and the need to build up the oil economy of the country to a self supporting level necessitated the identification of alternative sources, to bridge the gap between demand and supply of oil seeds and the cultivation of sunflower was one of the means sought for this purpose. In Andhra Pradesh, sunflower is cultivated largely in *kharif* as a rain fed crop and is mainly grown in Kurnool district of Rayalaseema region and Medak district, parts of Mahaboobnagar and Nalgonda districts of Telangana region. In this context it would be relevant to study the resource productivity and allocation efficiency in the production of sunflower and groundnut. Hence, this study was designed to evaluate empirically the relative profitability of sunflower and groundnut crops in Adoni taluk of Kurnool district and to analyse the constraints, if any, in the production of the sunflower. Data on quantity of inputs used and yields obtained with their respective value were recorded for both sunflower and groundnut production from the respondents, in 1999–2000. The functional analysis revealed that seeds, labour and fertilizers significantly influence production in both small and large farms. The ratio of MVP to MFC was more than unity for these factors indicating the possibilities for increased use of these resources to increase the yield of sunflower and groundnut. The results revealed that land was not significant in both small and large farms, which indicates that the allocation of additional area to sunflower and groundnut crops is not economical. The study indicated overall higher value of MVP to MFC ratios in sunflower production. It is inferred that the farmers would be better placed if they shift the resources (particularly seed to and fertilizers)/ sunflower production from groundnut to a limited extent. The same holds true in both the size groups. Hence, the concerned agencies should educate the farmers on scientific method of the cultivation of sunflower by providing other infrastructure facilities like supply of quality seeds, fertilizer, credit, etc., at right time. At the same time, we cannot afford to neglect the cultivation of groundnut. Hence, extension agencies should still intensify their educational activities to educate the farmers on scientific method of cultivation of groundnut also, which is an important oil seed crop in the country.

Introduction

Oil seeds occupy an important position in India's national economy, next to food grains, accounting for about 10 per cent of the cultivated area, as well as the value of all agricultural produce. They occupy very prominent position in the human dietary system. The bulk of vegetable oil production in India is derived from nine oil seeds; namely, groundnut, rape seed mustard, sesame, safflower, niger, soybean and sunflower, forming the edible group and linseed and castor, being the non-edible group. In addition, cottonseed, rice bran, seeds of some tree species etc., are also being exploited for vegetable oils. Both the area and production of oil seeds in India have been increasing (Anonymous, a). Both the area and production of oil seeds in India have been increasing (Anonymous, 1987a). The area has increased from an average area of 17.3 million hectares in 1974–1975 to 21.1 million hectares in 2002–2003 *i.e.*, by 22 per cent. The production has increased by about 63 per cent from 9.2 million tonnes in 1974–1975 to 15 million tonnes in 2002–2003.

The commercial cultivation of sunflower crop was started in India during 1972–1973, in the southern states *viz.*, Andhra Pradesh, Karnataka and Tamil Nadu. Among all the sunflower varieties tried, BC-68414, EC-68415 and modern were found adaptable to various agro-climatic conditions and different cropping systems. Varietal enovation programme, initiated in 1977, helped in stabilization of yield and oil content in these varieties. Due to the Intensive Research and Development Programme, the crop has now assumed importance as an oil seed in Indian agriculture. With its tremendous advantages over other oil seed crops, sunflower is poised to become a popular oil seed crop in India.

Sunflower is fast becoming popular with our farmers as an oil seed crop in view of its short duration, wide adaptability, photo insensitive nature and ability to with-stand drought. In India, sunflower was cultivated in 0.3 million hectares during 1975, as against an area of about 1.60 million hectares during 2002–2003; thereby recording an increase of 433 per cent and the production has increased nearly four fold times over the years, from a mere 0.2 million tonnes in 1974–1975 to nearly 0.9 million tonnes in 2002–2003.

The sunflower crop can be grown during all the seasons–*kharif*, *rabi* and summer. However, higher yields are obtained during summer. It is observed that sunflower crop has more income potential for the farmers and it is replacing some of the crops like *kharif* groundnut, *rabi* jowar and durum wheat, which are less remunerative. Hence, the area under sunflower is on the increase and it is gaining importance as one of the potential oil seed crops (Sindagi, 1985). Andhra Pradesh stands third in country's sunflower area occupying 29.40 thousand hectares with a production of 11.30 thousand tonnes.

Andhra Pradesh plays an important role in edible oil seed production of the country by contributing 13 per cent of total cultivated edible oil seeds production (1982–1983 to 1984–1985 average) and stands second next only to Gujarat, which contributes 16.6 per cent of total production of edible oil seeds (Anonymous, 1985).

In this context it would be relevant to study the resource productivity and allocation efficiency in the production of sunflower and groundnut. Hence, this study was designed to evaluate empirically the relative profitability of sunflower and groundnut crops in Adoni taluk of Kurnool district and to analyse the constraints, if any, in the production of the sunflower. It is sought to be accomplished by setting up the following objective. Study the resource productivity and allocation efficiency in the production of sunflower and groundnut.

Methodology

Kurnool district was chosen for the study as it ranks first in area under sunflower crop in Andhra Pradesh. The data relating to the area and production in the district were collected from District Agricultural Officer. Of all the taluks in Kurnool district, Adoni taluk was selected because it ranks first in area under sunflower. From the list of villages in Adoni taluk, the top 10 villages with the highest area under sunflower were selected. In each selected village, farmers growing both sunflower and groundnut were listed in consultation with Taluk Statistical Assistant. The farmers were categorized into "small" and "large" based on land size. Those with less than 7.50 acres of dry land were termed "small" and those with 7.50 acres or more of dry land were termed "large" in accordance with the classification followed in Andhra Pradesh.

From each of the 10 villages, 3 small farmers and 3 large farmers growing both sunflower and groundnut were randomly chosen. Thus, 30 small and 30 large farmers constitute the sample for the study.

Data Collection

Data was collected by personally interviewing the sample farmers with the help of pre-tested schedules specially designed for the purpose. Data on some selected socio-economic characteristics of the farmers, land holding, cropping pattern, farm assets were collected. Besides, data on quantity of inputs used and yields obtained with their respective value were recorded both for sunflower and groundnut production from the respondents. Data on marketing also were collected for both sunflower and groundnut crops. The data on opinions of respondents about constraints faced by them in production and marketing of sunflower were also elicited. The data pertaining to the year 1999–2000 *kharif* season were collected during February-March by survey method with assiduity.

Resource Productivity and Allocation Efficiency in Sunflower and Groundnut Production

The main objective of management is to coordinate and utilize the farm resources in the production of different crops in such a manner that they together yield the highest net returns. It is in this context the study of resource productivity assumes paramount importance.

The analysis of resource productivity and allocation efficiency of the involved inputs in sunflower and groundnut production was done to assess the extent to which these factors explained the variability in the yield of the crops, and to determine whether the inputs were optimally used in the production of these crops.

A modified Cobb-Douglas type of production function was fitted to estimate the elasticities {elasticity coefficients) of various factor inputs. Two Cobb-Douglas production functions were estimated, one for sunflower and the other for groundnut. For both the crops the dependent variable was yield (qtls/holding). The explanatory variables that were common to both the crops were:

x_1 = Land area under the sunflower/groundnut crop in acres

x_2 =Human labour in Mondays of eight hours (inclusive of family and hired labour) for the crop

x_3 =Value of seeds in rupees used for the crop

x_4 =Value of farm yard manure in rupees

x_s =Value of chemical fertilizers in rupees.

There was one more explanatory variable namely value of plant protection chemicals (kg) for groundnut. This variable is not included in the function for sunflower because the expenditure on plant protection chemicals in sunflower was negligible.

The equation of the production function for sunflower was:

$$y = a\, x_1^{b1}\, x_2^{b2}\, x_3^{b3}\, x_4^{b4}\, x_5^{b5} \qquad (1)$$

The equation of the production function for groundnut was:

$$y = a\, x^{b1}\, x_2^{b2}\, x_3^{b3}\, x_4^{b4}\, x_5^{b5} \qquad (2)$$

where,

y is the physical output (yield in qtls/holding), a is a constant term and x_1, x_2, x_3, x_4, x_5, x_6 are as explained above.

The production function (1) was fitted for:

1. The 30 small farms of sunflower
2. The 30 large farms of sunflower
3. The 60 all farms of sunflower.

Similarly, the production function (2) was fitted for:

1. The 30 small farms of groundnut
2. The 30 large farms of groundnut
3. The 60 all farms of groundnut.

In all six Cobb-Douglas functions were fitted.

Before fitting the function, the zero order correlation coefficients between the explanatory variables were calculated to rule out multi-collinearity.

Return to Scale

The return to scale can be estimated directly by getting the sum of 'b' coefficients. The return will be increasing, constant or diminishing as:

Σb_i is greater, equal or less than unity, respectively.

To see if the sum of 'b_i' coefficients were significantly different from unity, return to scale test (Rao and Miller, 1971) was conducted by using the test statistic of the form.

The resulting 't' computed values are compared with t (n-K)–critical values at 5 per cent level.

The marginal products were calculated at the geometric mean levels of the variables by using the formula:

$$\text{Marginal Product with Reference to } x_i^{th} \text{ resource} = b_i \frac{y}{x_i}$$

where,

y = Geometric mean yield from sunflower/groundnut

x_i = Geometric mean of ith independent variable

b_i = The regression coefficient (elasticity coefficient) of ith independent variable.

Marginal value product of each factor was calculated by multiplying the marginal product of each factor with the unit price of the output *i.e.,*

$$MVP = MPP \times P_y$$

where,

P_y = Price per unit of output.

Allocation Efficiency

Allocation efficiency of both sunflower and groundnut crops was studied by comparing the marginal value product of each factor to its marginal cost (Market input price) to determine whether the factors were used optimally in the production of crops and to estimate the possibility of shifting a resource if it can be more economical in production of other competing crop.

Results and Discussions

Resource Productivity and Allocation Efficiency in the Production of Sunflower and Groundnut

To study the productivities of resources involved in the production of sunflower and groundnut, analysis was done separately for small, large and all farms of sunflower and groundnut in order to determine whether or not the factors of production were used optimally. The efficiency of allocation of resources was studied by comparing the marginal value product with the marginal factor cost of each factor of production included in the analysis.

Further, an attempt is made to compare the marginal value products of sunflower with its competing groundnut for the resources in both the size groups and all farms and to estimate the possibility of shifting of resources from one crop to another. For this purpose the ratio between marginal value product and the respective marginal factor cost was calculated for the inputs involved in both the crops grown by different categories of respondents at their geometric mean levels.

A modified Cobb-Douglas type of production function was fitted to the data, since it provides a compromise between adequate fit to the data, computational feasibility and sufficient degrees of freedom to allow for statistical testing (Heady and Dillon, 1964).

Small Farms: Sunflower and Groundnut Crops

The results on elasticity coefficients in the production function of sunflower and groundnut for small farms are presented in Table 3.1. It could be observed from this table that in sunflower and groundnut crops the regression coefficients of seed (0.6606 and 0.4748) and fertilizers (0.0257 and 0.0180), were significant at one per cent level in case of sunflower and at five per cent level in case of groundnut. Similarly, the same for human labour (0.4022 and 0.7511) was significant at five per cent level in both the crops. On the other hand, the coefficients of variables like land (0.0286 and 0.0594), and farmyard manure (0.0123 and 0.0041) were non-significant in both the crops. The regression coefficient of plant protection chemicals (0.0123) was non-significant in case of groundnut.

Table 3.1: Factors Influencing Sunflower and Groundnut Production–Small Farms

Crop	*No. of Obser-vations*	*Intercept*	*Regression Coefficients (Production Elasticities)*							*Sum of Elasticities*
			Area Under Crop	*Human Labour*	*Seed*	*Farm-yard Manure*	*Fertilizer*	*Plant Protection Chemicals*	*R^2*	
Sunflower	30	– 3.0038	0.0286 NS (0.090)	0.4022* (1.732)	0.6606** (4.548)	0.0123 NS (1.67)	0.0257** (4.329)		0.8697**	1.294
Groundnut	30	– 4.9121	0.0594 KB (0.409)	0.7511* (2.241)	0.4748* (2.086)	0.0041 NS (0.778)	0.0180* (2.76)	0.0123 NS (2.76)	0.8821** (1.145)	1.3197

Note: * and ** indicate significant at five and one per cent level of significance respectively.

NS: Not Significant.

Figures in parentheses indicate t-values.

R^2: Significance of this was tested by 'F' test.

The results on allocation efficiency of important variables included in the analysis for sunflower and groundnut in small farms are presented in Table 3.2. In this table, the marginal factor cost of land was nothing but an imputed rental value of land. The marginal factor cost of human labour was the wage rate paid by the farmers per man-day. For the remaining resources such as seed, farmyard manure, fertilizers, plant protection chemicals the marginal factor cost of Re. 1 was considered.

As indicated earlier seed, fertilizer and human labour were significant variables, which would contribute in the production of both the crops. The marginal product of a rupee of seed in sunflower was 0.0161 qtls valued at Rs. 9.77 as compared to 0.0054 qtls valued at Rs. 3.01 in groundnut. Similarly the marginal product of a rupee worth of fertilizer was 0.0180 qtls (MVP = 10.93) and 0.0113 qtls (MVP = 6.29) in sunflower and groundnut respectively. The marginal product of one man-day of human labour was 0.0493 qtls (MVP = Rs. 29.92) in sunflower and 0.0935 qtls (MVP = Rs. 52.08) in groundnut. The marginal product of land however did not cover the cost of renting in additional land.

The ratio of marginal value product to marginal factor cost was less than unity in case of land in both the crops, indicating it's over use. The ratio for all other resources *viz.* human labour, seed, farm yard manure, fertilizer were greater than unity in both the crops indicating the possibilities of increasing the yield with increased doses of these resources baring farm yard manure where regression coefficient was not significant. The ratios were 2.99, 9.77, 2.73 and 10.93 in case of human labour, seed, farmyard manure and fertilizer, respectively, in a case of sunflower. The corresponding ratios for groundnut were 5.21, 3.01, 19.77 and 6.29. The ratio for plant protection chemicals, the variable included only for groundnut was 3.18. The sum of elasticities was 1.1294 for sunflower crop and this was not statistically different from unity indicating constant returns to scale. The sum of elasticities was 1.3197 for groundnut crop and this was statistically different from unity indicating increasing returns to scale.

Table 3.2: Allocation Efficiency in Sunflower and Groundnut Production–Small Farms

Variables	*Sunflower*					*Groundnut*				
	GM	*MP (Qtls.)*	*MVP (Rs.)*	*MFC (Rs.)*	*MVP/ MFC*	*GM*	*MP (Qtls.)*	*MVP (Rs.)*	*MFC (Rs.)*	*MVP/ MFC*
Area under crop (Acres)	2.34	0.0717	43.52	133.33	0.3263	2.58	0.1795	99.98	133.33	0.75
Human labour (man-days)	47.80	0.0493	29.92	10.00	2.992	70.88	0.0935	52.08	10.00	5.21
Seeds (Rs.)	240.09	0.0161	9.77	1.00	9.77	683.35	0.0054	3.01	1.00	3.01
Farmyard manure (Rs.)	15.94	0.0045	2.73	1.00	2.73	8.98	0.0036	2.00	1.00	2.00
Fertilizer (Rs.)	8.34	0.0180	10.93	1.00	10.93	12.39	0.0113	6.29	1.00	6.29
Plant protection chemicals* (Rs.)						16.74	0.0057	3.18	1.00	3.18
Yield (Quintals)	5.86					7.79				
Price per quintal** (Rs.)	606.92					557.00				

Note: * Variable included only for Groudnut.

** Estimated at arithmetic mean level.

The adjusted coefficient of Multiple Determination (R) was 0.8697 and 0.8821 in case of sunflower and groundnut crops, respectively which was significant at one per cent level. This implied that the variables selected in the function, together explained 86.97 per cent and 88.21 per cent of variation in yield significantly.

Large Farms: Sunflower and Groundnut Crops

The results on production elasticities of production functions for both sunflower crop (with five independent variables) and groundnut crop (with six independent variables) are presented in Table 3.3. It is observed from the table that regression coefficients of labour in sunflower and groundnut crops were 0.7104 and 0.6924 respectively and they were significant at one per cent level. The variable 'seed' was significant at one per cent level in case of sunflower (0.4992) and at five per cent level in case of groundnut (0.3442). In sunflower crop the variable farmyard manure (0.0241) was significant at one per cent level and the same in groundnut (0.0065) was non-significant. The use of 'fertilizer' (0.0186 and 0.0189) was significant at five per cent level in sunflower (0.0186) and at one per cent level in groundnut (0.0189). The plant protection chemicals (0.160), which are included only for groundnut was significant at five per cent level. The coefficient of land (0.0676 and 0.0493) though positive was not significant both in sunflower and groundnut.

Table 3.3: Factors Influencing Sunflower and Groundnut Production–Large Farms

Crop	No. of Observations	Intercept	Area Under Crop	Regression Coefficients (Production Elasticities)					R^2	Sum of Elasticities (Return to Scale)
				Human Labour	Seed	Farm-yard Manure	Fertilizer	Plant Protection Chemicals		
Sunflower	30	– 4.2691	0.0676 NS	0.7104**	0.4992**	0.0241**	0.0186*		0.9494**	1.3199
			(0.749)	(3.379)	(4.066)	(3.484)	(2.742)			
Groundnut	30	– 2.934	0.0493 NS	0.6924**	0.3484*	0.0065 NS	0.0189**	0.0160*	0.9372**	1.1315
			(0.877)	(4.299)	(2.692)	(1.718)	(2.922)	(2.591)		

Note: * and ** indicate significant at five and one per cent levels of significance respectively.

NS: Not Significant

Figures in parentheses indicate t-values

R^2: Significance of this was tested by 'F' test.

The results on allocation efficiency of variable included in the analysis for sunflower and groundnut crops in large farms are presented in Table 3.4. We have seen earlier that human labour was highly significant in both the crops. The marginal product of a man-day of human labour was 0.1023 qtls valued at Rs. 62.22 in case of sunflower production and in groundnut production it was 0.0759 qtls valued at Rs. 42.54. The marginal product of a unit (Rs.) of seed in sunflower was 0.0137 qtls (MVP = Rs. 8.33) whereas it was 0.0039 qtls (MVP = Rs. 2.19) in groundnut. Regarding farmyard manure and fertilizer the marginal products per unit (Re.) were 0.0032 qtls (MVP = Rs. 1.95) and 0.0031 qtls (MVP = Rs. 1.89) in sunflower and 0.0041 qtls (MVP = Rs. 2.30) and 0.0018 qtls (MVP = Rs. 1.01) on groundnut production. As in the small sized farms land did not significantly influence production.

Table 3.4: Allocation Efficiency in Sunflower and Groundnut Production–Large Farms

Variables	Sunflower					Groundnut				
	GM	MP (Qtls.)	MVP (Rs.)	MFC (Rs.)	MVP/ MFC	GM	MP (Qtls.)	MVP (Rs.)	MFC (Rs.)	MVP/ MFC
Area under crop (Acres)	9.97	0.2103	127.90	135.40	0.95	8.07	0.1535	86.03	135.40	0.64
Human labour (man-days)	215.29	0.1023	62.22	10.00	6.22	229.06	0.0759	42.54	10.00	4.25
Seed (Rs.)	1125.52	0.0137	8.33	1.00	8.33	2197.33	0.0039	2.19	1.00	2.19
Farmyard manure (Rs.)	237.70	0.0032	1.95	1.00	1.95	39.29	0.0041	2.30	1.00	2.30
Fertilizer (Rs.)	182.55	0.0031	1.89	1.00	1.89	260.60	0.0018	1.01	1.00	1.01
Plant protection chemicals*						79.99	0.0050	2.80	1.00	2.80
Yield (qtls.)	31.03					25.13				
Price per quintal** (Rs.)	608.17					560.48				

Note: * Variable included only for Groundnut.

**Estimated at arithmetic mean level.

The ratio of marginal value product to marginal factor cost was greater than unity in case of human labour, seed and farmyard manure in both the crops indicating the possibilities of increasing the yield with increased doses of these resources. The ratio for human labour was Rs. 6.22 and Rs. 4.25 in sunflower and groundnut respectively. The MVP to MFC ratio for seed was Rs. 8.33 in the case of sunflower, while it was only Rs. 2.19 in the case of groundnut. The ratio for farmyard manure in sunflower was Rs. 1.95 and the corresponding value for groundnut was Rs. 2.30. The ratio for fertilizer was Rs. 1.89 in case of sunflower whereas it was Rs. 1.01 in groundnut production. The ratio of marginal value product to marginal factor cost of land was less than unity in both the crops. Since in the regression the land variable was non-significant, the value marginal products have limited significance and only indicate a broad trend.

The sum of elasticities was 1.3199 for sunflower, and this was statistically different from unity indicating increasing returns to scale. The sum of elasticities was 1.1315 for groundnut and this was not statistically different from unity indicating constant returns to scale.

This means that if all the variables were increased together by one per cent the yield would increase by one per cent *Ceterus paribus*.

The adjusted R^2 values were 0.9494 and 0.9372 for sunflower and groundnut crops, respectively, which were highly significant at one per cent level. This meant that the variables included in the production function explained 94.94 per cent and 93.72 per cent of the variation in the yield of sunflower and groundnut crops.

All Farms: Sunflower and Groundnut Crops

Table 3.5 presents the results on regression coefficients of variables included in production functions for sunflower and groundnut crops.

Table 3.5: Factors Influencing Sunflower and Groundnut Production–All Farms

Crop	*No. of Obser-vations*	*Intercept*	*Regression Coefficients (Production Elasticities)*						*R^2*	*Sum of Elasticities*
			Area Under Crop	*Human Labour*	*Seed*	*Farm-yard Manure*	*Fertilizer*	*Plant Protection Chemicals*		
Sunflower	60	– 3.5909	0.0966 NS	0.5671**	0.6007	0.0177**	0.0253*#		0.9691**	1.307
			(1.005)	(2.791)	(6.136)	(3.4511)	(5.802)			
Groundnut	60	3.2267	0.0793 NS	0.4746**	0.5322*	0.0074*	0.0124*	0.0203**	0.9658**	1.126
			(1.216)	(2.684)	(2.135)	(2.135)	(2.547)	(3.475)		

Note: * and ** indicate significant at five and one per cent levels of significance respectively.

NS: Not Significant.

Figures in parentheses indicate t-values.

R^2: Significant of this was tested by 'F' test.

It is observed from the table that regression coefficients of seed (0.6007 and 0.5322), farmyard manure (0.0177 and 0.0074), fertilizers (0.0253 and 0.0124) were significant at one per cent level in sunflower production and at five per cent level in case of groundnut production. The elasticity coefficient of human labour (0.5671 and 0.4746) was significant at one per cent level in both the crops. In groundnut crop coefficient of plant protection chemicals (0.0203) was significant at one per cent level. The regression coefficients of land (0.0966 and 0.0793) were not significant in sunflower and groundnut production.

Table 3.6 presents the particulars regarding marginal product, marginal value product, marginal factor cost and the ratio of marginal value product to the marginal factor cost of each variable involved in sunflower and groundnut production in all farms.

Table 3.6: Allocation Efficiency in Sunflower and Groundnut Production–All Farms

Variables	*Sunflower*					*Groundnut*				
	GM	*MP (Qtls.)*	*MVP (Rs.)*	*MFC (Rs.)*	*MVP/ MFC*	*GM*	*MP (Qtls.)*	*MVP (Rs.)*	*MFC (Rs.)*	*MVP/ MFC*
Area under crop (Acres)	4.85	0.2697	163.86	134.37	1.22	4.56	0.2432	135.89	134.37	1.01
Human labour (man-days)	101.39	0.0753	45.80	10.00	4.58	127.49	0.0521	29.11	10.00	2.91
Seed (Rs.)	19.57	0.0155	9.42	1.00	9.42	1225.25	0.0114	6.27	1.00	6.37
Farmyard manure (Rs.)	61.56	0.0038	2.31	1.00	2.31	18.78	0.0055	3.07	1.00	3.07
Fertilizer (Rs.)	27.91	0.0122	7.41	1.00	7.41	56.83	0.0030	1.68	1.00	1.68
Plant protection chemicals*						36.60	0.0071	4.30	1.00	4.30
Yield (Quintals)	13.48					13.99				
Price per quintal** (Rs.)	607.55					558.74				

Note: * Variable included only for Groundnut.

**: Estimated at arithmetic mean level.

In both the crops, all the variables included in the analysis were significant except land. The marginal products of unit (man-days) of human labour were 0.0753 qtls and 0.0521 qtls in sunflower and groundnut production respectively. The marginal products of a rupee worth of seed, farmyard manure and fertilizer were 0.0155, 0.0038 and 0.0122 qtls respectively in sunflower production. The corresponding figures for groundnut were 0.0114, 0.0055 and 0.0030 qtls.

The marginal value product of human labour was Rs. 45.80 in sunflower production and groundnut production it was Rs. 29.11. The MVP of seed in sunflower production was Rs. 9.42 whereas groundnut production it was relatively lower at Rs. 6.37. Similarly, the marginal value product of fertilizer (Rs. 7.41) in sunflower was greater than of groundnut (Rs. 1.68).

The ratio of marginal value product to marginal factor cost was more than unity for all the resources in both the crops indicating the possibilities of increasing the yield with increased doses of these resources. The ratio of marginal value product of seed to its marginal cost was Rs. 9.42 in case of sunflower and Rs. 6.37 in case of groundnut. Similarly, the ratio for fertilizer in sunflower production was Rs. 7.41 and in groundnut production Rs. 1.68. The ratios for land, human labour, and farmyard manure were 1.22, 4.58, 2.31 in sunflower and 1.01, 2.91, and 3.07 in groundnut.

The sum of elasticities was 1.3074 for sunflower crop and this was statistically different from unity indicating increasing returns to scale. In groundnut, the sum of elasticities was 1.1262 and this was not statistically different from unity indicating constant returns to scale.

The values of adjusted Coefficient of Multiple Determination (R^2) were 0.9691 and 0.9658 in case of sunflower and groundnut, respectively, which were highly significant at one per cent level. Thus the variables chosen for the production function, together explained 96.91 per cent and 96.58 per cent of the variation in yield of sunflower and groundnut crops in the study area respectively.

Conclusion

The functional analysis revealed that seeds, labour and fertilizers significantly influence production in both small and large farms. The ratio of MVP to MFC was more than unity for these factors indicating the possibilities for increased use of these resources to increase the yield of sunflower and groundnut. The results revealed that land was not significant in both small and large farms, which indicates that the allocation of additional area to sunflower and groundnut crops is not economical. The study indicated overall higher value of MVP to MFC ratios in sunflower production. It is inferred that the farmers would be better placed if they shift the resources (particularly seed to fertilizers) sunflower production from groundnut to a limited extent. The same holds true in both the size groups.

References

Anonymous. 2000. *Bulletin on Food Statistics*. Directorate of Economics and Statistics, Department of Agriculture and Cooperation, Ministry of Agriculture, Govt. of India, pp. 185.

Anonymous. 2003a. *Manorama Year Book*. Malayala Manorama Press, Kottayam, Kerala, pp. 566.

Anonymous. 1987b. *The Europa Year Book*. Europa Publications Ltd., Love lane, Rochester, Kent. J., pp. 1354.

Anonymous. 1987c. *Sunflower Ray of Hope*, National Bank News Review. 2(3): 30.

Heady, E.G. and Dillon, J.L. 1964, *Agricultural Production Functions*. Iowa State Univ. Press, Iowa.

Sindagi, S.S. 1985. The current situation, constraints and prospects of sunflower crop in India. *Oil Seed Production: Constraints and Opportunities,* (Eds). Srivastava, H.C., Bhaskaran, S., Vatsya, B. and Menon, K.K.G. Oxford and IBH Publishing Co., New Delhi, pp. 144.

Chapter 4

Vr, Wr Graphical Analysis for Horticultural Traits in Cauliflower (*Brassica oleracea* var. *botrytis* L.)

Sanjeev Kumar*, U.K. Kohli** and Puja Rattan*

**Division of Olericulture and Floriculture, FOA, Main Campus, Chatha, SKUAST – 180 009, J&K*
***Department of Vegetable Crops, Dr. Y.S. Parmar University of Horticulture and Forestry, Nauni, Solan – 173 230, H.P.*
****Department of Vegetable Science and Floriculture, CSKHPKV, Palampur – 176 062, H.P.*

ABSTRACT

15 Hybrids and 6 parents were studied to investigate gene action through Vr, Wr graphical analysis. The regression coefficient (b) deviated significantly from unity for three characters namely stalk length, number of leaves per plant and curd depth depicting the presence of non-allelic interaction for them. The position of regression line passing through origin showed complete dominance for stalk length, position of regression line intersecting Wr axis above the point of origin depicted partial dominance for plant frame, leaf length, gross curd weight and days to marketable maturity. For characters like number of leaves per plant, leaf breadth, curd depth and harvest index the regression line passed the Wr axis below the point of origin showing over dominance, whereas for net curd weight the regression line did not intersect the Wr axis depicting no dominance.

Introduction

Although cauliflower has a very old history of cultivation and having wide range of available variability, yet not much work has been done on breeding and genetics of this crop. For studying the inheritance of metric traits, the diallele analysis is the most widely adopted biometrical approach. One of the important approaches of diallele analysis is the graphical representation of two statistics (Vr, Wr) which are derived from diallele tables used in present study of Vr, Wr graphical analysis of curd weight and other important horticultural traits in cauliflower involving six parents crossed in half diallele approach.

Materials and Methods

The present investigations were carried out at the research farm of Department of Vegetable Crops, Dr. Y.S. Parmar University of Horticulture and Forestry, Nauni, Solan (H.P.) during 1999–2002. Six genotypes of cauliflower, namely Olympus, Winter Cauliflower (Armel), Improved Majestic, Early Snowball, Autumn King and Snowflower were crossed in half diallele fashion (excluding reciprocals). Fifteen F_1 hybrids and 6 parents were grown in a randomized block design with 3 replications at spacing of 60 X 45 cm. Each replication comprised of fifty plants of parents and F_1's. The observations were recorded on 5 plants for various horticultural traits. Vr, Wr graphical analysis was based on procedure given by Jinks and Hayman (1953) and Jinks (1954 and 1955). Recent developments in this technique have been described in detail by Mather and Jinks (1981).

Results and Discussion

The values of regression coefficients of Wr on Vr and test of significance of linearity of regression coefficients for various horticultural traits are given in Table 4.1. 4 out of 10 cases gave significant 1-b Wr/Vr depicting the failure of some of the basic assumptions of the complementary type. In 2 cases the standard errors were larger the regression coefficients making them almost equal to zero. The regression line passed below origin for number of leaves per plant, leaf breadth, curd depth and harvest index expressing over dominance for these characters. The regression line passed well above the point of origin for characters like plant frame, leaf length, gross curd weight and days for marketable maturity indicating partial dominance for these characters. For stalk length the regression line intersected the Wr-axis at the point of origin indicating complete dominance whereas for net curd weight the regression line did not intersect the Wr-axis showing no dominance.

Table 4.1: Values of b wr/vr and Test of Significance of 1-b for Various Horticultural Traits in Cauliflower (*Brassica oleracea* var. *botrytis* L.)

	Stalk Length	*No. of Leaves per Plant*	*Plant Frame*	*Leaf Length*	*Leaf Breadth*	*Curd Depth*	*Gross Curd Weight*	*Net Curd Weight*	*Days to Marketable Maturity*	*Harvest Index*
bwr/vr	0.89 ±0.43	0.84 ±0.25	– 0.28 ±0.36	– 0.16 ±0.15	0.22 ±0.17	0.49 ±0.21	– 0.32 ±0.18	– 0.41 ±0.29	– 0.21 ±0.21	0.54 ±0.32
1-bwr/vr	*	*	–	ns	ns	*	ns	ns	–	*

The distribution of parental points along the regression line for stalk length and number of leaves per plant depicted wide genetic diversity for the characters because of scattering of array points. For stalk length the parent Snowflower and Early Snowball had excess of dominant genes whereas all other except Improved Majestic had more or less equal proportion of dominant and recessive genes. For number of leaves per plant Autumn King and Olympus had excess of dominant genes. For harvest index and plant frame most of the parents were having recessive genes.

For leaf length and leaf breadth parent Olympus was having dominant genes as it was lying near the point of origin, while remaining parents were having more or less equal number of dominant genes. The parent Snowflower had more of recessive genes for curd depth. For gross curd weight and net curd weight the parent Snowflower was having mostly recessive genes as it was lying far away from the point of origin, whereas remaining parents had more equal proportions of dominant and recessive genes. For days taken to marketable maturity the parent Olympus and Early Snowball were with more of dominant genes while Snowflower was having recessive genes.

For stalk length the findings in general revealed that both additive and non-additive effects can be exploited for the improvement of the character, as suggested by Sharma (1989) and Thakur (1998). The deviation from over dominance in component analysis to partial dominance in graphical analysis for stalk length and over dominance in component analysis for net curd weight to no dominance in graphical analysis for net curd weight could be due to epistatic effects.

References

Jinks, J.L. and Hayman, B.L. 1953. The analysis of diallele crosses. *Maize Genetics Coop. Newsletter*, 27: 48–54.

Jinks, J.L. 1954. The analysis of continuous variation in a diallele crosses of *Nicotiana rustica. Genetics,* 39: 767–788.

Jinks, J.L. 1955. A survey of genetical basis of heterosis in a variety of diallele crosses. *Heredity*, 9: 223–238.

Mather, K. and Jinks, J.L. 1981. *Inheritance of Characters: The Study of Continuous Variation*, 3rd edition, Chapman and Hall, London, pp. 394.

Sharma, H.R. 1989. Inheritance of horticultural traits and quality characters in cauliflower (*Brassica oleracea* var. *botrytis* L.). *Ph.D. Thesis*. Dr. Y.S. Parmar University of Horticulture and Forestry, Solan (H.P.).

Thakur, B.S. 1998. Breeding for resistance against black rot, stalk rot and heterosis studies in cauliflower (*Brassica oleracea* var. *botrytis* L.). *Ph.D. Thesis*, Dr. Y.S. Parmar University of Horticulture and Forestry, Solan (H.P.).

Chapter 5

Phyllosphere Studies in Sewage Water Irrigated Fodder Grass *Brachiaria mutica*

S.T. Girisha* and S. Umesha**

**Department of Botany, University of Mysore, Manasagangothri, Mysore – 570 006*
***Department of Applied Botany, Seed Pathology and Biotechnology, University of Mysore, Manasagangothri, Mysore – 570 006*

ABSTRACT

The effect of urban sewage water on the phyllosphere microflora of Paragrass was studied. The experiments were conducted in both summer and rainy seasons, irrigated with sewage water and normal water with dryland, irrigated and water logged conditions. The pathogenic bacterial population *viz.*, *Salmonella* sp., *Vibrio* sp., *Shigella* and *E. coli* were enumerated in all the type of irrigation in both sewage water irrigated and normal (control) experimental plots. The bacterial population was tremendously high with the plants irrigated with sewage water and nil with the normal irrigated plants except *E. coli*. The bacterial population was recorded in all the types of irrigation. Among the pathogenic bacteria *E. coli* population recorded highest followed by *Shigella*, *Salmonella* and *Vibrio* in water logged conditions, during summer season and the same was low in rainy season. The variation in phyllosphere microflora upon urban sewage water irrigation in Paragrass *Brachiaria mutica* was discussed in the present article.

Keywords: *Phyllosphere, Salmonella, Vibrio cholerae, Shigella, E. coil, Sewage water, Fodder grass, Brachiaria mutica.*

**Author for Correspondence.

Introduction

Sewage water irrigation leads changes in biological characteristics of our environment that may harmfully affect the human life. This problem have arises recently because of increasing pace of population explosion and urbanization. The sewage water has the capability of altering the ecological balance of ecosystems. *Brachiaria mutica* is one of the major fodder grasses grown in urban regions by using sewage water for cattle in the southern districts of Karnataka state. The harvested grass is transported to cattle farms. It is a versatile species growing in dry land, irrigated and water logged soils and in sewage waters. In the present study, an attempt has been made to ascertain and enhance knowledge about the relationship between this forage and sewage water pollution. The study also focuses on *B. mutica* and the relationship between transfer of sewage water bacteria to leaf surface.

Sewage water has been used as a source of plant nutrient to grow shrubs, trees, vegetables and flowers (Hirsch, 1969; Ciaccio, 1971; Davis, Graig and Kirkham, 1988). Though sewage water contains raw sludge and the soils are rich in nutrients, they lead to out break of several disease since it also contains a large number of pathogenic forms which may grow in effluent irrigated food crops. The effluent of many industries either harbour or act as vectors of several pathogens (Powell, 1974) was reported. The infectious disease caused by virus, bacteria, protozoa in sewage water has been reported (Burge and Marsh, 1978).

The effect of sewage on the physico-chemical and biological characteristics of inland fresh water bodies in India (Goel, Gopal and Trivedy, 1980). Most of the literature available is on various aspects of sewage farming (Feigin, Bielorai, Snalhevet, Kipnis and Dag, 1979).

The chance of transferring pathogenic bacteria particularly salmonella to live stock feed by sewage irrigated alfalfa are relatively more (Bell, 1976). The raw sewage contains a large number of Salmonella (Waite and Finch, 1986). Various type of bacteria isolated in domestic sewage of the urban areas, raw sewage was found to contain *E. coli, Enterobacter* (Shehada and Sallal, 1996) etc., even on the sewage pipes *Salmonella* species were recorded containing urban sewage (Papapetropoulou and Moschopoulos, 1996).

According (Khatri and Tamberkan, 2003) to ice goal contaminated with pathogenic microorganisms like *E. coli, Salmonella, Citrobacter* etc., act as a source of human infection. In the same thing (Chordia and Tamberkan, 2003) reported kulfi in a source of enteropathogenic infection which contains pseudomonas, staphylococcus, enterobacter and *E. coli.* Opined (Shilpa and Musaddiq, 2003) that drinking water contaminates by *Salmonella, E. coli* etc., due to the water from slum, market, slaughter house etc., and reported drinking water contaminated by coliform due to polluted water (Chetna and Saxena, 2003). The review presented in this paper with an objective to help the reader to a better understanding of the transfer of sewage water bacteria to the leaf surface of fodder grass and risks involved in public health during the use of fodder grass to feed cattles.

Materials and Methods

Mysore city wastewater from Hotels, Hospitals, Residential areas, Factories, Slum, Market etc., become unportable sewage water. This sewage water has been used for irrigation purposes. Growing fodder grasses in sewage irrigated land is practised. *B. mutica,* a commercial crop is one of the important fodder grasses grown in out skirts of Mysore and Bangalore cities in addition to other districts of Karnataka. Total area used for this purpose is estimated to be 170 ha. in Mysore and 470 ha. in Bangalore. Water is directly irrigated once a week without any treatment. The waste water flows from Mysore city to research site. This have been used for both sampling and population studies as treated and departmental garden as control. The leaves of *B. mutica* were analysed for the presence of pathogenic bacteria the bacterial species isolated during present studies are *Salmonella, Vibrio cholerae, Shigella* and *E. coli.*

Preparation of Saline Solutions

0.8g of NaCl was dissolved in 100 ml of distilled water and the solution was poured to test tubes, cotton plugged and sterilized these test tubes were used to collect the leaf materials from the experimental sites. Leaf material was collected in this saline solution, in dry land, irrigated and waterlogged sewage irrigated conditions, from the sewage farm site and from the control site. From this saline samples bacteria present were isolated and enumerated by serial dilution method by using specific medium from specific bacteria. Following are the compositions for various media.

Selentie-Cystine broth for *Salmonella spp.* Culture

Tryptone–5.0 g, Lactose–4.0 g, $Na_2 HPO_4$–10.0 g, Na-Acid selentie–4.0 g, L-Cystine–0.01 g, Agar–15.0 g, Water–1 ltr. The pH of the medium was adjusted to 7.0.

Monsur's Medium for *Vibrio cholerae* Culture

Tryptone–1.0 g, Nacl–1.0 g, Na-tauro cholate–0.5 g, $Na_2 CO_3$–0.1 g, Gelatin–3.0 g, Agar–1.5 g, Water–100 ml. The pH of the medium was adjusted to 8.5.

Deoxycholate Citrate Medium for *Shigella spp.* Culture

Base medium was first prepared as follows

Meat extract–5.0 g, Agar–12.0 g, Water–1000 ml.

The above mixture was heated in conical flask and peptone 5.0 g and Lactose 10.0 g were added, plugged and steamed. If the medium is too acidic, it becomes opaque, hence pH 7.6 of the medium was maintained.

Solution A

Trisodium citrate–17 g, Sodium thiosulphate–17 g, Water–1000 ml. The above ingredients were dissolved by heating and while the solution is warm 2 g of ferric citrate flakes were added.

Solution B

Sodium desoxycholate–10 g, Water–1000 ml–To maintain pH (7.0) NaoH was added to the solution B.

To 500 ml base medium at pH 7.7 (hot) 5 ml each of solution A and B and 0.8 ml of one per cent neutral red were added and then poured to the plates.

MaCconkey's Agar for *E. coli*

Agar–20 g, Sodium tauro glycocholate–5 g, Peptone–20 g, Sodium Chloride–20 g, Water–1000 ml. The pH of the medium was adjustcd to 7.6.

The solution was heat dissolved and 20 g of lactose and 10 ml of 1 per cent neutral red were added.

Results and Discussion

The results of the experiment reveals that the sewage water contains various types of microorganisms including bacteria, viruses, fungi, protozoan and algae. This sewage is a potential source of pathogenic microorganisms which cause dysentery, cholera, typhoid, polio, hepatitis etc.

Microorganisms like *E. coli,* species of *Shigella, Vibrio, Salmonella* etc., which are present in the sewage water contaminate the fodder grass and later may be transmitted to animals and man. Leaf samples from the research sites were analysed and microorganisms were isolated and the following are the findings.

Salmonella spp.

A causative agent of typhoid, the populations of *Salmonella* species have been investigated the colonies of *Salmonella* species ranging from 6.8 colonies/leaf to 90.25 colonies/leaf, during summer season in sewage water treated leaf samples, in control conditions the colonies were completely absent (Table 5.1).

Table 5.1: Effect of Sewage Irrigation on Phyllosphere of *B. mutica*

Sl.No.		*Sewage Irrigated Plants*			*Control Plants*		
		Dryland Condition	*Irrigation Condition*	*Waterlogged Condition*	*Dryland Condition*	*Irrigation Condition*	*Waterlogged Condition*
1.	*Salmonella* colonies/leaf (in 10^6)						
	Summer seasons	6.8	24.00	90.25	–	–	–
	Rainy seasons	1.35	1.97	3.24	–	–	–
2.	*Vibrio cholerae* colonies/leaf (in 10^6)						
	Summer seasons	2.60	26.00	48.00	–	–	–
	Rainy seasons	1.8	17.30	18.60	–	–	–
3.	*Shigella* colonies/leaf (in 10^6)						
	Summer seasons	4.6	64.50	94.16	–	–	–
	Rainy seasons	2.16	43.00	79.50	–	–	–
4.	*E. coli* colonies/leaf (in 10^6)						
	Summer seasons	8.0	78.00	130.00	–	3.50	8.00
	Rainy seasons	5.30	65.30	103.00	–	2.30	5.30

Values are means ± S.E. of four replicates each and two consecutive seasons. Data were subjected to analysis of variance and compared for significance according to DMRT (P = 0.05).

Vibrio cholerae

A causative agent of cholera, the population of this species growing on the surface of the leaves ranged from 2.6 colonies/leaf to 48 colonies/leaf during summer season on the leaves of control plant colonies were not detected (Table 5.1).

Shigella spp.

The population of dysentery causing organisms *Shigella* species was also carried higher on the leaves of sewage irrigated grasses. The number of colonies of this populations fluctuated from a minimum of 4.6 colonies/leaf to 94.16 colonies/leaf in control completely absent (Table 5.1).

E. coli

A potential enteropathogenic organisms which inhibits intestine of warm blooded animals. A maximum per cent of *E. coli* colonies were recorded on the leaf surface of sewage irrigated sites except in control. *E. coli* population ranged from 8.0 colonies/leaf to 130 colonies/leaf (Table 5.1).

Generally all microbial populations was slightly higher during summer reasons compared to raining seasons.

The phyllosphere study showed the percentage of pathogens of *Salmonella, V. cholarae, Shigella* and *E. coli,* transmitted from sewage water to leaf surface was more during summer seasons than in rainy seasons, however, during rainy seasons as the leaves get splashed with rain water, the number of colonies decreased. From the foregoing discussion it may be inferred that application of sewage water significantly changes phyllosphere microflora. Thus on one hand sewage water use helps to preserving scare water resources but at the same time its untreated and unplanned use can lead to disastrous effects on human health during cutting transporting and feeding this fodder grass to the cattles, to avoid these problems when using such water there has to be careful management is needed. Thus for using the sewage for a agricultural purpose or for discharge it into soil at a place where *B. mutica* is cultivated it should be either to be treated or diluted to minimize its harmful effects at the time of growth of crop and handling the crop.

References

Bell, R.G. 1976. Persistence of fecal coliform indicator bacteria on alfalfa irrigated with municipal sewage lagoon effluent. *J Environ. Qual.,* 5: 39–42.

Burge, W.D. and Marsh, P.B. 1978. Infectious diseases hazards of land spreading sewage waste. *J. Env. Qual.,* 7: 1– 9.

Chetna Chhabra and Saxena, M.M. 2003 A study on coliform contamination of drinking water in desert city Bikaner. *Poll. Res.,* 22(3): 439–441.

Chordia, P.S. and Tamberkan, D.H. 2003. Kulfi a source of enteropathogenic infection: A case report. *Poll. Res.,* 22(4): 615–618.

Ciaccio, L.L. 1971. *Water and Water Pollution Handbook, Vol. 1.* Marcel Dekker, Inc., New York.

Davis, T.L., Graig, J.K. and Kirkham, M.B. 1988. Waste water irrigation of vegetable crops. *Biocycle,* 29: 60–63.

Feigin, A., Bielorai, H., Snalhevet, J., Kipnis, T. and Dag, J. 1979. The effectiveness of some crops in remaining minerals from soils irrigated with sewage effluent. *Prog. Water Tech.,* 11: 151–162.

Goel, P.K., Gopal, B. and Trivedy, R.K. 1980. Impact of sewage on freshwater ecosystem I. General features of water bodies and sewage. *J. Ecol. Env. Sci.,* 6: 83–96.

Hirsch, L. 1969. Irrigation with reclaimed waste water. *Water Waste Engg.* 6: 58–60.

Khatri, R.M. and Tamberkan, R.M. 2003. Studies on bacteriological analysis of ice goal sold in Amaravati City. *Poll. Res.,* 22(4): 613–614.

Papapetropoulou, M. and Moschopoulos, H. 1996. Detection of *Salmonella* spp. in estuarine waters by using both the conventional culture and a probe technique. *Water, Air and Soil Poll.,* 89(1–2): 159–165.

Powell, L. 1974. Farm waste disposal and its effect on human health. *J. Public Hlth., (London)*, 88: 141–152.

Shehada, M.N.A. and Sallal, A.K.J. 1996. Preliminary studies on the bacterial and parasitic microflora and microfauna of domestic sewage in northern Jordan Inter. *J. Env. Hlth. Res.*, 6(1): 31–38.

Shilpa, D. Joshi and Musaddiq, M. 2003. Studies on distribution of *E. coli* and *Salmonella* in drinking water. *Poll. Res.*, 22(1): 121–124.

Waite, W.M. and Finch, P.E. 1986. Pathogens in sewage sludge. (iii) Heat inactivation of *Salmonella* in raw sewage sludge. *Water Poll. Control*, 85(1–4): 482–487.

Chapter 6

Studies on Seed Conservation in Cucumber

*C. Vanniarajan**, *Sanjeev Saxena*** *and T. Nepolean**

**Centre for Plant Breeding and Genetics, Tamil Nadu Agricultural University, Coimbatore – 3*

***National Bureau of Plant Genetic Resources, PUSA, New Delhi – 110 012, India*

ABSTRACT

A conservation study on seeds of cucumber was conducted. The seeds with moisture contents of 15, 10, 5 and 3 per cent were placed at storage temperatures of –20°C, 4°C, ambient temperature and 35°C. Samples were taken periodically at all storage temperatures and percentage germination count, seedling vigour and electrical conductivity were recorded. The results revealed that the seeds of cucumber (Pusa Safed) with low moisture contents (5 and 3 per cent) could be placed at –20°C, 4°C and ambient temperature for safe-storage for 80 days. These combinations maintained germination percentage and seedling vigour.

Keywords: *Seed storage, Cucumber, Seed moisture, Storage temperature, Safe storage.*

Introduction

Cucumis sativus is a member of the cucurbitaceae which comprises 90 genera and 750 species. *Cucumis* comprises of about 30 different species, which are distributed over two geographically different areas. Cucumber is a cross-pollinated crop. It is also self-compatible and expresses no significant inbreeding depression. The fruit of cucumber are eaten as vegetable and as pickles *i.e.,* the immature stage. Due to their low energy content, they are mainly regarded as a refreshing condiment. Seed conservation is an important aspect to prevent loss of biodiversity. Cucumber a minor important crop is not given much importance in conservation. Therefore the present study was formulated to study the behaviour pattern of cucumber seeds stored at different temperatures and seed moisture content.

Materials and Methods

Cucumber (*Cucumis sativus* L.) variety "Pusa Safed" was obtained from the National Seed Corporation, New Delhi, India. The seed storage experiments were conducted in the Physiology and Biochemistry Lab of the National Bureau of Plant Genetic Resources, New Delhi, India from February–April, 2000.

Seed moisture content was determined following the procedure prescribed by the International Seed Testing Association (Anonymous, 1985 and 1993) and expressed on a wet weight basis. The seeds were divided into four lots. Seed lots with the 15 and 10 per cent moisture contents were obtained using the formula of:

$$W_f = \frac{W_i \times 100i_m}{100 - f_m}$$

where,

W_f is the final weight of the sample, W_i is the initial weight of the sample, i_m is the initial moisture content of the sample and f_m is the final moisture content of the sample.

The two other lots of seeds were placed in dessicators over silica gel. Periodically samples were taken and tested for seed moistures of 5 and 3 per cent. Once the seed moistures of 15, 10, 5 and 3 per cent were achieved, the seeds were hermetically sealed in aluminium foil pouches.

The sealed seed lots with known moisture contents were placed at four different storage temperatures of –20°C, 4°, ambient temperature and 35°C. Seed samples were retrieved periodically from the respective storage conditions and subjected to various tests.

For the germination test, three replications of 25 seeds each were placed in a petridish lined with moist filter paper and placed between two sheets of paper towel (46 cm × 29 cm) which were then rolled carefully, wrapped in a sheet of butter. The petridishes with seeds were incubated in a germination chamber at 25°C. Germinated seeds were counted daily taking the protrusion of the radicle as an indication of germination.

Seedling vigour was calculated as germination percentage multiplied by seedling length. For the electrical conductivity tests, 25 seeds were weighed and soaked in 25 ml of decanted in a clean beaker and electrical conductivity was measured using digital conductivity meter and expressed in d Sm^{-1}.

The data obtained from the observations on the 7th day of the germination tests were statistically analysed according to a Factorial Completely Randomised Design using MSTAT: C package.

Results and Discussion

The mean values of germination of cucumber seeds after different storage periods are shown in the Table 6.1. The germination percentage was reduced in the treatments of high moisture contents (15 per cent and 10 per cent) as the age of the seeds advanced at all storage temperatures studied with more pronounced effect at higher temperatures (35°C and ambient). The seeds with low moisture contents of 5 and 3 per cent and storage temperatures of –20°C, 14°C and ambient maintained germination percentage in contrast to seeds stored at 35°C, which showed a trend of decreasing germination percentage over the storage period of time. A decrease in germination percentage in seeds with low moisture and stored at high temperatures has also been reported for other crops (Normah and Chin, 1991 and Mwasha *et al.*, 1997).

The mean values of seedling vigour after different aging periods are shown in Table 6.2. The vigour of cucumber and watermelon seeds are maintained in the treatments of low moisture contents (5 per cent and 3 per cent) with storage temperatures of –20°C, 4°C and ambient.

Table 6.1: Effect of Seed Storage Under Various Conditions on Germination Percentage of Cucumber

Storage Condition		*Period (Days)*									
Seed Moisture (%)	*Temperature (°C)*	*0*	*6*	*15*	*22*	*30*	*37*	*46*	*55*	*62*	*80*
15	– 20	88	85	83	80	80	79	75	75	75	75
15	4	88	83	80	76	75	75	73	70	65	60
15	Ambient	88	70	70	68	65	60	55	50	40	35
15	35	88	77	45	36	28	15	10	5	0	0
10	– 20	88	85	83	82	82	80	80	80	80	79
10	4	88	84	83	82	80	80	80	75	75	74
10	Ambient	88	88	88	86	84	80	75	70	65	63
10	35	88	70	70	70	70	68	66	66	65	65
5	– 20	88	85	85	85	83	83	83	83	80	80
5	4	88	88	88	88	87	87	87	87	87	87
5	Ambient	88	88	87	85	85	85	85	83	83	82
5	35	88	85	83	81	75	70	68	68	65	65
3	– 20	88	85	85	84	84	84	84	83	83	82
3	4	88	86	86	85	85	85	85	84	84	83
3	Ambient	88	85	85	85	85	83	83	83	83	82
3	35	88	85	80	76	75	75	72	70	68	68

The mean values of electrical conductivity of unaged and aged cucumber seeds are presented in Table 6.3. Seeds with higher moisture contents (15 per cent and 10 per cent) stored at all the temperatures registered higher values of electrical conductivity as compared to the seeds with moisture content of 5 and 3 per cent and stored at 20°C, 4°C and ambient. Koostra and Harrington (1969) and Halder and Gupta (1982) have reported that ultra dessicated seeds showed higher electrical conductivity values. It has also been reported that if seeds are dried to lower than the critical level, the lipid auto-oxidation is faster and consequent rapid seed deterioration (Hampton *et al.*, 1992 and 1994). In the present investigation no such deterioration in the seeds dried to low moisture content (3 per cent) was observed.

Table 6.4 shows the analysis of variance of the observations taken in the study. The results showed that all treatments differed significantly from each other. The interaction effects of moisture with temperatures are also highly significant for all the characteristics studied.

Table 6.2: Effect of Seed Storage Under Various Conditions on Seedling Vigour Index of Cucumber

Storage Condition		*Period (Days)*									
Seed Moisture (%)	*Temperature (°C)*	*0*	*6*	*15*	*22*	*30*	*37*	*46*	*55*	*62*	*80*
15	– 20	1487.2	1234.2	1060.7	1013.6	836.8	752.0	666.7	635.7	528.0	488.2
15	4	1487.2	1198.2	995.2	892.4	654.0	479.2	440.9	333.2	301.6	261.0
15	Ambient	1487.2	975.8	878.5	744.6	627.9	514.8	348.7	231.5	167.6	137.5
15	35	1487.2	1291.6	513.0	241.2	134.1	31.0	13.5	2.7	–	–
10	– 20	1487.2	1560.6	1173.6	1032.5	893.8	844.8	769.6	746.4	627.2	578.2
10	4	1487.2	1686.7	1537.1	1424.3	1000.0	832.0	779.2	654.0	626.2	540.2
10	Ambient	1487.2	1558.4	1275.1	1137.1	903.8	791.2	634.5	492.8	374.4	275.3
10	35	1487.2	847.3	779.1	709.1	435.8	253.6	190.0	165.6	96.2	42.9
5	– 20	1487.2	1760.3	1630.3	1645.6	1645.0	1834.3	1757.9	1816.8	1705.6	1660.8
5	4	1487.2	1692.2	1510.0	1942.1	1684.3	2001.8	1907.0	1801.7	1713.9	1763.4
5	Ambient	1487.2	1850.6	1740.0	1789.2	1778.2	1897.2	1612.4	1702.3	1689.0	1676.9
5	35	1487.2	1545.3	1190.2	1107.2	711.0	596.4	533.1	474.6	439.4	391.3
3	– 20	1487.2	2091.0	1932.2	2005.0	1918.5	1960.5	1933.6	1919.7	1918.9	1904.8
3	4	1487.2	1467.2	1511.0	1529.1	1681.3	1853.0	1842.8	1850.5	1782.4	1723.0
3	Ambient	1487.2	1541.9	1703.4	1774.8	1769.7	1768.7	1620.9	1770.3	1728.0	1727.7
3	35	1487.2	1530.0	1142.4	977.5	820.5	723.7	676.0	609.0	561.8	503.8

Table 6.3: Effect of Seed Storage Under Various Conditions on Electrical Conductivity (dSm^{-1}) of Cucumber Seeds

Storage Condition		*Period (Days)*									
Seed Moisture (%)	*Temperature (°C)*	*0*	*6*	*15*	*22*	*30*	*37*	*46*	*55*	*62*	*80*
15	– 20	18.2	32.4	50.4	62.6	70.8	77.2	79.0	80.0	150.6	204.4
15	4	18.2	66.4	62.8	78.3	82.8	90.0	105.8	200.2	212.4	219.5
15	Ambient	18.2	34.1	42.4	56.5	62.3	93.6	152.6	202.4	225.6	230.0
15	35	18.2	52.2	82.3	193.8	198.8	200.0	212.8	248.5	250.8	287.6
10	– 20	18.2	40.0	52.3	64.5	72.3	79.6	100.4	150.6	172.8	225.2
10	4	18.2	26.5	51.4	725	82.6	87.5	100.8	123.7	150.8	173.7
10	Ambient	18.2	52.0	40.6	48.2	56.2	76.0	121.4	152.7	177.4	260.0
10	35	18.2	41.9	62.8	182.0	185.0	190.0	200.4	225.2.	235.8	249.4
5	– 20	18.2	29.4	42.8	62.8	65.6	67.7	132.6	150.8	150.7	151.7
5	4	18.2	13.1	43.4	53.3	58.8	69.7	71.6	72.2	122.4	155.3
5	Ambient	18.2	27.6	38.2	43.6	68.9	92.7	94.8	100.0	124.8	156.3
5	35	18.2	24.3	58.6	114.0	112.6	119.0	120.0	123.2	132.8	162.0
3	– 20	18.2	20.4	32.2	45.5	81.5	89.1	92.6	95.6	126.7	146.3
3	4	18.2	25.9	40.6	66.6	92.6	100.4	100.8	102.2	101.4	131.5
3	Ambient	18.2	2.2	52.3	74.6	81.4	97.7	102.4	110.5	1120.4	138.2
3	35	18.2	24.9	42.6	94.1	99.2	101.4	114.8	126.2	135.8	171.0

Table 6.4: Analysis of Variance for Different Characteristics of Cucumber Seeds Stored After Two and Half Months at Different Temperature and Moisture

Source	*df*	*Germination*	*Vigour*	*Electrical Conductivity*
Treatment	15	1467.29**	1601921**	7017.03**
Moisture	3	3510.20**	5148859.7**	25736.90**
Temperature	3	2130.60**	213914.46**	5010.18**
Temperature × Moisture	9	565.22**	241943.73**	1446.02**
Error	32	1.63	86.38	72.61
CV %		1.89	1.09	4.45
S.E. (m)		0.74	5.37	4.92

From the study, it can be summarised that the seeds of cucumber (Pusa Safed) with low moisture contents (5 and 3 per cent) can be placed at 20°C, 4°C and ambient temperature for safe storage. These combinations maintained uniform germination percentage and seedling vigour.

Acknowledgements

We wish to express our sincere thanks to Dr. P.L. Gautam, Deputy Director General (National Agriculture Technology Project) for supporting us to carry out this research in the Physiology and Biochemistry Lab of National Bureau of Plant Genetic Resources, New Delhi, India.

References

Anonymous. 1985. International rules for seed testing. *Seed Sci. and Tech.,* 13: 356–513.

Anonymous. 1993. Intemational rules for seed testing. *Seed Sci. and Tech.,* 21: 1–288.

Halder, S. and Gupta, K. 1982. On the mechanism of sunflower seed deterioration under low and high RH. *Seed Sci. and Tech.,* 10: 276–376.

Hampton, J.G., Johnstone, K.A. and Eua-Umpon, V. 1992. Bulk conductivity test variables for mungbean, soybean and french bean seed lots. *Seed Sci. and Tech.,* 20: 677–686.

Hampton, J.G., Lungwangwa, A.L. and Hill, K.A. 1994. The bulk conductivity test for lotus seed lots. *Seed Sci. and Tech.,* 22: 171–180.

Koostra, P.T. and Harrington, J.F. 1969. Biochemical effects of age on membrane lipids of *Cucumis sativus. Seed Test Assoc.,* 34: 329–340.

Mwasha, A.J., Ellis, R.H. and Flon, T.D. 1997. The effect of dessication on the subsequent survival of seeds of cashew. *Seed Sci. and Tech.,* 25: 115–122.

Normah, M.N. and Chin, H.F. 1991. Changes in germination, respiration rate and leachate conductivity during storage of Hevea seeds. *Pertanika,* 14: 1–6.

Chapter 7

Integrated Weed Management in Soybean (*Glycine max.*)

Pardeep Kumar and Sat Paul Mehra

Department of Agronomy and Agrometereology, Punjab Agricultural University, Ludhiana

ABSTRACT

Soybean Cv. PK 416 was sown in six and nine rows of 45 cm and 30 cm and plant spacing of 5 cm and 7.5 cm, respectively using recommended seed rate of 87.5 kg/ha, keeping seed rate per plot constant and applying recommended dose of fertilizers. Sixteen treatment combinations of row spacings (30 cm and 45 cm) and eight weed control treatments *i.e.*, pendimethalin (pre-em.) and trifluralin (pre-plant) each at 0.5 and 0.75 kg/ha, both pendimethalin and trifluralin at 0.5 kg/ha (f.b.) one hand weeding (7 WAS) and hand weeding twice (3 and 6 WAS) and weedy check (control) were tested in a randomized block design with four replications. Both herbicides at applied doses proved safe to the crop. Application of pendimethalin/trifluralin each at 0.5 kg/ha f.b. one handing weeding 7 WAS effectively controlled weeds and yielded statistically at par with each other. Both herbicides applied at higher dose (0.75 kg/ha) gave better control of weeds than their lower dose (0.5 kg/ha). Alone application of both herbicides at lower dose proved less effective in controlling weeds and yielded significantly lower than hand weeding twice. Narrow row spacing (30 cm) proved significantly superior to wider row spacing (45 cm) due to higher growth parameter and yield attributes. The interactions between row spacing and weed control treatments in respect of growth parameters, yield attributes and seed yield were non-significant.

Introduction

Soybean is a high value pulse and oilseed crop with multiple food, feed and industrial uses. Soybean grown during the rainy season faces tremendous competition with weeds. Due to congenial temperature and moisture conditions, soybean is infested by a wide variety of both grass and broadleaf weeds. Wider row spacing and slow growth of the crop during its early growth period further aggravate the problem of weeds. Weeds cause significant reduction in the yield of this crop. Losses in yield to the tune of 50 to 71 per cent due to weed competition are quite usual (Bhan *et al.*, 1972; Muniyappa *et al.*, 1986 and Singh and Singh, 1987). Integrated weed management of soybean is important because no single method of weed control proves effective on long term basis. Through present study, an attempt was made to test the efficacy of trifluralin vis-à-vis pendimethalin and also the possibilities of doing away with requirement of one hand weeding by increasing the dose of herbicide integrated with closer row spacing to exploit the smothering potential of the crop.

Material and Methods

The field experiment was conducted at the Research Farm of the Department of Agronomy, Punjab Agricultural University, Ludhiana during *kharif* 1997 to test the efficacy of trifluralin vis-à-vis pendimethalin for weed control in soybean and to study the effect of soybean planting pattern integrated with herbicides and mechanical methods on weeds and yield of the crop. The soil was sandy loam, normal in pH (8.1), EC (0.30 mm hos/cm) low in organic carbon (0.12 per cent) and available nitrogen (61.4 kg/ha), high in available phosphorus (22.8 kg/ha) and potassium (165.0 kg/ha). A soybean Cv. PK 416 was sown in six and nine rows of 45 cm and 30 cm spacing at a plant spacing of 5 cm and 7.5 cm respectively using recommended seed rate of 87.5 kg/ha keeping seed rate per plot constant. Crop was given recommended dose of N (30 kg/ha) and P_2O_5 (80 kg/ha) in the form of urea and SSP. Sixteen treatment combinations of two row spacings (30 cm and 45 cm) and eight weed control treatments *i.e.*, both pendimethalin (pre-emergence) and trifluralin pre-plant incorporation (PPI) at 0.5 and 0.75 kg/ha, both pendimethalin and trifluralin at 0.5 kg/ha followed by (f.b.) one hand weeding 7 WAS, hand weeding twice (3 and 6 WAS) and weedy check (control) were tested in a randomized block design with four replications.

Results and Discussion

The weed flora of the experimental field comprised mainly of *Digitaria sanguinalis, Eleusine aegyptiacum, Trianthema portulacastrum* and *Euphorbia hirta.*

Effect on Crop and Weeds

The data showed that the effect of various weed control treatments on seedling emergence was non-significant indicating that at 0.5 and 0.75 kg/ha both pendimethalin and trifluralin proved safe to the crop as none of the herbicides showed any toxicity on seedling emergence. Pre-emergence application of pendimethalin 0.5 kg/ha combined with one hand weeding 7 WAS and PPI of trifluralin 0.5 kg/ha f.b. one hand weeding 7 WAS effectively controlled weeds and yielded statistically at par with each other (14.45 and 13.97 q/ha, respectively) (Table 7.2). These herbicides reduced weed biomass and provided conditions in favour of crop growth. Pendimethalin 0.5 kg/ha coupled with one hand weeding 7 WAS gave complete control of *Trianthema portulacastrum, Euphorbia hirta* and nearly complete control of *Eleusine aegyptiacum* and *Digitaria sanguinalis.* Similar effects of herbicides on weed in soybean crop were reported by Singh and Mani, 1977 and Jain *et al.*, 1985. Both herbicides applied at higher dose (0.75 kg/ha) gave better control of weeds than their lower dose (0.5 kg/ha) (Table 7.1). Pendimethalin 0.75 kg/ha (12.89 q/ha) and trifluralin 0.75 kg/ha (12.13 q/ha) though statistically at par with each other, but proved significantly inferior to hand weeding twice 3 and 6 WAS (13.59 q/ha). Alone application of pendimethalin and trifluralin at 0.5 kg/ha proved less effective in controlling weeds (Table 7.1) and yielded significantly lower than hand weeding twice 3 and 6 WAS. The comparatively lower yield was due to partial weed kill by herbicide at low dose (0.5 kg/ha).

Table 7.1: Effect of Row Spacing and Weed Control Treatments on Periodic Dry Matter Accumulation by Weeds (g m^{-2})

Treatment	*30 DAS*			*60 DAS*			*120 DAS*		
	S_1	S_2	*Mean*	S_1	S_2	*Mean*	S_1	S_2	*Mean*
Pendimethalin 0.5 kg/ha (PE)	65.96	79.10	72.52	51.68	69.52	60.60	31.85	42.49	37.17
	(8.12)	(8.82)	(8.47)	(7.12)	(8.35)	(7.74)	(5.64)	(6.53)	(6.09)
Pendimethalin 0.5 kg/ha (PE) f.b. 1 hoeing 7 WAS	61.86	76.85	69.36	10.93	21.43	16.18	7.12	15.10	11.11
	(7.86)	(8.69)	(8.28)	(3.34)	(4.63)	(3.98)	(2.79)	(3.98)	(3.38)
Pendimethalin 0.75 kg/ha (PE)	52.26	60.05	56.16	31.32	47.0	39.11	18.89	24.0	21.45
	(7.23)	(7.76)	(7.50)	85.60)	(6.88)	(6.24)	(4.15)	(4.95)	(4.55)
Trifluralin 0.5 kg/ha (PPI)	91.97	101.12	96.55	66.86	75.23	71.03	41.26	47.24	44.25
	(9.64)	(10.10)	(9.87)	(8.20)	(8.71)	(8.45)	(6.46)	(6.92)	(6.69)
Trifluralin 0.5 kg/ha (PPI) f.b. 1 hoeing 7 WAS	89.01	98.95	93.98	16.28	29.73	23.01	7.92	18.34	13.13
	(9.49)	(9.99)	(9.74)	(3.39)	(4.95)	(4.17)	(2.94)	(4.35)	(3.64)
Trifluralin 0.75 kg/ha (PPI)	56.85	66.19	61.52	35.89	56.94	46.42	17.44	28.11	22.78
	(7.59)	(8.18)	(7.88)	(5.87)	(7.52)	(6.69)	(4.20)	(5.17)	(4.68)
Hand weeding twice (3 and 6 WAS)	38.48	49.10	43.49	21.68	28.90	25.29	11.98	17.35	14.67
	(6.26)	(7.06)	(6.66)	(4.29)	(5.41)	(4.85)	(3.39)	(4.15)	(3.77)
Unweeded Control	237.09	301.35	269.23	141.0	176.72	158.86	98.16	125.58	111.90
	(15.42)	(17.37)	(16.40)	(11.88)	(13.30)	(11.90)	(13.32)	(12.61)	(10.60)
Mean	86.69	104.10	–	47.0	63.18	–	29.19	39.77	–
	(8.95)	(9.75)		(6.21)	(7.47)		(4.94)	(5.91)	
CD (P = 0.5) for									
Row spacing		0.36			0.72			0.42	
Weed control treatment		0.72			1.44			0.82	
Interaction		NS			NS			NS	

S_1: 30 cm; S_2: 45 cm.

Table 7.2: Seedling Emergence, No. of Pods/Plant, 100 Seed Weight and Seed Yield as Influenced by Various Treatments

Treatment	*Seedling Emergence per Metre Row Length*			*No. of Pods/ Plant (Mean of S_1 and S_2)*	*100 Seed Wt (g) (Mean of S_1 and S_2)*	*Seed Yield (q/ha) (Mean of S_1 and S_2)*
	S_1	*S_2*	*Mean*			
Pendimethalin 0.5 kg/ha (PE)	19.2	26.0	22.6	54.7	12.21	12.21
Pendimethalin 0.5 kg/ha (PE) f.b. 1 hoeing 7 WAS	18.8	25.9	22.4	63.6	12.79	14.45
Pendimethalin 0.75 kg/ha (PE)	17.6	27.0	22.3	57.5	12.64	12.89
Trifluralin 0.5 kg/ha (PPI)	18.5	25.3	21.9	51.6	11.82	11.87
Trifluralin 0.5 kg/ha (PPI) f.b. 1 hoeing 7 WAS	17.8	26.5	22.1	60.7	12.48	13.97
Trifluralin 0.75 kg/ha (PPI)	18.9	24.5	21.7	55.9	12.40	12.13
Hand weeding twice (3 and 6 WAS)	17.5	25.0	21.2	57.5	12.26	13.59
Unweeded Control	18.5	25.1	21.8	38.3	11.36	4.99
Mean	18.4	25.7	–			
CD (P = 0.5) for						
Row spacing		1.1		1.59	0.20	0.75
Weed control treatment		NS		3.18	0.41	1.51
Interaction		NS		NS	NS	NS

PE: Pre-emergence; PPI: Pre Plant Incorporation; WAS: Weeks After Sowing; DAS: Days After Sowing.

Narrow row spacing (30 cm) proved significantly superior to wider row spacing (45 cm) due to higher growth parameter and yield attributes (No. of pods per plant and hundred seed weight) and seed yield (Table 7.1) and weed suppression (Table 7.2).

Pendimethalin or trifluralin 0.5 kg f.b. one hand weeding 7 WAS gave effective control of weeds like *Trianthema portulacastrum, Euphorbia hirta, Eleusine aegyptiacum* and *Digitaria sanguinalis* and yielded significantly higher than weedy check, respectively. The integration of both pendimethalin and trifluralin 0.5 kg/ha with one hand weeding 7 WAS proved superior to pendimethalin or trifluralin alone (0.5 kg/ha). Herbicides applied at higher dose (0.75 kg/ha) provided better weed control than their lower dose (0.5 kg/ha) (Table 7.1). The interactions between row spacing and weed control treatments in respect of growth parameters, yield attributes and seed yield were non-significant.

References

Bhan, V.M., Singh, M. and Maurya, R.A. 1972. Studies on chemical weed control in soybean (*Glycine max* L. Merill). *Weed Sci.*, 4: 16–22.

Jain, H.C., Dubey, S.K., Koshta, L.D. and Raghu, J.S. 1985. Integrated weed management in soybean (*Glycine max.* L. Merill). *Ann. Conf. Indian Soc. Weed Sci.*, 67 (Abstr).

Muniyappa, T.V., Ramchandra, P.T.V. and Krishnamurthy. 1986. Critical stages of weed competition and chemical weed control in soybean. *Indian J. Weed Sci.,* 18: 34–38.

Singh, B.P. and Maini, V.S. 1977. *Investigations on Chemical Weed Control in Soybean*. Div. Agron., IARI New Delhi.

Singh, G. and Singh, D. 1987. Weed control efficiency of pendimethalin and methabenzthiazuron in soybean (*Glycine max* L. Merill). *Indian J. Weed Sci.,* 19: 230–232.

Chapter 8

Effect of Growth Regulators on Yield and Yield Components in Rice

P. Subbaramamma* and P.S.S. Murthy**

** Research Associate (Plant Physiology) at Regional Agricultural Research Station, Lam, Guntur – 522 034*

*** Associate Director of Research, Regional Agricultural Research Station, Chintapalle, Visakhapatnam, Professor and University Head, Department of Plant Physiology, A N G R Agricultural University*

ABSTRACT

Effect of foliar spray of plant growth regulators *viz.*, Cytokinin (10 ppm) and Abscisic acid (0.1 ppm) along with control (water spray) at 10 days after flowering (DAF) was studied on yield, yield attributes, harvest index and mobilization index in six rice cultivars *viz.*, Rasi, IR-64, Jaya, Vijeta, Mahsuri and Chaitanya. An increase in grain number panicle^{-1}, per cent filled spikelets panicle^{-1}, 1000-grain weight straw yield, grain yield, harvest index (HI) and mobilization index (MI) was observed in Kinetin treatment over control. Among the six rice cultivars, the cultivar Mahsuri recorded the highest grain yield by virtue of more number of grains panicle^{-1}.

Introduction

In rice, grain filling depends mostly on current assimilation under optimal growing conditions and the demand for assimilate and leaf senescence during grain filling phase are inter-related (Yoshida, 1981). Ray and Choudhuri (1981) observed that the function of growth hormones in improving grain filling in rice is due to increase in the longevity of leaves there by prolonging the period of mobilization of metabolites from leaves. Foliar application of growth regulators such as IAA and Kinetin at 5 and 10 ppm concentrations, respectively at the time of anthesis or one week after anthesis significantly increased the 1000 seed weight, filled grain number panicle^{-1} and yield of rice cultivars *viz.*, Jaya

Palmar, IR-8 and Jhona-349 (Singh and Kaur, 1987; Biswas and Choudhuri, 1981), whereas ABA and silicone oil decreased the yield tiller^{-1} in rice (Debata and Murthy, 1981). Bruinama (1982) reported that certain plant growth regulators are effective in the regulation of photosynthetic efficiency and chemical manipulation of internal distribution of photosynthate to seed. This study is taken up to evaluate the influence of growth regulators on popular rice varieties grown in Andhra Pradesh.

Materials and Methods

A field experiment was conducted at Agricultural College Farm, during *kharif* 1996 under irrigated conditions in split-plot design, with four replications. The soil was vertisol in nature. Soil samples were collected from the experimental field from 0-30 cm depth at random and composite samples were analysed as per the methods described by Jackson (1973). Soil is of clay loam with a pH of 7.6, EC-0.4 dSm^{-1}, Organic Carbon-0.45 per cent, available nitrogen–275 kgha^{-1}, available phosphorus-37 kgha^{-1} and available potassium-470 kgha^{-1}. Transplanting was done with 30-days-old seedlings by adopting a spacing of 20 × 15 cm in between rows and plants, respectively. Recommended package of practices were followed. Plant protection measures were taken up as and when necessary. The treatments *viz.*, foliar application of cytokinin @ 10 ppm, foliar application of Abscisic acid @ 0.1 ppm and control (water spray) were imposed in six rice cultivars at 10 DAF. Aqueous solutions of kinetin (10 ppm) and ABA (0.1 ppm) were sprayed @ 5 ml hill^{-1} day^{-1} for four consecutive days commencing from 10th day after flowering on 20 plants treatment^{-1} in each of the six rice cultivars, using 0.5 per cent (v/v) TEEPOL (pH 7.0) as a surfactant. On each occasion control plants were sprayed with water. The data on yield and yield components were recorded by adopting standard procedures.

Results and Discussion

The effect of foliar application of growth regulators on yield and yield components was furnished in Tables 8.1–8.3. Grain number panicle^{-1} was more in Mahsuri followed by Jaya. Foliar application of Cytokinin (10 ppm) increased the grain number panicle^{-1} (5.98 per cent) over control. The percentage of filled spikelets panicle^{-1} and 1000-grain weight were high in Jaya (87.4 per cent, 23.4 g, respectively). The grain number panicle^{-1} and percentage filled spikelets panicle^{-1} were low in IR-64. Foliar application of kinetin significantly increased the percentage of filled spikelets panicle^{-1} (6.1 per cent) and 1000-grain weight (6.76 per cent), where as AHA decreased (– 5.1 per cent, – 3.38 per cent, respectively) over control. Foliar application of kinetin increased the percentage of filled spikelets panicle^{-1}, individual grain size and 1000-grain weight. It could be attributed to an increase in the longevity of leaves thereby prolonging the period of mobilization of metabolites from source (leaves) to sink (panicles) by kinetin treatment. Similar results were reported by Debata and Murthy (1981); Ray *et al.* (1983); Singh and Kaur (1986) and Kaur and Singh (1987) in rice. The straw yield was high in Mahsuri and low in IR-64. The variation in straw yield among the cultivars could be due to genotypic variation specially the duration of the cultivars. The grain yield was more in Mahsuri due to more number of grains panicle^{-1}. The straw and grain yields were increased by foliar application of kinetin (7.08 per cent, and 13.2 per cent, respectively) over control. Foliar application of Abscisic acid (0.1 ppm) decreased the straw yield and grain yield (– 5.24 per cent and – 9.2 per cent, respectively). The interaction of cultivars and treatments was significant for grain yield only. The increase in straw and grain yields by kinetin may be attributed to delay in senescence of leaves, increase in photosynthetic capacity of leaves, and better translocation of metabolites from source to sink at the time of grain filling resulting in increase in individual grain size, percentage of filled spikelets panicle^{-1}, 1000-grain weight and finally the grain yield. Foliar application of ABA enhanced the leaf senescence, thereby decreased

Table 8.1: Effect of Foliar Application of Growth Regulators on Grain Number Panicle^{-1}, Percentage of Filled Spikelets Panicle^{-1} and 1000-grain Weight (g) in Different Rice Cultivars

Cultivars	*Grain Number Panicle^{-1}*				*Percentage of Filled Spikelets Panicle^{-1}*				*1000-grain Weight (g)*			
	Treatments				*Treatments*				*Treatments*			
	Foliar Application of Cytokinin (10 ppm) at 10 DAF	*Foliar Application of Abscisic Acid (0.1 ppm) at 10 DAF*	*Control*	*Mean*	*Foliar Application of Cytokinin (10 ppm) at 10 DAF*	*Foliar Application of Abscisic Acid (0.1 ppm) at 10 DAF*	*Control*	*Mean*	*Foliar Application of Cytokinin (10 ppm) at 10 DAF*	*Foliar Application of Abscisic Acid (0.1 ppm) at 10 DAF*	*Control*	*Mean*
Rasi	138.6	127.4	134.0	133.3	90.1	76.1	82.6	82.9	21.3	18.5	19.5	19.8
IR-64	117.8	104.5	112.8	111.7	81.6	69.6	77.5	76.2	24.5	21.9	22.6	23.0
Jaya	181.7	160.6	167.4	169.9	92.5	82.8	86.9	87.4	24.4	22.5	23.4	23.4
Vijeta	127.8	116.5	119.8	121.4	88.1	78.1	82.7	83.0	23.1	21.1	21.7	22.0
Mahsuri	188.6	172.3	178.0	179.6	87.7	75.9	79.1	80.9	18.0	16.4	16.7	17.0
Chaitanya	148.9	134.6	140.3	141.3	90.6	80.8	85.1	85.5	21.5	19.6	20.5	20.5
Mean	150.6	136.0	142.1	–	88.4	77.2	82.3	–	22.1	20.0	20.7	–
		SE±	CD (0.05)			SE±	CD (0.05)			SE±	CD (0.05)	
Cultivars (C)		1.1	3.4			1.0	3.1			0.15	0.45	
Treatments (T)		1.0	2.7			0.96	2.7			0.12	0.35	
Interactions												
C × T		2.22	NS			2.19	NS			0.29	NS	
T × C		2.34	NS			2.36	NS			0.31	NS	

NS: Non-significant.

the photosynthetic capacity of leaves, and translocation of metabolites from source to sink when the grain dry weight is still increasing. The percentage of filled spikelets panicle^{-1}, individual grain size, 1000-grain weight and grain yield were decreased by ABA. The results are in agreement with the earlier findings of Ray *et al.* (1983); Singh and Kaur (1987) and Khan and Choudhuri (1992) and Debata and Murthy (1981) in rice. The cultivar Jaya recorded the highest harvest index (41.4) and mobilization index (0.70) followed by Rasi while it was low in Mahsuri (Table 8.3). The variation in harvest and mobilization index could be due to genotypic variation in translocation of metabolites from source to sink. Exogenous application of kinetin (10 ppm) significantly increased the harvest and mobilization index by 4.3 per cent and 6.7 per cent respectively, while ABA decreased the same over control. It could be due to an increase in the number of vascular bundles and size of the phloem area in the transverse section of the young peduncle. This facilitated better translocation of metabolites from source to sink and increase in individual grain size, percentage of filled spikelets panicle^{-1}, 1000-grain weight, grain yield and finally the harvest and mobilization index. ABA enhanced the leaf senescence, decreased the synthetic capacity of leaves and metabolite translocation from source to sink finally decreased the grain yield, harvest and mobilization index. Similar results were reported by Biswas and Choudhuri (1981); Ray *et al.* (1983); Singh and Kaur (1986) and Debata and Murthy (1981) in rice.

Table 8.2: Effect of Foliar Application of Growth Regulators on Straw Yield (kgha^{-1}) and Grain Yield (kgha^{-1}) in Different Rice Cultivars

Cultivars	*Straw Yield (kgha^{-1})*				*Grain Yield (kgha^{-1})*			
	Treatments				*Treatments*			
	Foliar Application of Cytokinin (10 ppm) at 10 DAF	*Foliar Application of Abscisic Acid (0.1 ppm) at 10 DAF*	*Control*	*Mean*	*Foliar Application of Cytokinin (10 ppm) at 10 DAF*	*Foliar Application of Abscisic Acid (0.1 ppm) at 10 DAF*	*Control*	*Mean*
Rasi	54858.3	4585.3	4750.0	4931.2	3742.5	2736.0	3049.3	3175.9
IR-64	6157.5	5207.5	5412.3	5592.4	3675.8	2607.8	2904.3	3062.6
Jaya	7082.5	6481.0	6746.0	6769.8	5185.8	4415.8	4766.8	4789.5
Vijeta	7350.0	6791.0	7136.5	7092.5	4447.5	3736.8	4086.8	4090.4
Mahsuri	9730.0	8859.3	9327.3	9305.5	5435.8	4555.8	5002.5	4988.0
Chaitanya	7251.8	6154.0	6813.3	6739.7	4693.5	3749.3	4205.8	4616.2
Mean	7171.7	6346.4	6697.6	–	4530.2	3633.6	4002.6	–
		SE±	CD (0.05)			SE±	CD (0.05)	
Cultivars (C)		55.7	167.8			55.0	165.7	
Treatments (T)		69.3	192.2			18.3	50.7	
Interactions								
C × T		149.4	NS			66.0	194.2	
T × C		169.8	NS			44.8	124.1	

NS: Non-significant.

Table 8.3: Effect of Foliar Application of Growth Regulators on Harvest Index and Mobilization Index in Different Rice Cultivars

Cultivars	*Harvest Index*				*Mobilization Index*			
	Treatments				*Treatments*			
	Foliar Application of Cytokinin (10 ppm) at 10 DAF	*Foliar Application of Abscisic Acid (0.1 ppm) at 10 DAF*	*Control*	*Mean*	*Foliar Application of Cytokinin (10 ppm) at 10 DAF*	*Foliar Application of Abscisic Acid (0.1 ppm) at 10 DAF*	*Control*	*Mean*
Rasi	40.7	37.4	39.1	39.1	0.68	0.60	0.64	0.64
IR-64	37.5	33.6	35.0	35.4	0.60	0.51	0.54	0.55
Jaya	42.3	40.5	41.4	41.4	0.73	0.68	0.70	0.70
Vijeta	37.7	35.2	36.4	36.4	0.61	0.55	0.57	0.58
Mahsuri	35.8	34.0	34.7	34.8	0.56	0.52	0.54	0.54
Chaitanya	39.3	37.9	37.4	38.2	0.65	0.61	0.62	0.63
Mean	38.9	36.4	37.3	–	0.64	0.58	0.60	–
		SE±	CD (0.05)			SE±	CD (0.05)	
Cultivars (C)		0.25	0.75			0.006	0.02	
Treatments (T)		0.31	0.86			0.007	0.02	
Interactions								
C × T		0.67	NS			0.01	NS	
T × C.		0.76	NS			0.02	NS	

NS: Non-significant.

References

Biswas, A.K. and Choudhuri, M.A. 1981. Influence of plant senescence as modified by hormones and nutrients on yield of rice. *Indian Agriculturist*, 25: 77–90.

Bruinama, J. 1982. Plant growth regulators in field crops. In: *Chemical Manipulation of Crop Growth and Development*, (Ed.) McLaren, J.S. Butterworths, London, p. 3–12.

Debata, A. and Murthy, K.S. 1983. Endogenous ethylene content in rice leaves during senescence. *Indian J. Plant Physiol.*, 26: 425–427.

Jackson, M.L. 1973. *Soil Chemical Analysis.* Prentice Hall of India (P) Ltd., New Delhi.

Kaur, J. and Singh, G. 1986. Hormonal regulation of grain filling in relation to peduncle anatomy in rice cultivars. *Indian J. Exp. Biol.* (In Press).

Khan, R.I. and Choudhuri, M.A. 1992. Role of endogenous hormones in the regulation of whole plant senescence in rice *Indian J. Exp. Biol.*, 30: 131–134.

Ray, S. and Choudhuri, M.A. 1981. Effect of plant growth regulators on grain filling and yield of rice. *Ann. Bot.*, 42: 755–758.

Ray, S., Mondal, W.A. and Choudhrui, M.A. 1983. Regulation of leaf senescence, grain filling and yield of rice of kinetin and abscisic acid. *Physiologia Plantarum,* 59: 343–346.

Singh, G. and Kaur, J. 1987. Effect of growth regulators on grain filling in rice cultivars. *Oryza,* 24: 185–187.

Yoshida, S. 1981. Physiological analysis of rice yield. In: *Fundamentals of Rice Crop Science*. The International Rice Research Institute, Los Banos, Philippines, p. 231–251.

Chapter 9

Climatic Influences on Water Use-Efficiencies in Irrigated Wheat in India

S. Venkataraman

Retired Director, India Meteorological Department and Former WMO/FAO Agromet Expert, 59/19, Nav Sahyadri Society, Pune – 411 052, India

ABSTRACT

Factors affecting and importance of regional variations in water use-efficiencies of irrigated crops are outlined. Relative Evapotranspiration, RET and Relative Water Use-Efficiency, RWUE are defined. Agroclimatically determined data on water-needs, life-duration, RET and RWUE of (a) normal and late sown wheat cultivars and (b) common variety of wheat sown on normal dates are presented for representative stations covering major regions of the wheat belt. The emerging indications are discussed.

Keywords: *Yield per day, Accumulated heat units, Wheat-climate zones, Reference station.*

Introduction

The term water use-efficiency is employed by irrigation engineers to indicate the fraction of water let out from the reservoirs that is used in crop production at the field level. Agronomists refer to the economic yield produced per unit amount of water used in crop production as Water Use-Efficiency, WUE, to which this communication refers. Thus, both yield and water requirements of a crop determine WUE. Yield of a given crop cultivar is affected by its field life-duration (Swaminathan, 1968; Venkataraman and Muchinda, 1988). Crop water needs are affected by (*a*) both life-duration and developmental-rhythm of the crop which in turn are influenced by climatic factors and (*b*) the

Evaporative Power of Air, EPA, which is the resultant of factors like solar radiation, dryness of air and wind speed. The Evapotranspiration loss from a cropped field, ET constitutes the essential, minimum crop water requirement. Thus in evaluating WUE only actual ET losses from a field must be considered.

There is no physiological basis for relating crop yields to their water requirement. When both a low and a high yielding variety of a crop of nearly same life-duration and growth rhythm are exposed to the same weather there will be no significant differences in the crop water needs. Thus water use-efficiency of a crop at a place will be governed solely by the yielding capacity of the varieties. However, across locations the WUE of a crop can be influenced by climatic factors through variations in (*i*) EPA, (*ii*) field-life duration and (*iii*) developmental-rhythm of a crop. Therefore, in assessment of variations in WUE of an irrigated crop consideration has to be given to both direct and indirect influences of climate. Frequently reference is being made to lower unit area yields and lower water use-efficiencies of irrigated, clear season crops in India, compared to countries in the temperate zone. Similar inter-regional differences in unit area yields of wheat in India are explainable on the basis of the climatically influenced factor of yield per day of field crop-life (Swaminathan, 1968). It was decided to examine if a similar climatic influence, if any, would prevail on water use-efficiency of irrigated wheat in India.

Materials and Methods

Standard mesh-covered class a pan is used in India as a network instrument to measure Pan Evaporation EP. Factors affecting ET also influence EP. The ratio of ET to EP is called Relative Evapotranspiration, RET. Use of the ratio RET takes account of the variations in crop water needs due to the climatic factors. For irrigated wheat, Venkataraman (2004) has presented agroclimatically determined values of ET, field-life duration and RET ratios for climatically appropriate wheat cultivars sown on optimal dates at representative stations covering important, homogenous wheat-climate zones (Venkataraman and Rahi, 1983). It was seen that ET was minimal and field life-duration maximal at Ludhiana. Thus, it was decided to take the WUE of optimal wheat crop at Ludhiana as 1.0. Since yield is proportional to field-life duration a reduction in life duration at other stations would mean a reduction in yield proportional to the reduction in field-life of crop at Ludhiana. If reduction in yield was matched by a reduction in ET there will be no change in WUE. On the other hand variations in ET will influence WUE. In view of the foregoing, using Ludhiana data for an optimal crop as a standard, Relative Water Use-Efficiencies, RWUEs at other stations were calculated by dividing ET per day of crop life at Ludhiana by the ET per day of crop life at the given station. Data of Venkataraman and Rahi (1983) show that a common wheat variety optimally sown and with requirement of Accumulated Heat Units, AHUs of 1950 above a base degree of 4.5 degrees centigrade can be grown at all stations. Delay in sowings of wheat due to temperature vagaries leading to use of shorter duration cultivars is not infrequent. In view of the above it was decided to examine the RWUE values at representative stations in the wheat-climate zones (Venkataraman, 2004) for wheat cultivars (*i*) sown on climatically optimal date and with climatically appropriate variety, (*ii*) sown 15 days later than optimal at a station with variety of correspondingly shorter life-duration and (*iii*) sown on optimal date with common wheat cultivar of 1950 AHUs accumulated above a base mean temperature of 4.5 degrees centigrade. For the above scenarios the method, as outlined above, was used to calculate RWUE.

Results and Discussion

Station-wise data on (*i*) optimal sowing dates and (*ii*) ET, RET and RWUE at 18 stations covering relevant wheat-climate regimes for 3 types of cultivars, as enumerated above, are presented in Table 9.1. One may postulate, *a priori*, that higher temperatures reducing crop-life duration will also increase EP and if the increase in EP is compensated by a reduction in crop-life duration total ET may remain nearly remain the same at all locations. Even then reductions in yield due to reduced crop-life duration will decrease water use-efficiency. However, Table 9.1 shows that ET can vary significantly even with same duration of crop-life and RET values and that despite differences in crop-life duration and RET nearly same value of ET may obtain. For example though crop-life at Parbhani is only 75 per cent of that at Ludhiana and Delhi total ET at Parbhani is nearly 50 per cent higher than at Ludhiana but only 20 per cent higher than at Delhi. The above features are due to absence of any relationship between heat units and EP.

Table 9.1: Relative Water Use-efficiencies of Normal and Late Sown Wheat cultivars

Station	*Zone*	*Optimal Date of Sowing*	*Duration Days*			*Total ET mm*			*RET*			*RWUE*		
			O	*L*	*C*	*O*	*L*	*C*	*O*	*L*	*C*	*O*	*L*	*C*
Ludhiana	A	15 Nov	150	134	159	399	361	470	0.96	0.97	1.00	1.00	0.99	0.90
Delhi	B	15 Nov	150	134	153	504	466	528	0.96	0.97	0.96	0.79	0.77	0.77
Hissar	B	15 Nov	150	134	154	486	446	531	0.95	0.96	0.94	0.82	0.80	0.77
Jobner	C	15 Nov	142	128	158	521	493	596	0.95	0.99	0.96	0.72	0.69	0.71
Agra	D	15 Nov	142	126	145	472	431	496	0.95	0.96	0.97	0.80	0.78	0.78
Kota	E	1 Nov	135	121	140	501	445	534	0.95	0.95	0.95	0.72	0.72	0.70
Jhansi	E	1 Nov	135	121	147	387	328	463	0.95	0.91	0.95	0.93	0.98	0.85
Adhartal	E	1 Nov	135	121	135	428	412	428	0.97	1.05	0.97	0.84	0.78	0.84
Patna	F	1 Nov	135	121	139	406	366	432	1.00	1.00	1.03	0.89	0.88	0.86
Sabour	F	1 Nov	135	121	141	398	358	437	1.00	1.00	1.0	0.90	0.90	0.86
Chinsurah	G	15 Nov	120	105	124	465	412	490	1.02	1.00	1.02	0.69	0.68	0.67
Cuttack	H	1 Nov	135	121	108	495	428	374	1.04	1.00	1.05	0.73	0.75	0.77
Labhandi	H	1 Nov	135	121	121	535	485	450	1.03	1.03	1.03	0.67	0.66	0.71
Anand	IW	15 Nov	112	96	117	434	465	371	1.00	1.00	1.00	0.69	0.69	0.67
Vasad	IW	15 Nov	112	96	112	508	456	508	1.03	1.03	1.03	0.59	0.50	0.59
Pune	IS	15 Nov	112	96	117	516	452	553	0.99	0.99	0.99	0.58	0.57	0.56
Akola	IS	15 Nov	112	96	118	568	502	623	0.97	0.98	0.97	0.53	0.51	0.50
Parbhani	IS	15 Nov	112	96	112	595	515	595	1.0	0.99	1.0	0.50	0.50	0.50

O: Sown on optimum date with appropriate variety; L: Sown 15 days later than optimal date with variety of 15 days shorter duration; C: Sown on optimal date with variety with accumulated heat units above 4.5 degrees centigrade equal to 1950.

It is seen from Table 9.1 that for an optimally sown crop both life duration and RET are the same at:

1. Ludhiana, Hissar and Delhi but RWUE at Hissar and Delhi is only 80 per cent of that of Ludhiana,
2. Jobner and Agra but RWUE is 70 per cent and 80 per cent respectively at Jobner and Agra,
3. Kota, Jhansi and Adhartal but RWUE is 70 per cent at Kota, 85 per cent at Adhartal and 90 per cent at Jhansi, and
4. Patna and Sabour in zone F and Cut tack and Labhandi in zone H, but RWUE is 90 per cent at Patna and Sabour and 70 per cent at Cuttack and Labhandi.

Though life duration at Chinsurah is only 90 per cent as at zone H, values of RET and RWUE are the same as in zone H. At the five southern stations life duration and RET are the same but RWUE shows a progressive decrease from North to South.

From Table 9.1 it is seen that (*i*) at a given location for all the three types of cultivars both RET and RWUE are nearly the same. Thus while changes in crop agronomy at a place will have little influence on water use-efficiency they can affect total crop water needs through temporal changes in level of EPA. Again, while RET shows a range of 0.99 plus or minus 0.04. RWUE ranges from 0.50 to 1.00. The practical implication is that for raising a bushel of wheat the water needed compared to Punjab will be (*a*) 10 per cent higher in Bihar (*b*) 25 per cent higher in Delhi, Haryana and West U.P. (*c*) 40 per cent higher in East Rajasthan, West Bengal, Orissa and North Gujarat (*d*) 70 per cent higher in western regions of peninsular wheat belt and (*e*) 100 per cent higher in Central Maharashtra. Madhya Pradesh shows wide variations in relative crop water need, the increase being 10 per cent in North, 20 per cent in central areas and 50 per cent in the eastern region.

Pointers for Further Work

Additional information relating to phasic durations of wheat for three types wheat culture can enable (*i*) climatic planning of irrigation scheduling and (*ii*) issue of operational irrigation advisories based on evaporimetry. More corroborative studies of above nature for more stations will help in re-examination of cropping patterns in the wheat tract and in more rational allocation of irrigation water to meet the demands of existing acreage of wheat under cultivation in different areas. However, replacement of wheat by another grain crop in regions of low WUE will need very careful consideration because the water needs of a replacement grain crop is likely to be higher than that for wheat since (*i*) maturity durations of grain crops are far less variable than their vegetative phase and (*ii*) most of the hybrid varieties of maize, sorghum and millets show no physiological senescence during maturity (Venkataraman, 1995) such that the peak period of water consumption continues up to physiological maturity.

References

Swaminathan, M.S. 1968. Genetic manipulation of productivity per day. Special Lecture, *ICAR Symposium on Cropping Patterns in India.*

Venkataraman, S. 1995. Agrometeorological determination of the optimal distribution of total water requirements of crops. *Internat. J. Ecol. & Env. Sci.*, 21: 251–261.

Venkataraman, S. 2004. Agroclimatic determination of water needs of crops: A case study for wheat in India. *Indian J. Env. & Ecoplan.*

Venkataraman, S. and Muchinda, M.R. 1988. Influence of temperature and moisture on maturation periods of rainfed maize. *Productive Farming*, p. 20–25.

Venkataraman, S. and Rahi, A.K. 1983. Influence of temperature climatology on productivity of wheat crop in India. *Mausam*, 34: 81–84.

Chapter 10

Genetic Divergence in Mungbean (*Vigna radiata* L. Wilczek)

Ch. Mallikarjuna Rao and Y. Koteswara Rao

Regional Agricultural Research Station, Lam Farm, Guntur – 522 034

ABSTRACT

Sixty genotypes of mungbean were evaluated for 13 different characters and mean values were worked for variability, heritability, correlation, genetic advance and genetic diversity by Mahalanobis D^2 statistic. The character total dry matter, plant height, number of pods per plant and yield per plant exhibited high variability and heritability coupled with genetic advance indicating the influence of additive gene action. Correlation studies indicated that the total dry matter, number of pods per plant, number of clusters per plant, number of branches per plant and days to 50 per cent flowering are positive and significantly associated with seed yield. Total dry matter and number of pods per plant had also direct positive effect on seed yield while, the plant height had negative effect on seed yield. The results of multivariate analysis (D^2) indicated the presence of considerable genetic divergence among the 60 genotypes studied. The genotypes were grouped into 8 clusters. Days to maturity, 100 seed weight, number of pods per plant and total dry matter contributed maximum towards diversity. Crosses can be effective between the genotypes of cluster V and VII followed by cluster I and VII where the maximum intercluster distance was exhibited for getting desirable segregants.

Keywords: *Mahalanobis D^2 statistic, Genetic diversity, Heritability, Correlation.*

Introduction

Mungbean is one of widely grown short duration grain legume in India fitting well in all cropping systems. It is grown as *Kharif* (both as sole and intercrop), *rabi* and summer crop in Andhra Pradesh.

The genetic variability present in the base population for desired characters plays a vital role in development of new varieties. As the variability present in the cultivated mungbean lines is less, the identification of divergent parents is need of hour. Hence the study on variability, heritability and genetic divergence in mungbean germplasm lines is attempted.

Materials and Methods

The experimental material consists of 60 exotic and indeginous Mungbean germplasm lines grown in Randomized block design with two replications during *rabi* 2002 at Regional Agricultural Research Station, Lam, Guntur. Each genotype was sown in 4 m length plot of 2 rows with inter and intra row spacing of 30 × 10 cm respectively. Observations were recorded on five randomly selected plants for each genotype in two replications for days to 50 per cent flowering, days to maturity, plant height (cm), number of branches per plant, number of clusters per plant, number of pods per plant, pod length (g), number of seeds per pod, total dry matter per plant (g), 100 seed weight (g), harvest index, seed yield (g) and protein content. The variability, heritability and expected genetic advance were estimated as cited by Johnson *et al.* (1955). Character association and path coefficient analysis are estimated as per the procedure of Dewey and Lu (1959). The data were subjected to multivariate analysis as suggested by Mahalanobis and genotypes were grouped into different clusters following Tochers method as described by Rao (1952).

Results and Discussion

The range, coefficient of variation, heritability in broad sense and genetic advance for yield and its components are presented in Table 10.1.

Table 10.1: Estimates of Variability Parameters for Yield and Yield Component Characters in Mungbean

Character	*Range*		*Mean*	*GCV*	*PCV*	*h^2 (b)*	*GA*
	Max	*Min*					
Days to 50% flowering	42.50	25.50	33.17	9.04	9.19	96.7	6.07
Days to maturity	83.50	56.50	65.25	7.37	7.44	98.1	9.86
Plant height	62.15	22.16	41.17	16.00	16.27	96.6	13.34
Number of branches per plant	5.00	0.40	2.31	41.52	43.28	93.4	1.93
Number of clusters per plant	12.20	5.00	7.58	17.44	25.13	48.2	1.89
Number of pods per plant	64.75	15.64	28.78	34.66	36.69	89.2	19.42
Pod length	9.70	5.60	7.05	10.66	11.19	90.7	1.48
Number of seeds per pod	12.40	5.00	10.28	12.04	12.43	93.9	2.47
100 seed weight	5.30	2.95	4.08	13.16	13.42	96.2	1.09
Seed yield per plant	24.17	6.11	12.18	36.65	39.22	87.3	8.54
Total dry matter per plant	73.32	13.74	35.39	37.18	38.12	95.1	26.45
Harvest index	40.72	25.69	34.07	4.80	9.15	27.5	1.77
Protein content	25.75	21.31	22.80	1.24	5.19	5.70	0.14

PCV: Phenotypic coefficient of variation; h^2(b): Heritability in broad sense; GCV: Genotypic coefficient of variation; GA: Genetic advance.

The genotype Lam M2 was found very late in flowering (42.50) and maturity (83.50) with tall plant habit (62.15) from the genotypes studied, while the genotype M 1450 recorded maximum number of pods per plant (64.75), seed yield per plant (24.17) and total dry matter per plant (73.32). However, early flowering (25.50) and early maturity (56.50) were recorded in UPM-98-1. The genotypic coefficient of variation and phenotypic coefficient of variation are higher for number of branches per plant, seed yield per plant and number of pods per plant indicating considerable variation in these traits. The effect of environment is found to be low for all other traits except for number of clusters per plant.

The heritability is high for all characters studied except number of clusters per plant, harvest index and protein content. The environment interaction with genotype makes us to select traits with high heritability coupled with high genetic advance. The traits like plant height, total dry matter per plant and number of pods per plant possess high heritability with high genetic advance. These results are in accordance with findings of Yadav *et al.* (2001). Therefore, direct selection of these characters would be effective (Panse, 1957).

The association between seed yield and all other characters except plant height, 100 seed weight and protein content was positive and highly significant (Table 10.3). The highest association was with total dry matter per plant (0.933**) followed by number of pods per plant (0.907**). These results are in accordance with the findings of Sandhu *et al.* (1988). However, seed yield per plant recorded negative correlation with plant height which was also reported by Ahuja and Chowdhury (1991).

The direct and indirect effects of yield and its components are presented in Table 10.4. The traits like total dry matter per plant, number of pods per plant and 100 seed weight exhibited high direct effects on seed yield. The positive indirect effect of total dry matter per plant on seed yield is through number of pods per plant indicating that these traits are to be takes care of while formulating breeding strategies of Mungbean. The residual effect is very low (0.0607) indicating that the characters selected for the study are appropriate.

The multivariate analysis of 60 genotypes grouped them into 8 clusters (Table 10.2). The cluster I and IV has 14 genotypes each, while cluster VII has 1 genotype (Lam M2) and cluster VIII has 3 genotypes (M 501, M 638, M 1450). The pattern of distribution of genotypes from different eco-graphical regions into 8 clusters was at random indicating that geographical diversity and genetic diversity are not related. The maximum intercluster distance (1588.34) was found between cluster V and cluster VII (Table 10.5). The cluster V includes genotypes which flower and mature early. The cluster VII comprises one genotype (Lam M2) which is late in flowering and maturity. It is noted that cluster VIII posses the genotypes with more number of pods, high seed yield and with high total dry matter per plant. The crosses can be effective between the genotypes of cluster V and VII followed by cluster I and VII for obtaining desirable and useful segregants.

Table 10.2: Distribution of 60 Genotypes of Mungbean into Different Clusters

Cluster	*No. of Genotypes*	*Genotypes*
I	15	AKM 92401, Sipai, HUM-1, ML-803, PDM 84-139, K-851, M 1319B, LGG 460, Ganga-8, PS-16, Pusa baisaki, LGG 446, Pusa 102, ML-955
II	3	BDYR-3, TM-99-15, UPM-79-3-4
III	8	AKM 9243, Pusa 105, Pusa bold, M 1643, Pusa 9072, TM-96-2, M 1218, Pusa Visal
IV	14	LGG 407, LGG 450, LGG 444, WGG-37, VGG-14, LGG-521, MGG 347 Neelalu, M 1475, ML-613, TARM-1, ML-267, ML-33, PDM-54
V	9	Basanti, BDYR-1, BDYR-4, NM-1, VC6173A, NM-92, VC 3960-88, UPM-98-1, VC 6370
VI	7	WGG-2, LGG-410, CO-4, LGG 456, M 59, LGG 494, LGG 489
VII	1	Lam M2
VIII	3	M 501, M 638, M 1450

Table 10.3: Phenotypic Correlation Between Yield and Yield Components of Mungbean

Character	*Days to 50% Flowering*	*Days to Maturity*	*Plant Height*	*Branches per Plant*	*Clusters per Plant*	*Pods per Plant*	*Pod Length*	*Number of Seeds per Pod*	*100-Seed Weight*	*Yield per Plant*	*Total Dry Matter*	*Harvest Index*	*Protein Content*
Days to 50% flowering	1.000	0.860**	0.165	0.316**	0.237**	0.306**	−0.008	0.115	−0.357	0.236**	0.283**	−0.061	0.007
Days to maturity		1.000	0.228*	0.270**	0.207*	0.257**	0.004	0.167	−0.302	0.215*	0.278**	−0.088	0.022
Plant height			1.000	−0.045	0.082	−0.085	0.178	0.020	0.036	−0.069	−0.042	−0.049	0.068
Branches per plant				1.000	0.471**	0.513**	0.147	0.033	−0.269	0.417**	0.417**	0.045	−0.074
Clusters per plant					1.000	0.724**	−0.032	0.112	−0.133	0.683**	0.549**	0.498**	0.072
Pods per plant						1.000	0.126	0.120	−0.163	0.907**	0.841**	0.365**	0.047
Pod length							1.000	0.224*	0.272**	0.248**	0.282**	−0.031	−0.132
Number of seeds per pod								1.000	0.145	0.402**	0.451**	−0.016	−0.066
100-seed weight									1.000	0.167	0.109	0.199*	−0.106
Yield per plant										1.000	0.933**	0.390**	0.000
Total dry matter											1.000	0.058	−0.036
Harvest index												1.000	0.126
Protein content													1.000

*: Significant at 5 per cent level; **: Significant at 1 per cent level.

Table 10.4: Direct (Diagonal) and Indirect Effects of Yield and its Components in Mungbean

Character	*Days to 50% Flowering*	*Days to Maturity*	*Plant Height*	*Branches per Plant*	*Clusters per Plant*	*Pods per Plant*	*Pod Length*	*Number of Seeds per Pod*	*100-Seed Weight*	*Total Dry Matter*	*Harvest Index*	*Protein Content*
Days to 50% flowering	0.0188	0.0166	0.0033	0.0063	0.0053	0.0060	−0.0002	0.0021	−0.0069	0.0055	−0.0024	−0.0001
Days to maturity	−0.0060	−0.0068	−0.0017	−0.0019	−0.0018	0.0019	−0.0001	−0.0012	0.0022	−0.0020	0.0009	−0.0003
Plant height	−0.0038	−0.0054	−0.0217	0.0009	−0.0009	0.0028	−0.0043	−0.0004	−0.0010	0.0011	0.0043	0.0013
Branches per plant	0.0024	0.0020	−0.0003	0.0070	0.0043	0.0039	0.0011	0.003	−0.0021	0.0031	0.0003	−0.0003
Clusters per plant	0.0125	0.0118	0.0019	0.0273	0.0447	0.0317	0.0001	0.0059	−0.0090	0.0274	0.0109	−0.0022
Pods per plant	0.1275	0.1115	−0.0514	0.2238	0.2849	0.4014	0.0656	0.0484	−0.0759	0.3498	0.0993	−0.0101
Pod length	−0.0001	0.0001	0.0015	0.0011	0.0000	0.0012	0.0073	0.0017	0.0022	0.0023	0.0000	−0.0010
Number of seeds per pod	0.0122	0.0195	0.0022	0.0043	0.0146	0.0133	0.0263	0.1101	0.0170	0.0501	−0.0064	−0.0136
100-seed weight	−0.0544	−0.0471	0.0071	0.0435	−0.0299	−0.0280	0.0443	0.0228	0.1481	0.0142	0.0437	−0.0158
Total dry matter	0.1488	0.1457	−0.0263	0.2257	0.3109	0.4419	0.1604	0.2308	0.0487	0.5070	0.0282	−0.0303
Harvest index	−0.0149	−0.0164	−0.0232	0.0042	0.0289	0.0293	0.0004	−0.0068	0.0349	0.0066	0.1183	−0.0028
Protein content	0.0000	−0.0002	0.0003	0.0002	0.0002	0.0001	0.0006	0.0005	0.0005	0.0003	0.0001	−0.0043
r_{iy}	0.2340	0.2314	−0.1083	0.4554	0.6611	0.9016	0.3014	0.4142	0.1585	0.9655	0.2972	−0.0794

Residual effect: 0.0607.

Table 10.5: Intra and Inter Cluster (D^2) Values in Mungbean

	1 Cluster	*2 Cluster*	*3 Cluster*	*4 Cluster*	*5 Cluster*	*6 Cluster*	*7 Cluster*	*8 Cluster*
1 Cluster	66.08	135.93	223.59	193.04	176.46	373.03	1122.92	823.12
	(8.12)	(1.65)	(14.9)	(13.89)	(13.28)	(19.31)	(33.5)	(28.69)
2 Cluster		36.31	301.96	232.83	282.03	320.33	907.15	980.94
		(6.02)	(17.3)	(15.25)	(16.79)	(17.89)	(30.11)	(31.32)
3 Cluster			172.52	274.29	388.72	288.93	957.46	446.05
			(13.13)	(16.56)	(19.72)	(16.99)	(30.94)	(21.12)
4 Cluster				184.19	435.76	283.45	816.07	723.07
				(13.57)	(20.88)	(16.83)	(28.56)	(26.89)
5 Cluster					103.53	678.28	1588.34	1066.02
					(10.17)	(26.04)	(39.85)	(32.65)
6 Cluster						174.39	417.30	492.84
						(13.20)	(20.42)	(22.20)
7 Cluster							0	920.51
							(0)	(30.34)
8 Cluster								212.58
								(14.58)

Square root values of D^2 are given in parenthesis.

References

Ahuja, S.L. and Chowdhury, R.K. 1991. Correlation and path coefficient analysis in greengram. *Agril. Sci. Dig.*, 11: 223–226.

Dewey, D.R. and Lu, K.H. 1959. A correlation and path coefficient analysis of components of crested wheat grass seed production. *Agronomy J.*, 51(9): 515–518.

Johnson, H.W., Robinson, H.F. and Comstock, R.E. 1955. Estimates of genetic and environmental variability in soybean. *Agronomy J.*, 47: 314–318.

Mahalanobis, P.C. 1936. On the generalized distance in statistic. *Proceedings of National Institute of Science* (Indian), 2: 49–55.

Rao, C.R. 1952. *Advanced Statistical Methods in Biometrical Research.* John Wiley and Sons, New York, p. 390–395.

Yadav, G.C., Singh, P.K., Singh, B.B. and Rituverma. 2001. Genetic variability and path coefficients in urdbean. *Indian J. Pulses Res.*, 14: 143–144.

Panse, V.G. 1957. Genetics of quantitative characters in relation to plant breeding. *Indian J. Genet. & Plant Breed.*, 17: 318–328.

Sandhu, T.S., Singh, K. and Singh, B. 1988. Mungbean germplasm resourses: Evaluation and Utilization in AVRDC, Mungbean. *Proceedings 2nd International Symposium Shanhua, Taiwan*, p. 29–34.

Chapter 11

Effect of Different Growing Media on Cut Flower Production of Gerbera (*Gerbera jamesonii*) Under Polyhouse Conditions

Lalita Bhangare, A.S. Jadhav, Madhuri Shirole, T.K. Tiwari and Subodhini Chavan
Mahatma Phule Krishi Vidyapeeth, Rahuri – 413 722, District Ahmednagar, Maharashtra, India

ABSTRACT

The experiment was conducted at Hi-Tech Floriculture and Vegetable Project, College of Agriculture, Pune (Maharashtra) during the year 2002–2003 on effect of different growing media on cut flower production of gerbera. The days required for flower bud initiation were significantly more under soil + FYM (1 : 1) and sand + cocopeat (1 : 1) growing media and significantly lower under soil + cocopeat + FYM (1 : 1 : 1) and pure perlite growing media. The mean stalk length, flower head diameter and stalk thickness of gerbera flower was significantly more under growing media of soil + cocopeat + FYM (1 : 1 : 1) and soil + perlite + FYM (1 : 1 : 1) than pure perlite media of cocopeat, perlite, sand and combination of soil + FYM (1 : 1) and soil + sawdust (1 : 1). Significantly the highest flower yield (25.67 per plant) was observed under pure media of perlite. It was followed by that observed under soil + cocopeat + FYM (1 : 1 : 1) and cocopeat + perlite (1 : 1).

The net monetory returns were comparatively more (Rs. 435.29 m^{-2}) under soil + cocopeat + FYM (1 : 1) followed by pure cocopeat (Rs. 363.54 m^{-2}) growing media indicating suitability of these media for cultivation of gerbera under protected cultivation.

Keywords: *Gerbera, Polyhouse, Perlite, Growing media, Yield.*

Introduction

In India, at present only 10,000 ha area is under modern cut flowers. Among the floriculture products exported, cut flowers are the single most important items in terms of value and growth in recent years. In Maharashtra, there are 188 projects with an area of 33.5 acres under gerbera with average daily production of 1,19,800 flowers (Mahalle, 2002).

The use of soilless culture has substantially increased during the last decades as it contributes to the intensification of horticultural production and provides high crop yield even in areas with adverse growing culture solutions. Gerbera grown in perlite media had higher number of flowers per plant with longest stem and bigger flower diameter than in the soil base substrate (Mascarini, 1998). Use of cocopeat as growing media for roses and gerbera ensures uniform growth and superior quality due to excellent air-water balance and ideal medium for roots to breath and grow (Jadhav *et al.*, 2002). Various workers have mixed number of available substrates in different proportions to provide good media mixtures. Hence proposed experiment was undertaken to study the effect of different growing media on cut flower production of gerbera under polyhouse conditions.

Materials and Methods

The present investigation was carried out at Hi-Tech Floriculture and Vegetable Project, College of Agriculture, Pune (India) during the year 2002–2003 in polyhouse (erected for research purpose). The experiment was laid out in Completely Randomised Design, (CRD) with ten treatments and three replications. The treatment details are furnished in Table 11.1. Tissue culture seedlings of 5 weeks' age of 'Sangria' variety were planted on 04–02–2002 in different growing media as per treatments. Growing media used were cocopeat, perlite, sand soil, FYM and sawdust and their combinations in different proportion. First flower bud was removed to develop plant further and next flower was harvested. In gerbera, the plants produced marketable flowers, 3 to 4 months after planting. The flowers were harvested when outer first ring of disc florets was in full bloom and ray florets were fully opened or perpendicular to petals.

Table 11.1: Growth, Yield and Net Monetory Returns as Influenced by Different Growing Media in Gerbera

Treatment	*Number of Suckers per Plant*	*Days to Flower Bud Initiation*	*Number of Flowers per Plant in Six Months*	*Number of Flowers per m^2 in Six Months*	*Net Monetory Returns (Rs. m^{-2})*
T1–Cocopeat (pure)	4.50	68.67	19.93	132.73	363.54
T2–Perlite (pure)	4.58	59.75	25.67	159.73	234.19
T3–Sand (pure)	4.50	61.25	16.65	115.57	332.71
T4–Soil + FYM (1 : 1)	4.42	79.42	15.53	102.59	347.12
T5–Soil + Sawdust (1 : 1)	4.50	66.25	17.53	110.69	288.67
T6–Soil + Cocopeat (1 : 1)	5.67	72.42	15.98	109.07	301.31
T7–Sand + Cocopeat (1 : 1)	4.42	77.92	18.67	120.04	335.62
T8–Cocopeat + Perlite (1 : 1)	4.33	73.09	20.60	142.98	288.94
T9–Soil + Cocopeat + FYM (1 : 1 : 1)	3.08	58.75	21.53	151.98	435.29
T10–Soil + Perlite + FYM (1 : 1 : 1)	4.33	73.25	18.98	125.69	245.47
CD at 5%	NS	6.23	1.53	2.64	–

Grading was done on the quality of bloom, length and strength of stem. Flowers with stalk length more than 50 cm and diameter 10 cm were graded as grade–I. Subsequent grades were the stalk length 30 to 50 cm and flower diameter of 8 to 10 cm as grade–II and flowers having stalk length less than 30 cm as grade–III. Vase life of fresh gerbera flower was recorded at room temperature until petals dropped and stem bending was observed. All other operations differing from treatment allocations were uniformly carried out.

Results and Discussion

Effect on Growth Characters

Number of suckers per plant were not significantly influenced by the different growing media under study, as these are genetically controlled to some extent (Table 11.1). The days required for flower bud initiation were significantly more under growing media of soil + FYM (1 : 1) and sand + cocopeat (1 : 1) whereas significantly less number of days for flower bud initiation were required in growing media of soil + cocopeat + FYM (1 : 1 : 1) and pure perlite (Table 11.1). In former growing media, improper aeration, nutrient and moisture supply might have delayed flower bud initiation. Whereas, proper aeration, adequate nutrient and moisture supply under latter growing media might have resulted in early flower bud initiation. These findings were similar to those reported by Pivot (1985).

Effect on Yield

Number of flowers per plant/six months as well as number of flowers/m^2/six months were significantly the highest (25.67 and 159.73 respectively) in pure media of perlite (Table 11.1). This might have been ascribed to improved aeration, drainage, optimum moisture retention and nutrient availability under this media compared to other substrates. Similar findings were also reported by Maluopa and Papodopoulos (1997) and Grillas *et al.* (2002). Further it was noticed that soil + cocopeat + FYM (1 : 1 : 1) and cocopeat + perlite (1 : 1) growing media were on par but significantly produced more flowers than other growing media.

Effect on Quality

Stalk length, flower head diameter and flower stalk thickness were significantly more in growing media of soil + cocopeat + FYM (1 : 1 : 1) than pure growing media of cocopeat, perlite, sand, soil + sawdust (1 : 1) and soil + FYM (1 : 1) (Table 11.2). The increase in stalk length, flower head diameter and flower stalk thickness might be due to ideal soil environment for crop growth with adequate supply of moisture and nutrients resulting in improved fertilizer use efficiency. Whereas, in pure media, due to more drainage, most of the macro and micronutrients might have been lost in leachate.

In case of soil FYM (1 : 1) and soil + sawdust (1 : 1) media, there was no proper aeration and wide C : N ratio because of compactness. This may be resulted in shorter stalk length, decreased flower head diameter and thinner flower stalk. The results are in conformity with the findings of Boztok (1995), and Georget (2000).

The percentage of grade–I flowers was significantly the highest (94.42) when gerbera was raised on soil + cocopeat + FYM (1 : 1 : 1) growing media (Table 11.2). It was further noticed that growing or gerbera under soil + perlite + FYM (1 : 1 : 1), cocopeat + perlite (1 : 1) and pure perlite media also recorded significantly more percentage of grade–I flowers than other growing media indicating that these media are also good for growing gerbera. The percentage of grade–II flowers was significantly highest under soil + FYM (1 : 1) (20.03 per cent) owing to improper aeration due to compactness of media, and grade–III flowers were significantly more (9.57 per cent) under soil + cocopeat (1 : 1) media.

Table 11.2: Quality of Gerbera Flowers as Influenced by Different Growing Media

Treatment	Flower Stalk Length (cm)	Flower Head Diameter (cm)	Flower Stalk Thickness (cm)	Grading (%)			Vase Life (Days)
				Grade I	Grade II	Grade III	
T1–Cocopeat (pure)	57.72	9.89	0.71	85.31	14.32	0.36	7.67
T2–Perlite (pure)	60.04	10.83	0.75	92.70	6.83	0.45	8.33
T3–Sand (pure)	58.71	10.58	0.73	80.40	17.47	2.11	7.67
T4–Soil + FYM (1 : 1)	54.03	9.69	0.65	76.69	20.03	2.48	6.67
T5–Soil + Sawdust (1 : 1)	57.17	10.04	0.67	82.25	17.43	0.38	6.33
T6–Soil+Cocopeat (1 : 1)	61.43	11.00	0.78	79.81	8.68	9.57	6.67
T7–Sand + Cocopeat (1 : 1)	63.15	10.70	0.80	90.98	8.40	0.98	7.33
T8–Cocopeat + Perlite (1 : 1)	64.48	11.38	0.80	92.96	6.48	0.64	7.33
T9–Soil + Cocopeat + FYM (1 : 1 : 1)	69.14	11.78	0.85	94.42	5.32	0.25	7.67
T10–Soil + Perlite + FYM (1 : 1 : 1)	66.40	11.62	0.83	93.36	6.13	0.49	7.33
CD at 5%	3.87	0.47	0.029	1.01	0.30	0.05	0.98

The vase life of gerbera flower was significantly more (8.33 days) under pure perlite medium than other media except, pure media of cocopeat and sand and combination of soil + cocopeat + FYM (1 : 1 : 1) (Table 11.2). The more vase life under perlite media may be ascribed to the strong thick stem which was less likely to fall under vase due to turgor pressure maintained and it also showed rapid recovery of turgor when placed in water after harvest. This confirms the findings of De Jong (1978). It was further noticed that gerbera grown in pure media of cocopeat and sand and combination of soil + cocopeat + FYM (1 : 1 : 1) also have a significantly more vase life than gerbera grown under soil + FYM (1 : 1), soil + sawdust (1 : 1) and soil + cocopeat (1 : 1) growing media owing to comparatively less transpiration losses and consequently increased vase life. Whereas, thick and long stalk under soil + cocopeat + FYM (1 : 1 : 1) media might have increased the vase life of flower. Similar findings were reported by Yoon *et al.* (1996).

Net Monetary Returns

The net monetary returns realized were comparatively more under soil + cocopeat + FYM (1 : 1 : 1) (Rs. 435.29 per m^2) followed by pure cocopeat (Rs. 363.54 per m^2) (Table 11.1) growing media indicating suitability of these media for cultivation of gerbera under protective environment.

It can therefore, be suggested to use soil + cocopeat + FYM (1 : 1 : 1) or cocopeat (pure) as a media for growing of gerbera under protective cultivation.

References

Boztok, S. 1995. Effect of different substrates on quality and production of Gerbera. *Ege–Universitesi–Ziraaat–Fakultesi–Dergi*, 32(3): 57–64.

De Jong, J. 1978. Dry storage and subsequent recovery of cut gerbera flowers as an aid in selection for longevity. *Scientia Hort.*, 9: 389–397.

Georget, M. 2000. Research for new substrates (for use) in nurseries and cut flowers. PHM-Revue. *Horticole*, 417: 20–24.

Grillas, S., Lucas, M. and Bardopouplou. 2002. Perlite based soilless culture systems: Current commercial applications and prospects. *Acta Hort.*, 548: 1–2.

Jadhav, A.S., Patil, M.T. and Patil, P.V. 2002. *Protected cultivation*, p. 51–58.

Mahalle, J.P. 2002. Floriculture development in Maharashtra. *Floric. Today*, 6(7): 25–30.

Maloupa, E. and Papadopoulos, A. 1997. Seasonal crop coefficient of gerbera soilless culture. International seminar on soilless culture. Technology protected crops. *Acta Hort.*, 480: 498–503.

Mascarini, I. 1998. Gerbera cultivation in growing media. *Horticultural International*, 6(19): 86–88.

Pivot, D. 1985. Cultivation of gerbera on softwood bark of *Picea alba* and *Abies alba*: Physical characteristics of these substrates. *Acta Hort.*, 172: 211–222.

Yoon, H., Sang, C. and Choi, B.J. 1996. The relationship between morphological characteristics and scape deformation of cut Gerberas. *J. Korean Soc. Hort. Sci.*, 37(4): 593–597.

Chapter 12

Correlation and Path Analysis for Yield and Other Economic Traits in White x Colour Linted Crosses of American Cotton (*G. hirsutum* L.)

B. Subbareddy* and N. Nadarajan**

**Research Associate, NATP RCPS, 8 Regional Agricultural Research Station, Lam, Guntur – 522 034*
***Professor, Department of Genetics and Plant Breeding, AC&RI Madurai*

ABSTRACT

For character association and path analysis for yield and economic traits, in 8 × 8 diallel mating design involving colour linted and high seed oil content genotypes of cotton were under taken. Seed cotton yield per plant exhibited significant positive association with number of bolls per plant, plant height, number of sympodia per plant, boll weight and lint index. Number of bolls per plant, plant height and boll weight exhibited positive high direct effects on seed cotton yield whereas, lint index had moderate positive direct effect on seed cotton yield. Hence number of bolls per plant is the major determinant for improving seed cotton yield.

Keywords: *Gossypium hirsutum, Correlation, Path coefficient analysis, Yield attributes.*

Introduction

Cotton, a crop of prosperity having a profound influence on men and matter, is an industrial commodity of world wide importance. Cotton cultivation in India encompasses economic, political and social affairs of the world. Besides being a money spinner, it is also an employment generator. Higher yield is the main breeding objective in all the crops and yield is regarded as a complex character or super character which is influenced by many component traits.

Selection for seed cotton yield will be effective only when it lies along with the yield components rather than relying on yield alone (Grafius, 1960). Hence, the knowledge about relationship between yield and its contributing characters is needed for an efficient breeding programme. Correlation provides information on the nature and extent of relationship among characters. Therefore, this technique aids in the isolation of superior yielding genotypes from genetically variable population. The direction and degree of association of each component trait with seed cotton yield is ascertained by correlation coefficients between these traits. However, path coefficient analysis (Wright, 1921) provides an effective means of finding direct and indirect causes of association.

Materials and Methods

The experimental material consisted of eight genotypes of which included two colour linted genotypes *viz.*, Arkansas green, Algerian brown, and six white linted genotypes *viz.*, SVPR 2, MCU 11, MCU 5, MCU 12, MCU 7, Sahana. Among these MCU 7, Sahana had high seed oil content. These eight parents were crossed in half diallel fashion (excluding reciprocals) during summer 2001–2002 at AC and RI Madurai. Twenty eight resultant crosses (F_1) were evaluated along with their parents in randomized block design with two replications. Recommended package of practices and plant protection measures were followed to raise good crop. The biometrical observations were recorded on five randomly selected competitive plants for the characters *viz.*, days to 50 per cent of flowering, plant height (cm), number of sympodia per plant, number of bolls per plant, boll weight (g), seed cotton yield, ginning outturn, lint index, seed index and oil content. Genetic correlation coefficients were estimated as per the procedures of Johnson and Path coefficient analysis was made following the procedure suggested by Dewey and Lu (1959).

Results and Discussion

The analysis of variance exhibited that high significant differences was observed among the genotypes for all the ten traits studied. Genotypic correlation coefficients between yield and nine other component traits and also inter correlation among the component traits are furnished in Table 12.1. In the present investigation, the association between seed cotton yield and its component traits *viz.*, plant height, number of sympodia per plant, number of bolls per plant, boll weight and lint index were positive and significant. The positive association between seed cotton yield and plant height was already reported by Sumathi and Nadarajan (1995) and Sambamurthy (1999). Positive association of number of sympodia per plant, number of bolls per plant and boll weight with seed cotton yield was earlier confirmed by many researchers (Shanti and Selvaraj, 1993; Basha, 1997 and Mandoli *et al.*, 1998). Positive contribution of lint index to seed cotton yield was in accordance with the results obtained by Shanti and Selvaraj (1993) and Sambamurthy (1999). This indicated that high yielding genotypes possessed more boll number per plant, number of sympodia per plant and plant height were rather tall in habit.

Knowledge on inter correlation between yield attributes revealed the strength and direction of association with each other. This could facilitate effective selection for simultaneous improvement of one or more yield contributing characters. The intercorrelation among the yield contributing traits *viz.*, plant height, number of sympodia per plant and number of bolls per plant were positive and significant. This was in accordance with the results obtained by Shanti and Selvaraj (1993) and Gnana Arul

Samuel Rajan (1997). Thus, selection of anyone of these traits might be useful to improve the seed cotton yield substantially. Besides this, the interrelationship of lint index with seed index was significantly positive. Hence simultaneous selection for lint index or seed index will improve the seed cotton yield substantially. These results are in conformity with the results obtained by Tyagi (1994), Sumathi and Nadarajan (1995) and Basha (1997).

Table 12.1: Genotypic Correlation Coefficients for Yield and Yield Components

Character	*Plant Height*	*Number of Sym-podia per Plant*	*Number of Bolls per Plant*	*Boll Weight*	*Ginning Outturn*	*Lint Index*	*Seed Index*	*Oil Content*	*Seed Cotton Yield per Plant*
Days to 50% flowering	– 0.37*	– 0.017	– 0.045	– 0.057	– 0.396*	– 0.452*	0.017	0.081	0.022
Plant height		0.508*	0.448*	0.090	– 0.091	0.326*	0.074	– 0.090	0.460*
Number of sympodia per plant			0.369*	0.085	– 0.061	0.262	0.026	– 0.074	0.433*
Number of bolls per plant				– 0.362*	0.148	0.229	0.024	0.790	0.670*
Boll weight					0.063	0.134	0.140	– 0.196	0.409*
Ginning outturn						0.223	– 0.081	0.330*	– 0.185
Lint index							0.721*	0.246	0.314*
Seed index								0.074	0.160
Oil content									0.042

* Significant at 5 per cent level.

The genotypic path coefficient analysis of different yield contributing characters traits are furnished in Table 12.2. In the present investigation number of bolls per plant had high direct effect on seed cotton yield. Similar results were reported by Sumathi and Nadarajan (1995) and Rao *et al.* (2001). The direct effect of boll weight on seed cotton yield was high which was in accordance with the results obtained by Basha (1997) and Sangeetha (1998). Lint index showed positively moderate direct effect on seed cotton yield, which was in accordance with the results obtained by Sumathi and Nadarajan (1995). None of the quality traits including lint colour showed direct effect on seed cotton yield.

The indirect effect of plant height through number of bolls per plant was positive and high. Similar findings were quoted by Sumathi and Nadarajan (1995). Number of sympodia per plant showed positive indirect effect through number of bolls per plant. Similar findings were also reported by Sangeetha (1998). Indirect effect of lint index through number of bolls per plant was positive and moderate. Similar findings were quoted by Rao and Mary (1996). Hence, number of bolls per plant is the important character on which selection pressure is to be applied for improving the seed cotton yield.

Correlation studies inferred that intensive selection for plant height, number of sympodia per plant and number of bolls per plant will enhance the seed cotton yield substantially. Path analysis indicated that number of bolls per plant has a prime role in determining seed cotton yield per plant involving colour linted genotypes of cotton which might be considered as a selection criteria including *per se* in a selection of a selection of a breeding programme. Some of the parents identified having superior performance for the important component characters for increasing yield were Algerian brown, SVPR 2, MCU 7 and Sahana.

Table 12.2: Genotypic Path Coefficients for Yield and Yield Components

Character	*Days to 50% Flowering*	*Plant Height*	*Number of Sympodia per Plant*	*Number of Bolls per Plant*	*Boll Weight*	*Ginning Outturn*	*Lint Index*	*Seed Index*	*Oil Content*	*Correlation with Seed Cotton Yield per Plant*
Days to 50% flowering	0.216	0.003	0.000	−0.041	−0.040	−0.030	−0.034	−0.002	−0.050	0.022
Plant height	−0.008	−0.086	−0.004	0.435	0.062	−0.007	0.083	−0.010	−0.005	0.460*
Number of sympodia per plant	−0.004	−0.044	−0.007	0.388	−0.059	−0.005	0.097	−0.003	0.070	0.433*
Number of bolls per plant	−0.010	−0.039	−0.003	0.917	−0.250	0.051	0.018	−0.003	−0.011	0.670*
Boll weight	−0.012	−0.020	−0.001	−0.332	0.691	0.055	0.034	−0.018	0.012	0.409*
Ginning outturn	−0.095	0.051	−0.006	0.136	−0.043	−0.077	0.052	−0.011	−0.090	−0.185
Lint colour	−0.098	−0.028	−0.002	0.210	0.093	0.017	0.232	−0.095	−0.015	0.314*
Seed index	0.004	−0.006	0.000	0.074	0.047	−0.006	0.183	−0.132	−0.004	0.160
Oil content	0.018	0.008	0.001	0.132	−0.135	0.025	0.062	−0.010	−0.059	0.042

Residual effect: 0.04.

Diagonal values are direct effects.

References

Dewey, D.R. and H.K. Lu. 1959. A correlation and path coefficient analysis of components of crested wheat grass seed production. *Agron. J.*, 51: 515–516.

Basha, M. 1997. Genetic analysis of yield. and yield components of intraspecific hybrids of cotton (*Gossypium hirsutum* L.) *Ph.D. (Ag.) Thesis.* ANGR Agric. Univ., Rajendranagar, Hyderabad.

Grafius, J.E. 1960. Does over dominance exist for yield in corn. *Agron. J.*, 52: 361.

Gnana Arul Samuel Rajan. 1997. Studies on heterosis and combining ability in cotton (*Gossypium hirsutum* L.). *M.Sc. (Ag.) Thesis,* Tamil Nadu Agric. Univ., Coimbatore.

Mandoli, K.C., G.K. Koutu, U.S. Mishra, R. Julka, S. Holkar and S.K. Pandey. 1998. Character association in coloured cotton. *J. Indian Soc. Cotton Improv.*, 23(2): 172–175.

Rao, K.V.K. and T.N. Mary. 1996. Variability, correlation and path analysis of yield and fibre traits in upland cotton. *J. Res. Andhra Pradesh Agric. Univ.*, 24(3–4): 66–70.

Sambamurthy, J.S.V. 1999. Character association and component analysis in upland cotton. *Madras Agric. J.*, 86(1–3): 39–42.

Sangeetha, K. 1998. Character association of quality traits on yield and yield components in cotton (*Gossypium hirsutum* L.). *M.Sc. (Ag.) Thesis* ANGR Agric. Univ. Rajendranagar, Hyderabad.

Shanti, N. and U. Selvaraj. 1993. Association of yield related characters in cotton (*G. hirsutum* L.). *J. Indian Soc. Cotton Improv.*, 18(2): 152–154.

Sumathi, P. and N. Nadarajan. 1995. Character association and component analysis in upland cotton. *Madras Agric. J.*, 82(4): 225–228.

Tyagi, A.P. 1994. Correlation coefficients and selection indices in upland cotton (*Gossypium hirsutum* L.). *Indian J. Agric. Res.*, 28(3): 189–196.

Wright. 1921. Correlation and causation. *J. Agric. Res.*, 20: 557–587.

Chapter 13

Allelopathic Effect of *Chenopodium murale* Towards *Lens culinaris*

K. Lavanya, Daizy R. Batish**#, H.P. Singh** and R.K. Kohli***

**Department of Botany, Punjab University, Chandigarh – 160 014, India*

***Center for Environment and Vocational Studies, Punjab University, Chandigarh, India*

ABSTRACT

Laboratory studies were conducted to determine the phytotoxicity of *Chenopodium murale* towards *Lens culinaris*, a common winter season crop of India. In a laboratory bioassay, effect of different concentrations of leachates from stem, leaves, roots and inflorescence was studied on early growth of *L. culinaris.* A significant reduction both in seedling length and dry weight of *L. culinaris* was observed when treated with *C. murale* leachates. However, the level of phytotoxicity was dependent on concentration as well as on the part used. The order of decreasing phytotoxicity was inflorescence >stem>leaves >roots in terms of seedling length and in respect of biomass accumulation the trend was stem>roots>leaf>inflorescence. The study indicates that leachates of *C. murale* contain phytotoxic chemicals that upon release affect the growth and establishment of *L. culinaris.* In *C. murale* leachates significant amount of phenolics was found in each part under use. The amount increased with increasing concentration. Based on these observations it could be concluded that different parts of *C. murale* exert phytotoxic influence towards *L. culinaris* mediated through water-soluble phenolics.

Keywords: *Phytotoxicity, Leachate, Stem, Leaves, Roots, Inflorescence, Phenolics, Chenopodium murale, Lens culinaris.*

**Author for Correspondence; E-mail: daizybatish@yahoo.com.

Introduction

Weeds are unwanted plants which interfere with the utilization of land and water resources. In agricultural fields, weeds compete with beneficial and desired vegetation reducing the yield and quality of produce through their allelopathic effect (Kohli *et al.*, 1998; Qasem and Foy, 2001). Allelopathy is an important ecological phenomenon reported in a number of weeds and other plants (Rice, 1984). However, under natural conditions it is very difficult to segregate allelopathy from other ecological phenomena. Because of this reason, the term interference has been used especially in weed communities (Muller, 1969). Allelochemicals responsible for bringing about allelopathic effects are released through many modes such as leachation, volatilization, root exudation or decomposition (Rovira, 1969). However, leachation is the most common method of escape of allelochemicals.

Chenopodium murale L. (commonly known as Green fat hen) is an annual exotic weed commonly found in agricultural fields and wastelands. In India, the weed is common in the states of Kashmir, Himachal Pradesh, Punjab, Haryana, Rajasthan and West Bengal and extends up to peninsular India (Anonymous, 1988). In these parts, it appears during the winter season and infests the fields of wheat, gram, potato, and lentil. Consequently the growth of crops is poor and stunted in such fields. It appears that *C. murale* releases inhibitory chemicals in soil that adversely affect the growth of other plant. In order to check this, a study was planned to test the allelopathic nature of *C. murale.* For this purpose a common winter crop *Lens culinaris* was selected as a target plant.

Material and Methods

Study Material

For the study, the weed was collected from the surrounding areas around Chandigarh. Its various parts were separated and were used afresh for leachate preparation. Certified seeds of lentil (*Lens culinaris* Medik. var. T-9) were procured from Punjab Agricultural University, Ludhiana.

Bioassay Studies

Preparation of Leachates

To study the phytotoxicity of leachable allelochemicals aqueous leachates of fresh parts of *C. murale* were prepared. For this 8 g each of fresh leaves, stem, roots and inflorescence were dipped in 100 ml of distilled water for 18 h at 24°C and filtered through triple layer of muslin cloth followed by Whatman no. 1 filter paper. This solution was further diluted to obtain 4 and 2 per cent solutions. The pH of leachates varied between 5.65 and 6.60 whereas Osmotic Potential (O.P) ranged from – 0.190 to – 0.632 bars.

In laboratory bioassays, effect of different concentrations of leachates was studied on early seedling growth and biomass of *L. culinaris.* For this 10 seeds of *L. culinaris* were placed in 9 cm Petri dishes lined with a Whatman no. 1 filter circle moistened with 5 ml of different concentrations of leachates and distilled water to serve as control. For each treatment, 5 replicates were maintained in randomized block design and after a week the length and dry weight of seedlings was determined.

Determination of Phenolics

The amount of total water soluble phenolics in leachates prepared from different parts of *C. murale* was estimated by the methods of Swain and Hillis (1959) using Folinciocalteu reagent. Their amount was determined spectrophotometrically at 700 nm against standard of Ferulic acid. Five replicates were maintained per estimation.

Statistical Analysis

For each experiment 5 replicates were maintained and repeated. The data were presented as mean and statistically analyzed using one way analysis of variance (ANOVA) followed by Duncan's multiple range test. Regression analysis was used in case of amount of phenolics.

Results and Discussion

It is clear from the results that the leachates prepared from various parts of *C. murale* significantly reduce the early seedling growth as well as dry matter accumulation of *L. culinaris.* The seedling length of *L. culinaris* was significantly reduced in response to different concentrations of aqueous leachates prepared from different parts of *C. murale* (Table 13.1). With increasing concentration of leachates, a steady decrease in seedling length was observed. In response to treatment of 2 per cent stem leachates, it was reduced by 29 per cent whereas at higher concentrations of 4 and 8 per cent, the reduction was 48.8 and 71 per cent, respectively (Table 13.1). Almost similar inhibitory effect was observed in response to the root, leaf and inflorescence leachates, though of different magnitude. In general, the leachates prepared from inflorescence were found to be most phytotoxic followed by stem, 7 leaves and roots.

Table 13.1: Effect of Aqueous Leachates of Different Parts of *C. murale* on Seedling Length (cm) of *L. culinaris*

Concentration (%)	*Stem*	*Leaf*	*Roots*	*Inflorescence*
0	8.38±0.65a	8.38±0.65a	8.38±0.65a	8.38±0.65a
2	5.95±0.28b	6.06±0.59b	6.17±0.55b	6.03±0.37b
4	4.29±0.45c	3.99±0.47c	4.54±0.54c	4.20±0.44c
8	2.43±0.44d	2.47±0.52d	2.63±0.62d	2.32±0.32d

Different alphabets in a column represent significance at $P < 0.05$.

Like seedling length, a similar trend of decrease was observed in dry weight of *L. culinaris* in response to leachates from different parts of *C. murale* (Table 13.2). The leachates prepared from stem were found to be most inhibitory followed by those from roots. In response to 2 per cent stem leachates reduction in dry weight was nearly 46 per cent, whereas at 8 per cent concentration the reduction was nearly 76 per cent (Table 13.2). The inhibitory effect of different parts of *C. murale* with respect to effect on dry weight was in the order of stem > root > leaf > inflorescence.

Table 13.2: Effect of Aqueous Leachates of Different Parts of *C. murale* on Biomass Accumulation (mg/seedling) of *L. culinaris*

Concentration (%)	*Stem*	*Leaf*	*Roots*	*Inflorescence*
0	6.22±1.09a	6.22±1.09a	6.22±1.09a	6.22±1.09a
2	3.37±0.27b	3.67±0.39b	3.34±0.21b	4.58±0.56b
4	2.82±0.22c	2.70±0.30c	2.51±0.12c	3.35±0.22c
8	1.46±0.25d	1.98±0.15d	1.82±0.12d	2.29±0.44d

Different alphabets in a column represent significance at $P < 0.05$.

Leachation is an important mechanism of release of allelochemicals from various parts such as foliage, flower, stem, fruits and roots of plants through the natural agencies of rain, dew water and mist (Rice, 1984). Their phytotoxicity, however, is dependent upon various biotic and abiotic factors (Einhellig, 1996). Several studies have revealed that phytotoxic substances leach out from allelopathic plants and adversely affect the growth of other plants (Rice, 1984; Tawaha and Turk, 2003; Singh *et al.*, 2003). In *C. murale* also, leachation might be playing an important role in releasing inhibitory metabolites in the environment. The chemical nature of these metabolites was determined to be phenolics experimentally. The amount of the phenolics, however, varied from tissue to tissue and was found to be the maximum in leaves followed by inflorescence and least in roots (Figure 13.1). Allelopathy of *C. murale* is thus tissue specific. Higher amount of phenolics in the leaves could be due to their larger biomass and greater metabolic activity. These phenolics are known to play a significant role in allelopathic interactions and their amount varies with the age, metabolic stage, season and influx rate (Appel, 1993; Einhellig, 1996; Mizutani, 1999).

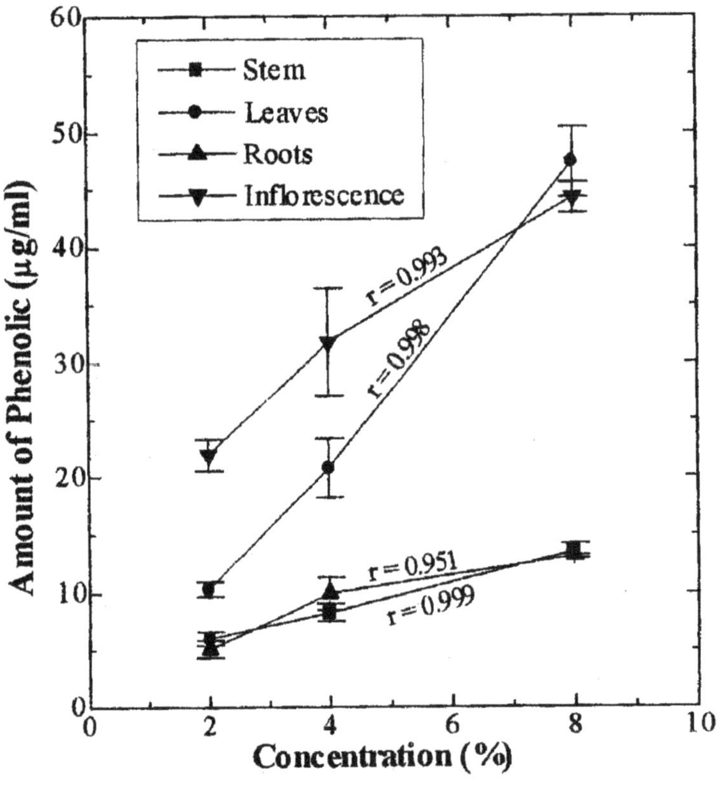

Figure 13.1: Amount of Phenolics in Leachates of Different Parts of *C. murale*.

Though the amount of phenolics was found to be maximum in leaves and lesser in stem, yet the stem imparted greater phytotoxic effect towards *L. culinaris*. This indicates that possibly some other types of water-soluble inhibitors are also present in stem of *C. murale*.

Based on the above observations, it could therefore be concluded that *C. murale* exerts phytotoxicity by releasing growth inhibitory chemicals (phenolics) in the soil.

Acknowledgement

K. Lavanya is thankful to C.S.I.R., New Delhi for financial help in the form of Senior Research Fellowship.

References

Anonymous. 1988. *The Wealth of India: Raw Materials, Vol. 2* (revised). CSIR Publications, New Delhi, India, pp. 471

Appel, H.M. 1993. Phenolics in ecological interactions: The importance of oxidation. *J. Chem. Ecol.,* 19: 1521–1552.

Einhellig, F.A. 1996. Interactions involving allelopathy in cropping systems. *Agronom. J.,* 88: 886–893.

Kohli, R.K., Batish, D. and Singh, H.P. 1998. Allelopathy and its implications in agroecosystems. *J. Crop Prod.,* 1: 169–202.

Mizutani, J. 1999. Selected Allelochemicals. *Crit. Rev. Plant Sci.,* 18: 653–671.

Muller, C.H. 1969. Allelopathy as a factor in ecological process. *Vegetatio.,* 18: 348– 357.

Qasem, J.R. and Foy, C.L. 2001. Weed allelopathy, its ecological impacts and future prospects: A review. *J. Crop Prod.,* 4(2): 43–119.

Rice, E.L. 1984. *Allelopathy*, 2nd Edition. Academic Press, Orlando, Florida.

Rovira, A.D. 1969. Plant root exudates. *Bot. Rev.,* 35: 35–59.

Singh, H.P., Batish, D.R., Kaur, S. and Kohli, R.K. 2003. Phytotoxic interference of *Ageratum conyzoides* with wheat (*Triticum aestivum*). *J. Agron. Crop Sci.,* 189: 341–346.

Swain, T. and Hillis, W.E. 1959. The phenolic constituents of *Prunus domestica* I. The qualitative analysis of constituents. *J. Sci. Food Agric.,* 10: 63–68.

Tawaha, A.M. and Turk, M.A. 2003. Allelopathic effects of black mustard (*Brassica nigra*) on germination and growth of wild barley (*Hordeum spontaneum*). *J. Agron. Crop Sci.,* 189: 298–303.

Chapter 14

Effect of Sulphur Nutrition on Dry Matter Accumulation, Sugar Yield and Sulphur Uptake in *suru* Sugarcane

A.S. Bhosale, T.K. Tiwari**, C.M. Thakre**, P.V. Mahatale** and P.G. Ingole**

**Sugarcane Research Centre, College of Agriculture, Dr. Panjabrao Deshmukh Krishi Vidyapeeth, Akola, M.S.*
***Sugarcane Research Centre, PGI, Mahatma Phule Krishi Vidyapeeth, Rahuri, M.S.*

ABSTRACT

A field experiment was conducted during 2001–2002 at Sugarcane Research Centre, Akola (M.S.) to study the response of sugarcane (*Saccharum officinarum* L.) to the graded levels and sources of sulphur. Number of millable canes, dry matter accumulation, sugar yield and sulphur uptake remained unaffected due to both the sources of sulphur (Gypsum and elemental sulphur). Incremental sulphur levels failed to show any remarkable effect on number of tillers per plant. The dry matter accumulation was significantly increased with 80 kg S ha^{-1} over control and was at par with the level of 60 kg S ha^{-1}. Whereas, significantly the highest number of millable canes, sugar yield and sulphur uptake (95.6 thousands, 19.01 t ha^{-1} and 43.22 kg S ha^{-1}, respectively) were recorded with application of 80 kg S ha^{-1} which was at par with that of 60 kg S ha^{-1}.

Keywords: *Sulphur, Sugarcane, Sugar yield, Uptake, Dry matter.*

Introduction

Intensive agriculture, increasing prevalence of high yielding varieties, multiple cropping system and increased shift from so called low analysis sulphur containing fertilizers (Ammonium sulphate and single super phosphate) to high analysis fertilizers (Urea and DAP) have combinely created gap between soil sulphur supply and crop requirements which led to marked increase in the incidence of sulphur deficiency in soils and crops (Tandon, 1986). Sugarcane, being a heavy feeder crop and yielding 100 t ha^{-1}, it removes about 30 kg S ha^{-1} (Shanmugam, 1995). Presently, about 30 per cent of crop land (120 districts in India) is estimated to suffer from various degree of sulphur deficiency. Hence, it is imperative to explore the possibilities of sulphur application in sugarcane for better uptake, dry matter accumulation and thereby increasing sugar yield for fulfilling daily human needs.

Materials and Methods

A field experiment was carried out during 2001–2002 at Akola, Maharashtra (India) on clay loam soil with pH 7.5, electrical conductivity 0.62 dSm^{-1}, organic carbon 0.57 per cent, available N-280 kg ha^{-1}, available P_2O_5–20.10 kg ha^{-1}, available K_2O–325 kg ha^{-1} and available S–12.40 ppm. Ten treatment combinations consisted of two sulphur sources (Gypsum and elemental sulphur) and 5 levels of sulphur (0, 20, 40, 60 and 80 kg S ha^{-1}) were laid out in Factorial Randomised Block Design with three replications (Table 14.1). Sugarcane variety CO-86032 (Nira) was planted in rows 90 cm apart on 3 February, 2001. The crop matured in 373 days. The crop was fertilized with 175 kg N ha^{-1} through urea, 100 kg each P_2O_5 and K_2O ha^{-1} through single sulphur phosphate and murate of potash. Nitrogen was applied in 3 splits as 25 kg ha^{-1} basal dose, 100 kg N ha^{-1} eight weeks after planting and 50 kg N ha^{-1} at earthing up. Phosphorus and potash were applied in two equal splits at planting and earthing up and sulphur was applied at the time of planting. Sulphur content in plant was estimated from tri acid extract by turbidimetric method (Chesnin and Yien, 1951) and available sulphur in soil by turbid metric method (Jackson, 1967). The total sulphur uptake was calculated in kg ha^{-1} using the following formulae:

Total dry matter ha^{-1} = Dry matter per plant × No. of millable canes per ha

Total sulphur uptake = % sulphur in plant × Total dry matter ha^{-1}

Table 14.1: Dry Matter Accumulation Sugar Yield, Sulphur Uptake and Soil Sulphur Content as Affected by Sulphur Nutrition in *suru* Sugarcane

Treatment	*Dry Matter Weight per Cane (g)*	*No. of Tillers per Cane*	*No. of Millable Canes ('000) ha^{-1}*	*Sugar Yield (t ha^{-1})*	*Sulphur Uptake (kg ha^{-1})*	*Soil Sulphur Content (ppm)*
Sulphur Sources (S)						
S_1–Gypsum	461.40	1.88	89.38	16.87	37.12	13.11
S_2–Elemental sulphur	457.06	1.87	88.23	16.81	36.83	13.85
CD at 5 %	NS	NS	NS	NS	NS	NS
Sulphur Levels (kg ha^{-1})						
0	434.66	1.83	82.17	14.68	30.81	8.38
20	448.83	1.85	85.10	15.56	33.93	12.08
40	462.33	1.88	87.80	16.59	35.82	13.81
60	472.66	1.90	93.36	18.34	41.11	15.81
80	477.66	1.92	95.60	19.01	43.22	17.31
CD at 5%	23.95	NS	6.59	1.70	4.14	1.74

Results and Discussion

Growth Attributes

Both the sources of sulphur *i.e.*, gypsum and elemental sulphur were equally effective in increasing number of millable canes, number of tillers per cane and dry matter accumulation (Table 14.1). This confirms the findings of Yaduvanshi (1998).

Number of tillers per plant remained unaffected due to increasing sulphur levels from 0 to 80 kg S ha^{-1}. Whereas, significantly more dry matter accumulation and number of millable canes (477.66 and 95,600 respectively) were observed with application of 80 kg S ha^{-1} and was at par with those recorded with application of 60 kg S ha^{-1} (Table 14.1). Application of 40 kg S ha^{-1} was also found at par with these two rates of application *i.e.*, 60 and 80 kg S ha^{-1} in respect of dry matter accumulation. This increase in dry matter accumulation might be due to the increase availability of nitrogen in soil because of sulphur application. As nitrogen and sulphur are the constituents of protein and involved in chlorophyll formation, that might have caused increased in chlorophyll content in growing region, meristematic tissue ultimately enhancing cell division and thereby increased the crop growth and finally growth attributes. Similar results were obtained by Ghosh *et al.* (1990). Singh *et al.* (1995) observed the increased number of millable canes with increment in sulphur levels.

Sugar Yield

Sugar yield was not significantly affected due to application of either gypsum or elemental sulphur. Application of 80 kg S ha^{-1} significantly increased the sugar yield by 29.5 per cent over control and was at par with that observed with application of 60 kg S ha^{-1} (Table 14.1). The increase in sugar yield was primarily due to the beneficial effect of sulphur in raising cane yield per unit area. Similar results were reported by Saroha and Singh (1979), Rahman *et al.* (1987) and Satisha *et al.* (1996). Significantly the lowest sugar yield was recorded under control and was at par with application of 20 kg S ha^{-1} indicating its failure in increasing sugar yield.

Sulphur Uptake

Sulphur uptake and residual sulphur content were not significantly affected due to either gypsum or elemental sulphur application. Whereas, significantly more uptake of sulphur and sulphur content in soil were observed due to application of 80 kg S ha^{-1} and was at par with 60 kg S ha^{-1} (Table 14.1). The increase in sulphur uptake might be due to deficient levels of sulphur in soil supporting growth. These results are in close conformity with the findings of Yaduvanshi and Yadav (1997), Choudhari and Sinha (2001).

Thus, it can be concluded that, application of 60 kg S ha^{-1} through gypsum or elemental sulphur was found as effective as 80 kg S ha^{-1} for better sugar yield and sulphur uptake by crop.

References

Chesnin, L. and Yien, C.H. 1951. Turbidimetric determination of available sulphate. *Proc. Soil Sci. Ame.*, 15: 149–151.

Choudhari, C.N. and Sinha, U.P. 2001. Dry matter production, sugar yield and nitrogen sulphur, recovery by sugarcane as influenced by concentrated organic manure, nitrogen and sulphur fertilization. *Indian J. Agric. Sci.*, 71(1): 31–34.

Ghosh, A.K., Rai, R.K., Saxena, Y.R. and Srivastava, A.K. 1990. Effect of sulphur application on the nutritional status, yield and quality of sugarcane. *J. Indian Soc. Soil Sci.*, 38: 73–76.

Jackson, M.L. 1967. *Soil Chemical Analysis*. Prentice Hall of India Pvt. Ltd., New Delhi.

Rahman, M.H., Choudhary, M.K.A. and Azad, A.K. 1987. Effect of sulphur on growth and yield of sugarcane in old Himalayas piedmont soils. *Indian J. Sugarcane Tech.*, 4(2): 104–107.

Saroha, M.S. and Singh, H.G. 1979. Effect of prevention of iron chlorosis on quality of sugarcane growth in vertisols. *Plant and Soil*, 52: 467–473.

Satisha, G.C., Krishnappa, M. and Srikanth, K. 1996. Influence of sulphur on yield and quality of sugarcane. *Indian Sugar*, 46(6): 397–401.

Shanmugam, K.S. 1995. Sulphur nutrition of sugarcane. *Fert. News*, 40(4): 237–238.

Singh, A., Dua, S.P. and Singh, G.P. 1995. Effect of sulphur on yield and quality of sugarcane. *Indian Sugar*, 45(4): 237–238.

Tandon, H.L.S. 1986. *Sulphur Research and Agricultural Production in India*, 2nd Edition. Fertilizer Development and Consultation Organization, New Delhi.

Yaduvanshi, N.P.S. 1998. Effect of different sulphur fertilizers on content and balance of sulphur and yield of sugarcane. *J. Indian Soc. Soil Sci.*, 46: 701–703.

Yaduvanshi, N.P.S. and Yadav, D.V. 1997. Effect of foliar application of sulphur on yield and uptake by sugarcane. *Indian J. Agric. Sci.*, 67(12): 599–600.

Chapter 15

Dry Matter Accumulation and Nitrogen Uptake of Basmati Rice Varieties as Influenced by Nitrogen Application and Lodging Management

Harmandeep Singh, M.S. Sidhu* and Virender Sardana***

**Department of Agronomy and Agrometeorology, Punjab Agricultural University, Ludhiana – 141 004, Punjab*

***Department of Plant Breeding, Genetics and Biotechnology, Punjab Agricultural University, Ludhiana, Punjab*

ABSTRACT

A field study was conducted at Punjab Agricultural University, Ludhiana during *kharif* 2002 to study the effect of variety, nitrogen and lodging management on nitrogen content and uptake in different plant parts at different growth stages of basmati rice in sandy loam texture. Dry matter in stem was significantly influenced by varieties and lodging management at all growth stages and panicles and grain at harvest. Different nitrogen levels significantly influenced the dry matter of leaves, stem, panicle and grains at all growth stages. Nitrogen content in leaves at 60 DAT and at harvest was significantly influenced by varieties. Nitrogen levels significantly affected the nitrogen content at different growth stages *i.e.*, 30 and 60 days after transplanting (DAT) and at harvest. But effect was non-significant in lodging management treatments. Nitrogen uptake was also affected by different treatments at different growth stages.

Keywords: *Super Basmati, Basmati 386, Dry matter, Nitrogen content, Nitrogen uptake, Lodging management.*

Introduction

Basmati rice assumes a special status in rice cultivation in Punjab owing to its excellent cooking and eating qualities. It has tremendous export potential and commands a premium price in the domestic and international market due to the specific ambrosial and culinary qualities. Nitrogen is an integral part of all the important function of plants and has profound affect on growth and yield of rice. However, heavy application of nitrogen is known to promote lodging in basmati due to its week stem. Moreover nitrogen requirement of different varieties of basmati rice varies. There is scope to increase response of nitrogen to higher doses and improve productivity if the lodging is controlled in basmati rice. Little is known about the dry matter accumulation and nitrogen uptake pattern particularly of recently released varieties of Super basmati of basmati rice which has shown higher yield potential. An attempt was therefore made to study the dry matter partitioning and nitrogen uptake in two basmati rice varieties as influenced by doses of nitrogen and lodging management.

Materials and Methods

The field experiment was carried out at Punjab Agricultural University, Ludhiana during *kharif* 2002 on sandy loam soil with low organic carbon (0.36 per cent), available nitrogen (125.4 kg/ha) and available K (130 kg/ha) but high in available P (29.1 kg/ha). The treatments comprised combination of two varieties (Super basmati and Basmati 386) and three nitrogen levels (20, 40 and 60 kg/ha) in main plots and four lodging management treatments (no lopping, lopping at 45 days after transplanting (DAT), trinexapacethyl 30 g/ha at 45 DAT and lopping at 45 DAT followed by (fb) trinexapacethyl 30 g/ha at 60 DAT in sub-plots. The experiment was conducted in split-plot design with three replications. Thirty days old seedlings of test varieties were transplanted at 20 × 15 cm spacing. Application of nitrogen was applied in two equal splits *i.e.,* half at 3 weeks after transplanting and remaining at 6 weeks after transplanting. The dry matter of plants was recorded at 30, 60 DAT and at harvest. The dry matter was partitioned into leaves and stem at 30 and 60 DAT and into leaves, stem, panicle (without grains) and grains at harvest. The nitrogen content in different plant parts was determined by using the standard Kjeldahl method. The nitrogen uptake was computed by multiplying nitrogen concentration with respective dry matter.

Results and Discussion

Dry Matter Partitioning

The two varieties differed non-significantly for leaf dry matter at different growth stages (Table 15.1). However differences in dry matter of stem were significant in test varieties. While at 30 DAT, Basmati 386 recorded significantly higher stem dry matter (11.6 q/ha) than Super basmati (10.5 q/ha). At 60 DAT and harvest, stem dry matter of Super basmati was significantly higher (30.2 and 31.0 q/ha) than Basmati 386 (28.3 and 29.2 q/ha). This may be due to more number of tillers and leaves per plant in Super basmati as compared to Basmati 386. Similar trend was observed in panicle and grains for Super basmati. Nitrogen application significantly influenced the dry matter in different plant parts (Table 15.1). In general there was increase in dry matter with increasing dose of nitrogen up to 60 kg N/ha. At 30 DAT, application of 60 kg N/ha significantly increased the dry matter in leaves(13.2 q/ha) over lower levels whereas in stem such an increase over 20 kg N/ha was significant. At 60 DAT and at harvest, application of 60 kg N/ha significantly increased the dry matter accumulation in leaves and stems as compared to lower doses of nitrogen.

Table 15.1: Effect of Variety, Nitrogen and Lodging Management Treatments on Dry Matter Partitioning (DMP) at Different Growth Stages of Basmati Rice

Treatment	*DMP (q/ha)*							
	30 DAT		*60 DAT*		*Harvest*			
	Leaves	*Stem*	*Leaves*	*Stem*	*Leaves*	*Stem*	*Panicle*	*Grain*
Variety								
Super basmati	12.1	10.5	24.4	30.2	34.2	31.0	3.6	30.5
Basmati 386	12.8	11.6	25.4	28.3	33.4	29.2	3.4	28.4
CD (P = 0.05)	NS	0.6	NS	1.0	NS	0.6	0.1	0.8
Nitrogen (kg/ha)								
20	11.6	9.5	22.6	25.8	31.0	26.9	3.3	27.9
40	12.4	11.4	25.0	29.6	33.0	30.0	3.4	29.3
60	13.2	12.2	27.1	32.2	37.2	33.3	3.7	31.2
CD (P = 0.05)	0.6	1.0	1.2	1.6	1.4	0.7	0.2	0.9
Lodging Management								
No lopping	12.7	11.1	25.6	32.1	33.9	31.9	3.6	30.0
Lopping at 45 DAT	12.5	10.9	24.8	27.8	33.7	29.8	3.4	29.0
Trinexapacethyl spray 30 g/ha at 45 DAT	12.3	10.9	24.7	29.4	34.3	29.2	3.4	29.7
Lopping at 45 DAT fb Trinexapacethyl spray 30 g/ha at 60 DAT	12.4	11.2	24.5	27.4	33.3	29.4	3.5	29.4
CD (P = 0.05)	NS	NS	NS	1.1	NS	1.1	0.1	0.7

Similarly differences in dry matter of panicle and grains at harvest with 60 kg N/ha were significant as compared to lower doses of nitrogen.

Lodging management treatments did not affect the dry matter at different growth stages except in stem at 60 DAT and stem, panicles and grains at harvest (Table 15.1). No lopping resulted in significantly more dry matter in stem at 60 DA T (32.1 q/ha) and in stem, panicle and grains at harvest (31.9, 3.6 and 30.0 q/ha) as compared to other treatments.

Nitrogen Content

Nitrogen content in both and leave and stem decreased with increasing age of plant from 30 DAT upto harvest (Table 15.2). Difference in nitrogen content between two varieties was significant only in leaves at 60 DAT and in leaves and grains at harvest. At 60 DAT, leaves of Super basmati contained significantly more nitrogen (1.71 per cent) than that of Basmati 386 (1.64 per cent). However, at harvest Basmati 386 accumulated significantly more nitrogen content in leaves and grains (0.84 and 1.25 per cent) as compared to Super basmati (0.72 and 1.23 per cent).

Application of nitrogen significantly influenced the nitrogen content in leaves and stem at different growth stages (Table 15.2). The nitrogen content of leaves with 60 kg N/ha was significantly higher (1.99, 1.74 and 0.83 per cent) as compared to 20 kg N/ha (1.79, 1.62 and 0.74 per cent) and 40 kg N/ha (1.85, 1.66 and 0.77 per cent) at all the growth stages. In case of stem 60 kg N/ha showed significantly higher nitrogen content at 60 DAT and at harvest. Dubey and Bisen (1989) also reported that the nitrogen content increased significantly upto 50 DAT with increased nitrogen levels from 0–114 kg/ha. But after 50 DAT the nitrogen content in plants decreased due to age as nitrogen is absorbed vigorously at early stages of growth due to greater cell region and accumulation of protein at panicle primodial stage and after flowering the rate of absorption decreases (Lal and Mahapatra 1975). At harvest with increase in nitrogen levels nitrogen content increased in panicle and grains. At 60 kg N/ha nitrogen content in panicles (0.92 per cent) and grains (1.26 per cent) was significantly lower than 20 kg N/ha.

Table 15.2: Effect of Variety, Nitrogen and Lodging Management Treatments on N Content (per cent) at Different Growth Stages of Basmati Rice

Treatment	*N Content (%)*							
	30 DAT		*60 DAT*		*Harvest*			
	Leaves	*Stem*	*Leaves*	*Stem*	*Leaves*	*Stem*	*Panicle*	*Grain*
Variety								
Super basmati	1.87	1.17	1.71	1.02	0.72	0.78	0.84	1.23
Basmati 386	1.89	1.18	1.64	1.03	0.84	0.80	0.84	1.25
CD (P = 0.05)	NS	NS	0.04	NS	0.02	NS	NS	0.01
Nitrogen (kg/ha)								
20	1.79	1.13	1.62	0.97	0.74	0.72	0.75	1.21
40	1.85	1.17	1.66	1.02	0.7	0.7	0.83	1.24
60	1.99	1.22	1.74	1.08	0.83	0.86	0.92	1.26
CD (P = 0.05)	0.04	0.05	0.05	0.05	0.02	0.03	0.11	0.02
Lodging Management								
No lopping	1.93	1.20	1.67	1.01	0.75	0.78	0.86	1.25
Lopping at 45 DAT	1.92	1.15	1.70	1.03	0.78	0.77	0.81	1.23
Trinexapacethyl spray 30 g/ha at 45 DAT	1.81	1.17	1.66	1.02	0.80	0.79	0.80	1.25
Lopping at 45 DAT fb Trinexapacethyl spray 30 g/ha at 60 DAT	1.85	1.17	1.68	1.04	0.76	0.80	0.86	1.24
CD (P = 0.05)	0.02	NS	NS	NS	0.03	NS	NS	NS

Lodging management treatments significantly influenced the nitrogen content in only leaves at 30 DAT and at harvest (Table 15.2). At 30 DAT no lopping (1.93 per cent) resulted in significantly more nitrogen content than trinexapacethyl spray (1.81 per cent) and lopping at 45 DAT fb trinexapacethyl spray at 60 DAT (1.85 per cent) but statistically at par with lopping at 45 DAT (1.92 per cent). This may be due to adverse effect of these treatments at harvest trinexapacethyl spray at 45 DAT (0.80 per cent) resulted in significantly more nitrogen content than other treatments except lopping at 45 DAT.

Nitrogen Uptake

Nitrogen uptake in leaves and stems continued to increase upto 60 DAT after which it decreased upto harvest in both the varieties (Table 15.3). The uptake in leaves at harvest was significantly more in Basmati 386 (27.7 kg/ha) than Super basmati (24.7 kg/ha) mainly due to significant increase in its nitrogen content as compared to Super basmati. Application of nitrogen significantly increased the uptake with each increment of nitrogen at all the growth stages. Nitrogen uptake in leaves at harvest with 20 kg N /ha was 22.8 kg/ha which increased to 30.5 kg/ha with 60 kg/ha due to increase in both dry matter accumulation and nitrogen content. Lodging management treatments failed to influence the nitrogen uptake in leaves.

Table 15.3: Effect of Variety, Nitrogen and Lodging Management Treatments on N-uptake (kg/ha) at Different Growth Stages of Basmati Rice

Treatment	*N-uptake (kg/ha)*							
	30 DAT		*60 DAT*		*Harvest*			
	Leaves	*Stem*	*Leaves*	*Stem*	*Leaves*	*Stem*	*Panicle*	*Grain*
Variety								
Super basmati	22.4	12.3	41.8	30.9	24.7	24.1	3.1	37.9
Basmati 386	24.3	13.6	41.7	29.1	27.7	23.4	2.8	35.4
CD (P = 0.05)	1.1	0.7	NS	NS	1.3	NS	NS	1.0
Nitrogen (kg/ha)								
20	20.8	10.7	36.5	25.1	22.8	19.2	2.5	34.0
40	22.8	13.4	41.5	30.1	25.4	23.3	2.9	36.4
60	26.3	14.9	47.1	34.8	30.5	28.8	3.4	39.5
CD (P = 0.05)	1.4	0.8	2.6	2.3	1.6	0.9	0.4	1.3
Lodging Management								
No lopping	24.6	13.3	42.8	32.5	25.7	25.1	3.0	37.4
Lopping at 45 DAT	23.6	12.7	42.2	29.0	26.6	23.2	2.8	35.8
Trinexapacethyl spray 30 g/ha at 45 DAT	22.2	12.9	40.9	30.2	27.1	23.1	2.7	37.0
Lopping at 45 DAT fb	23.0	13.1	41.1	28.2	25.6	23.6	3.0	36.3
Trinexapacethyl spray 30 g/ha at 60 DAT								
CD (P = 0.05)	1.3	NS	1.5	1.9	NS	NS	NS	0.8

Basmati 386 accumulated significantly more nitrogen uptake (13.6 kg/ha) than Super basmati (12.3 kg/ha) at harvest (Table 15.3). Application of nitrogen affected the N uptake in stem significantly at all growth stages. At harvest the nitrogen uptake in stem was maximum (28.8 kg/ha) with 60 kg N/ha which was 9.6 and 5.5 kg higher than 20 and 40 kg N/ha respectively. Lodging management did not influence the nitrogen uptake in stem at 30 DAT and at harvest though it was highest in low lopping treatment in all the growth stages (Table 15.3). However, at 60 DAT increase in nitrogen uptake in lodging management treatments over other treatments and trinexapacethyl spray application at 30 DAT over lopping at 45 DAT and lopping fb trinexapacethyl application was significant. Similarly nitrogen uptake in panicle was not significantly influenced by varieties and lodging management where it significantly increased with increasing dose of nitrogen (Table 15.3). Nitrogen uptake in grains was significantly influenced by different treatments. Super basmati has significantly more N uptake in grains (37.9 kg/ha) than Basmati 386 (35.4 kg/ha). Increasing levels of nitrogen significantly increased the N uptake by grains up to the harvest level. The increase with 40 and 60 kg N/ha over 20 kg N/ha was 7.1 and 16.2 per cent respectively. Dubey and Bisan (1989) also reported that nitrogen uptake by grains in rice significantly increased with increasing levels of nitrogen from control to 114 kg/ha. No lopping resulted in higher uptake of nitrogen in grains and it was at par with trinexapacethyl spray at 45 DAT and significantly better than lopping and lopping fb trinexapacethyl spray.

This study indicated uptake of nitrogen in different plant parts *i.e.*, leaves, stem, panicle and grain in relation to dry matter accumulation. Nitrogen uptake was 27.7, 23.4, 2.8 and 35.4 kg/ha at harvest in Basmati 386 and 24.7, 24.1, 3.1 and 37.9 kg/ha in Super basmati. This shows that nitrogen accumulation by leaves more in Basmati 386. There is need to further study for its translocation to the sink.

References

Dubey, S.K. and Bisen, C.R. 1989. Nitrogen uptake by rice as influenced by different levels: Sources and methods of nitrogen application. *Oryza*, 26: 37–42.

Lal, M. and Mahapatra, L.C. 1975. Effect of phosphorus carrier varying in water-soluble and citrate soluble phosphorus contents on uptake of major nutrients (NPK) by rice. *Riso*, 25: 45–48.

Chapter 16

Role of Copper and Manganese Application of Nitrate Reductase and Protease Enzyme Activities of *Zingiber officinale* Rosc. L. Var-1

A. Ksheroda Devi and P.K. Singh
Department of Life Sciences, Manipur University, Canchipur, Imphal – 795 003, Manipur

ABSTRACT

Field experiments were carried out during the period of April to December 2002 and 2003 on ginger (*Zingiber officinale* Rosc. L.) VAR-1, with soil and foliar application of copper ($CuSO_4$), manganese ($MnSO_4$) and total micronutrients (TM) in a randomized block design (RBD). The response of micronutrients was determined 90, 120, 150 and 180 days after plantation (DAP). Among the treatments, maximum efficiency of Nitrate Reductase (NR) was found in T_5 (Mn in foliar sprayed plants) at 90 DAP. The highest concentration of protease was recorded in T_4 ($CuSO_4$) in foliar sprayed plants) at 150 DAP. The lowest concentration of NR and Protease were found in T_5 and T_3. The applications of micronutrients to plant become important for their efficient utilization and better performance of the crop.

Keywords: *Ginger, Enzymes, Micronutrients, Soil and Foliar application.*

Introduction

Micronutrients play an important role in maintaining the plant growth, yield and quality of vegetables. It also influences the physiological and biochemical parameters of plants. Manganese

(Mn) is absorbed by plants as manganese ion (Mn^{++}) and plays an important role in nitrate assimilation, evolution of oxygen during photosynthesis and occurs in growing tissues. Mn deficiency causes inhibition of photosynthesis and nitrogen metabolism. Copper (Cu) acts as a prosthetic group of the enzymes. It is also a component of plastocyanin, which takes important role in electron transport system in photosynthesis. Cu deficiency causes curling of young leaves. The application of micronutrients to plant becomes important for their efficient utilization and better performance of the crop.

Materials and Methods

The experiment was conducted at the Life Sciences Experimental Field, Manipur University Campus, during the period of April to December 2002 and 2003. The plot was laid out in a randomized block design (RBD) with four replications each having a plastic bag (size 15 cm diameter, 8 cm height and 2 kg soil capacity) and spacing of 15 cm. The size of the experimental area is 6.3 m length and 1.75 m breadth. Number of plots was 24 with 6 treatments and a population of 240 plants. Application of micronutrients was performed on soil and foliar (40 and 70 DAP). The treatments are T_1 (control without micronutrients), T_2 (with copper–$CuSO_4$ 0.012 g/bag), T_3 (with manganese–$MnSO_4$ 0.12 g/bag). Foliar application was done with 0.05 per cent of spraying reagents Cu, Mn and total micronutrients (0.05 per cent : 0.05 per cent); they are T_4, T_5 and T_6 respectively. Estimation of N.R. and Protease enzyme activities were conducted with modification of the methods by Sadasivam and Manickam (1996). The plants were transplanted and arranged in a RBD. Representation and statistical analysis of experimental data were also worked out (Cochran and Cox, 1965; Gomez and Gomez, 1976).

Results and Discussion

The NR is the first enzyme, which converts NO_3 into NO_2. NR is an inducible enzyme and located in the cytosol, hence it is very likely that in the process of the, absorption, the agrochemical comes in contact with the NR present in the cytosol (Devi *et al.*, 2003).

In the present study, the response of micronutrients on ginger was observed at 90, 120, 150 and 180 DAP. The level of NR were found in an increasing trend from 90 to 120 DAP which were followed by a decline at 150 DAP and then increase at 180 DAP. The maximum values were obtained in T_5 (2.04 mg/g) with Mn in foliar sprayed plants 90 DAP. The same trend was observed at 120 and 180 DAP *i.e.*, T_5 (2.03 mg/g) at 120 DAP, T_5 (1.40 mg/g) at 180 DAP. At 150 DAP maximum values were obtained in T_4 (0.68 mg/g) with Cu in foliar sprayed plants. The minimum value was recorded in the control plant in all stages.

The present finding indicates the indispensable role of Mn in enhancing the production of NR in the early growth stages of the plant from 90 DAP to 120 DAP (Table 16.1). However, the activity of the Mn depleted as the test plant advances in age. It was observed in a decreasing trend of NR enzyme production in case of 150 DAP and 180 DAP (Table 16.1). The foliar application of Mn (T_5) on test plant as a key role in NR enzyme production. The results are in confirmative with the application of Zn in NR enzyme production (Ghildiyal *et al.*, 1981; Devi, *et al.*, 2003).

Protease are a complex group of enzymes varying quality in their physiological and catalytic properties and playing important roles in the catalytic and regulatory processes of eukaryotes and prokaryotes alike (North, 1982; Kalisz, 1988).

In the present investigation, the maximum protease enzyme activity was observed on T_4 (7.35 mg/g)-Cu in foliar application at 150 DAP. However, at 180 DAP, almost similar maximum value of protease activity was observed at T_2 (7.10 mg/g)-Cu in soil-applied plant (Table 16.2).

Table 16.1: Effect of Cu, Mn and TM on NR Enzyme Activities at Different Stages of *Zingiber officinale* Rosc. L. Var-1

Treatments	*Levels of Nitrate Reductase Activities (mg/g Leaf Fresh Weight)*			
	90 DAP	*120 DAP*	*150 DAP*	*180 DAP*
T_1	0.76±0.06	1.59±0.01	0.44±0.00	0.72±0.00
T_2	0.80±0.04	1.94±0.09	0.48±0.02	1.00±0.04
T_3	1.10±0.03	1.57±0.06	0.52±0.12	1.12±0.00
T_4	0.96±0.16	2.02±0.09	0.68±0.04	1.10±0.02
T_5	2.04±0.84	2.03±0.09	0.48±0.02	1.40±0.08
T_6	1.56±0.07	1.73±0.03	0.54±0.02	1.28±0.16
CD at 5%	0.237	0.070	0.040	0.056
CD at 1%	0.386	0.117	0.067	0.094

Table 16.2: Effect of Cu, Mn and TM on NR Protease Enzyme Activities at Different Stages of *Zingiber officinale* Rosc. L. Var-1

Treatments	*Levels of Protease Enzyme Activities (mg/g Leaf Fresh Weight)*		
	120 DAP	*150 DAP*	*180 DAP*
T_1	5.25±0.15	5.40±0.10	6.25±0.14
T_2	5.05±0.14	4.45±0.45	7.10±0.20
T_3	5.75±0.01	3.90±1.00	3.95±0.05
T_4	5.20±0.10	7.35±0.44	6.35±0.25
T_5	4.75±0.05	5.50±1.00	4.50±0.00
T_6	4.70±0.09	5.50±1.10	5.40±0.10
CD at 5%	0.102	0.771	0.143
CD at 1%	0.171	1.288	0.238

From the present findings, it has been observed that in the mature stages Cu is required for protease enzyme by the plants. The minimum activity of protease was observed on T_3 (3.90 mg/g)-Mn in soil application at 150 DAP. The result shows that Mn has a little role in protease enzyme activity.

In the present investigation, the role of micronutrients application on NR and protease activities on the test plant *Zingiber officinale* plays a very important role, selectively on various growth stages of the plant.

Acknowledgements

The authors are grateful to the Head, Department of Life Sciences, Manipur University, Canchipur, Imphal, for laboratory facilities. Thanks goes to my colleagues, S. Padmabati, Y. Nanda and S. Sangeeta for their valuable helps during the tenure of the programme.

References

Cochran, W.G. and Cox, G.M. 1965. *Experimental Design*. John Wiley, New York.

Devi, W.I. 1997. Effect of some micronutrients on the physiological and biochemical studies of *Brassica oleraceae* var. *capitata* L. *Ph.D. Thesis.* Manipur University.

Devi, W.I., Singh, P.K. and Devi, G.A.S. 2003. Role of micronutrients application on nitrate reductase, amylase, total free amino acids and sugar content of *Brassica oleraceae* var. capitata L. *Agric. Biol. Res.,* 19: 59–70.

Ghildiyal, M.C., Sirohi, G.S. and Pandey, R.M. 1981. Growth and nitrate reductase activity in mustard variety as affected by zinc nutrition. *Indian J.Pl. Physiol.,* 24: 170–175.

Gomez, K.A. and Gomez, A.A. 1976. *Statistical Procedures for Agricultural Research with Emphasis on Rice.* IARI. LOS Banos. Philippines.

Jones, B.G. 1980. *Annual Review of Plant Physiology,* pp. 257.

Kaliz, H.M. 1988. Microbial proteinases. *Advanced in Biochemical Engineering,* Vol. 36.

Lal, R.K., Ram, P.C. and Chaturvedi, G.S. 1977. *Plant Physiology Laboratory Manual.* Centre of Advanced Studies, Department of Crop Physiology, NDU of Agri. and Tech, Kumarganj, Farizabad, U.P.

North, M.J. 1982. Comparative. Biochemistry of the Proteinases of eukaryotic microorganisms. *Microbiological Reviews,* 46: 308–340.

Sadasivam, S. and Manickam, A. 1996. *Biochemical methods.* Centre for Plant Molecular Biology, TNAU, Coimbatore.

Wallace, A. and Romney, E.M. 1977. Synergistic trace metal effects in plants Commun. *Soil Sci. Plant Anal.,* 8: 699–707.

Chapter 17

Reaction of Rice Cultivars Against Gall Midge (*Orseolia oryzae* Wood Mason) Population of Sambalpur, Orissa Under Natural Infestation Conditions

L. Behera, S.C. Sahu, S. Rajamani, H.N. Subudhi and L.K. Bose
Central Rice Research Institute, Cuttack – 753 006, Orissa, India

ABSTRACT

Thirty-two donors and eighty-nine released varieties were evaluated under natural infestation conditions against gall midge at Sambalpur. Nine donors showed moderately resistant reaction, eleven were moderately susceptible while twelve donors showed susceptible reaction. Twelve released varieties showed moderately resistant reaction, seventeen were moderately susceptible while sixty showed susceptible reaction. None of the donors and released varieties showed resistant reaction. Sambalpur population is found to be different from biotype 1 as evidenced from host reaction studies.

Keywords: *Rice gall midge (Orseolia oryzae), Resistance, Biotype.*

Introduction

The Asian rice gall midge, *Orseolia oryzae* Wood-Mason (Cecidomyiidae: Diptera), is a serious pest of rice in many states of India, Sri Lanka, China and Thialand (Reddy *et al.*, 1967; Lai *et al.*, 1984;

Tan *et al.*, 1993). The larvae of the pest feed on the apical meristem, causing the formation of a tubular leaf sheath called 'silver shoot' in place of the normal panicle. The loss in grain yield varies from 3 per cent to 70 per cent in India (Chatterji *et al.*, 1976). Chemical and cultural methods for control of this pest are neither very effective nor environmentally safe. Therefore, the use of resistant varieties seems to be the best alternative. The variable reaction of the resistant donors due to variability in gall midge population, limits the use of resistant donors. Six biotypes of the rice gall midge are presently known to exist in India (Pasalu and Huang, 1998). This has necessitated systematic studies to identify donors with non-allelic resistance genes against gall midge biotypes and pyramid them into high yielding varieties for sustainable resistance over large areas and many years. Several national/state high yielding varieties having resistance to one or the other pest have been released. However, their resistance status after passage of time under different agro-climatic zones need verification in order to monitor the pest status as well as to recommend suitable varieties to the farmers from time to time. The present study was undertaken to evaluate donors and released varieties of different states under natural infestation conditions at Sambalpur, Orissa to identify resistant cultivars.

Materials and Methods

The screening of 32 donors and 89 released rice varieties was carried in field conditions against gall midge at Regional Research and Training Station, OUAT, Sambalpur during 2002 wet season. The known susceptible cultivar, TN1 was used as check. Twenty five-day-old seedlings of each variety were transplanted in two rows at 20 × 15 cm spacing with three replications. To ensure adequate insect pressure, susceptible cultivar, TN1 was transplanted in the border and after every 20 rows. Transplanting was delayed till first week of August to coincide maximum growth of the crop with high emergence level of gall midge. Standard agronomic package of practices every 20 rows. Transplanting was delayed till first week of August to coincide maximum growth of the crop with high emergence level of gall midge. Standard agronomic package of practices were followed for raising the seedlings in nursery and the crop in the main field. No pesticide was applied to the crop at any of its growth stage. Gall midge infestation was recorded using 0-9 scale as per the standard evaluation system (IRRI, 1996), when all the plants of the susceptible check showed more than 15 per cent silver shoot.

Results and Discussion

The reaction of 32 donors indicated that nine cultivars *viz.*, ARC 6605, INDCR 140, INDCR 210, INDCR 232, Phalguna, Samridhi, T 1477, Velluthacherra and W 1263 were moderately resistant (SES score 3), eleven cultivars were moderately susceptible (score 5) while 12 were susceptible (score 7) (Table 17.1). Out of 89 released varieties, twelve varieties *viz.*, Bharathi, CSR 27, Lalat, Kakatiya, Panidhan, Rudra, Samalei, Sarala, Sarasa, Sarathi, Shaktiman and Triguna showed moderate level of resistant reaction while 17 varieties showed moderate level of susceptible reaction (score 7 and 9) (Table 17.1). Sixty varieties showed susceptible reaction. None of the donors and released varieties are found to be resistant.

The moderate level of susceptibility of Abhaya, ARC 5984, PTB 10, RP2068-18-3 and Kranti; and susceptibility of Aggani, Bangalei, Leaung 152, NHTA 8, RP 2333 and Siam 29, which are known to resistant against biotype 1 (Kumar, 2001), indicated that gall midge population of Sambalpur is different from Raipur population of gall midge. Two varieties, Samalei and Sarasa were reported to be resistant to biotype 2, gall midge both at tillering and flowering stage (Singh *et al.*, 2001; Behera *et al.*, 2004). These varieties are also tolerant to Sambalpur population of gall midge. Thus, these varieties should be grown in Orissa for gall midge resistance.

Table 17.1: Reaction of Different Rice Cultivars to Gall Midge at Sambalpur, Orissa During 2002 Wet Season

% of Silver Shoot Damage	*SES Score*	*Donors/ Popular Varieties*	*Name of the Cultivars*	*Remark**
> 1–5	3	Donors	ARC 6605, INDCR 140, INDCR 210, INDCR 232, Phalguna, Samridhi, T 1477, Velluthacherra and W 1263	MR
		Popular varieties	Bharathi, CSR 27, Lalat, Kakatiya, Panidhan, Rudra, Samalei, Sarala, Sarasa, Sarathi, Shaktiman and Triguna	
> 5–10	5	Donors	Abhaya, ARC 5984, INDCR 967, INDCR 2139, MNP 152, PTB 10, PTB 18, PTB 31, PTB 32, RP 2068-18-3 and Tadukan	MS
		Popular varieties	ADT 39, Biraj, Chandrama, Jalaprava, Karuna, Kranti, Lunishree, Meher, Mohan, PR 114, PY 3, Rohini, Saket 4, Samanta, Surekha, Tara and Vytilla 2	
> 10–25	7	Donors	Aggani, ARC 6650, Bangalei, Bhumansan, INDCR 199, INDCR 273, INDCR 1351, Leaung 152, Madhuri A9, NHT A8, RP 2333 and Siam 29	S
		Popular varieties	Annapurna, ADT 33, Bhanja, Dharitri, Durga, Getu, Gouri, Gour 3, Heera, Indira, IR 36, IR 64, Jajati, Kalinga I, Konark, Moti, Narendra I, Neela, Padmini, Pathara, Pooja, PR 108, Pragati, Prasad, Punshi, Purnendu, Radhi, Rajashree, Ratna, Satabdi, Savitri, Seema, Supriya, Surendra, Tripti, Udaya, Utkalprva, Vandana, VL Dhan 61, Vivekdhan 62 and Vijeta	
> 25	9	Popular varieties	Daya, Dhalaheera, Gayatri, Hanseswari, Himalaya, Himdhan, Jaya, Kalinga II, Pokkali, Poorva, Ratna, Sabita, Saraswati, Sattari, Shyamla, Sneha, Swarna, TN 1 and Vanaprava	S

* MR: Moderately resistant; MS: Moderately susceptible; S: Susceptible.

References

Anonymous 1996. *Standard Evaluation System for Rice*. International Rice Research Institute, Manila, Philippines.

Behera, L., Sahu, S.C., Rajamani, S., Subudhi, H.N., Bose, L.K. and Singh, B.N. 2003. Screening of rice cultivars against rice gall midge (*Orseolia oryzae* Wood Mason) under glass house conditions. *Oryza* (Accepted).

Chatterji, S.M., Kulshrestha, J.P., Rajamani, S. and Prakash Rao, P.S. 1976. Insect pest of rice and control. *Pesticide Information*, 2: 33–36.

Lai, K., Tan, Y. and Pan, Y. 1984. Rice gall midge biotypes in Guandong province. *Int. Rice Res. Newsl.* 9(3): 17–18.

Kumar, A. 2001. Reaction of new rice gall midge (*Orseolia oryzae* Wood Mason) donors against biotype 1 in Madhya Pradesh, India. *Oryza*, 38(3&4): 158–159.

Pasalu, I.C. and Huang, B. 1998. Current status of gall midge biotypes in India and China. In: *International Workshop on New Approaches to Gall Midge Resistance in Rice*. November 22–24, Hyderabad, India, pp. 30.

Reddy, D.B. 1967. The rice gall midge *Pachidiplosis oryzae* (Wood Mason). In: *Major Insect Pests of the Rice Plant*. John Hopkins Press, Baltimore, p. 457–491.

Singh, B.N., Rajamani, S., Behera, L., Panda, A.R. and Subudhi, H.N. 2001. Panicle gall midge at CRRI farm. *CRRI News Lett.*, 22(1): 11–12.

Tan, Y., Pan, Y. and Zang, Y. 1993. Resistance to gall midge (GM), *Orseolia oryzae* in Chinese rice varieties compared with varieties from other countries. *Int. Rice Res. Notes*, 18(4): 13–14.

Chapter 18

Influence of Carbon Sources on *In vitro* Seed Germination, Protocorm and Shoot Formation in *Vanilla planifolia*

*M. C. Gayatri and R. Kavyashree**

Plant Biotechnology Unit, Annexe-Department of Botany, Bangalore University, Bangalore – 560 056, Karnataka
**E-mail: cavya@yahoo.com*

ABSTRACT

The influence of sucrose, sorbitol or mannitol on *in vitro* seed germination, protocorm and shoot formation of scandent epiphytic orchid *Vanilla planifolia* Andrews was compared. All the three carbon sources, sucrose, sorbitol or mannitol were equally suitable for seed germination and PLB's formation on VW medium fortified with Kn (4 mgl^{-1}), CH (4 mgl^{-1}) and CM (10 per cent). However, when the PLB's obtained from *in vitro* germinated seeds were subcultured on VW medium having sucrose, sorbitol or mannitol for retrieval of shoots, then the polyols promoted significantly more growth of the shoots than sucrose. The shoots were rooted *ex vitro* and complete plantlets were successfully transferred to field with varying percentage of survival frequency. Statistical analysis revealed that mannitol is significantly superior to sorbitol and sucrose for seed germination and shoot formation.

Keywords: *Vanilla planifolia, Carbon source, Seed germination, Protocorm like bodies.*

Abbreviations

VW:	Vacin and Went medium	Kn:	Kinetin
CM:	Coconut Milk	CH:	Casein Hydrolysate

Introduction

Orchids, comprise one of the most popular ornamental crops for their exquisite and occasionally fragrant flowers. Depending upon the variety, they are commercially utilized for their cut flower production or as pot plants (Roy and Bannerjee, 2002). Propagation of orchids through conventional methods is not conducive to their mass scale production because they are slow growers and seed germination is rather poor under natural conditions (Rao, 1977). However, the orchid seeds germinate only by mycorrhizal association (Frank, 1892; Bernard, 1909; Downie, 1957; Burgeff, 1959; Campbell, 1964; Arditti, 1966). First successful asymbiotic germination of orchid seed was obtained by Knudson (1922) on sucrose enriched nutrient medium. Since then *in vitro* culture techniques has been used as an alternative method to symbiotic germination for obtaining large number of orchids (Knudson, 1929; Tanaka, 1957; Livingston, 1962; Rosa and Laneri, 1977; Arditti and Ernst, 1984; Suttle, 1996).

In vitro cultured seeds and tissues need an exogenous supply of carbohydrate as a source of carbon because such plants are not fully autotropic (Garcia *et al.*, 2002). In general, sucrose is chosen as the carbon source for establishment of *in vitro* cultures because it is the major transport-sugar of many plants (Murashige and Skoog's, 1962; Vacin and Went, 1949; Thorpe, 1982; Thompson and Thorpe, 1987). However, there are many reports available at present that number of plant species can be grown *in vitro* on different carbon sources other than sucrose (Pua and Chong, 1984; Nadel, 1989; Marchal *et al.*, 1992; Vu *et al.*, 1993). In majority of the cases these alternative carbon sources are sugar alcohols such as sorbitol or mannitol. These polyols are primary photosynthetic products that fulfill the functions of reduced carbon translocation and storage in a wide range of plant species (Steinitz, 1999).

The prime objective of the present work is to compare the influence of sugar–sucrose and sugar alcohols–sorbitol, mannitol on *in vitro* seed germination, protocorm formation and retrieval of shoots of *V. planifolia* to obtain large number of plants to disseminate into countryside for commercial exploitation.

Material and Methods

Vanilla planifolia Andrews an epiphytic orchid maintained in the Orchidarium, Plant Biotechnology Unit, Annexe-Department of Botany, Bangalore University, Bangalore – 560 056 was used as source material. It is widely cultivated throughout the tropics to extract commercial Vanilla essence. *V. planifolia* although flower regularly, fail to set fruits due to lack of pollinating insect. To overcome this constraint artificial pollination technique was employed for fruit setting. The fruit of *V. planifolia* is a capsule commercially called as a 'Bean' measuring upto 15 cm in length.

The mature bean of *V. planifolia* were harvested and washed thoroughly in tap water, surface sterilized using liquid detergent Labolene – 5 per cent (v/v), given a rinse in 70 per cent ethanol for 30 secs, disinfected in 0.1 per cent Mercuric chloride ($HgCl_2$) for 3 mins and rinsed several times in sterile water to remove the traces of Mercuric chloride ($HgCl_2$).

The sterilized beans were excised aseptically and tiny seeds were cultured on Vacin and Went basal medium (VWBM) fortified with Kn (4 mgl^{-1}), CH (4 mgl^{-1}) and CM (10 per cent) containing different concentrations ranging from 1-5 per cent of carbon sources such as sugar–sucrose and sugar alcohols–sorbitol, mannitol separately to study their influence on *in vitro* seed germination, protocorm and shoot formation. The cultures were maintained in the culture room at a temperature 25±2°C with photon flux density of 30–50 $\mu Em^{-2}s^{-1}$. The shoots were transferred to culture bottles containing horticultural grade soilrite mix–peat : perlite : vermiculate (1 : 1 : 1 v/v) with half strength MS salts supplemented NAA (1 mgl^{-1}) for *ex vitro* rooting. The *ex vitro* rooted plants were then transferred to field and their survival frequency was recorded.

Results

In the present investigation, the influence of different concentrations of sucrose, sorbitol or mannitol on the germination of seeds was studied (Table 18.1). Initial swelling of seeds was influenced by the presence of sugar sucrose in VW medium fortified with Kn (4 mgl^{-1}), CH (4 mgl^{-1}) and CM (10 per cent) in all the concentrations tried with 21.8–77.0 per cent germination response during the first phase of 79–114 days of culture (Table 18.1). But, further development of the seedlings depended on the optimum concentration of sucrose. It was noticed that when there is no supply of sugar to the medium, the seeds failed to germinate. Further, the swollen seeds differentiated into Protocorm Like Bodies (PLB's) after 84–145 days of culture during second phase (Table 18.2). However, in the third phase, optimum concentration of 3 per cent sucrose allowed maximum number of shoot formation 46.6 from PLB's after two passages of sub-culture (Table 18.2). Although, the seeds germinated in the increased concentration of sucrose (5 per cent), it did not differentiate into PLB's.

Table 18.1: Influence of Different Concentration of Various Carbon Sources on *in vitro* Seed Germination of *V. planifolia*

Type of Carbon Source	*Concentration of Carbon Source (%)*	*Seed Germination (%) X*±SD*	*No. of Days for Swelling of Seeds*
Sucrose	1	39.3±3.98	91
	2	63.8±2.78	82
	3	77.0±2.60	79
	4	53.5±1.87	103
	5	21.8±2.31	114
Sorbitol	1	47.5±1.87	68
	2	67.5±1.87	58
	3	88.6±2.16	45
	4	70.3±2.16	61
	5	59.3±2.80	86
Mannitol	1	63.3±1.86	62
	2	78.8±2.48	42
	3	98.1±1.94	30
	4	70.0±2.36	49
	5	59.0±2.00	70

*: Mean of 6 replications.

When sorbitol was used for the seed germination of this orchid, the results were almost similar to that of sucrose. In all the concentrations of sorbitol initial swelling of seed was observed with 47.5–88.6 per cent germination response during the first phase of 45–86 days of culture (Table 18.1). But, further development and proliferation of seeds into PLBs and retrieval of large number of shoots depended on the suitable concentration of sorbitol. In the present experiment, 3 per cent sorbitol was found to be the best concentration of carbon source to obtain maximum number of PLB's after 60 days of culture during the second phase and also for the development of maximum number of shoots (58.8) during third phase when compared to other concentrations tried (Table 18.2).

Table 18.2: Influence of Different Concentration of Various Carbon Sources on Protocorm and Shoot Formation of *V. planifolia*

Type of Carbon Source	*Concentration of Carbon Source (%)*	*No. of Days for Protocorm Formation*	*No. of Shoots/Explant X*±SD*	*Survival Frequency (%)*
Sucrose	1	126	23.8±2.32	33
	2	102	31.3±1.63	35
	3	84	46.6±1.63	55
	4	133	27.5±1.87	36
	5	145	0	0
Sorbitol	1	98	30.5±1.87	70
	2	63	42.3±2.16	69
	3	60	58.8±1.47	73
	4	76	47.5±1.87	42
	5	106	32.0±1.41	50
Mannitol	1	80	42.6±1.75	60
	2	57	63.1±1.47	70
	3	42	89.1±2.48	78
	4	77	60.0±2.61	63
	5	102	33.3±2.16	68

*: Mean of 6 replications.

Further, when mannitol was used for the seed germination of *Vanilla planifolia* the results were superior to those with sucrose and sorbitol. The highest 98.1 per cent seed germination frequency was noticed during the first phase at 3 per cent concentration of mannitol after 30 days of culture (Table 18.1). During second phase, the seeds further differentiated into PLB's after 42 days of culture. These PLB's developed into maximum number of shoots 89.1 during third phase (Table 18.2).

The shoots obtained from all the types of carbon sources were successfully rooted *ex vitro* and transferred to field, which established in soil with varying percentage of survival frequency. However, the highest survival frequency (78 and 73 per cent) was noticed in 3 per cent polyol (mannitol and sorbitol) enriched nutrient medium (Table 18.2).

The analysis of the results revealed that the *in vitro* seed germination, protocorm and shoot formation is strongly influenced by the carbon source employed. Mannitol was much more efficient followed by sorbitol and sucrose in promoting highest percentage of seed germination and maximum number of shoot formation. Further, the increased or decreased concentration of carbon sources did not promote the increased seed germination, protocorm and shoot formation.

Discussion

The *in vitro* seed germination has become a powerful tool to overcome the problems derived from the low ability of the seeds to germinate in the natural conditions (Garcia *et al.*, 1994, 2002; Sarmiento *et al.*, 1994; Acebedo *et al.*, 1997) and thus to speed up the process of obtaining large number of *in vitro* plants. The analysis of the results indicates that the *in vitro* seed germination is highly influenced by the presence of carbon source in the nutrient medium. This does not coincide with the findings of Garcia *et al.* (2002) in Olive who have reported that the *in vitro* germination of Olive zygotic embryos are independent of the presence of carbon source in the medium. In the present study, the source of carbon is required for the germinating seeds at initial stage without any specific preference for sucrose, sorbitol or mannitol to grow properly. Further, the use of mannitol as carbon source proved to be efficient than sorbitol and sucrose for protocorm and shoot formation. Moreover, the positive effect of mannitol on the development of shoots was persistent in further sub-cultures as suggested by Garcia *et al.* (2002). Further, in the present study it was found that sugar alcohols–mannitol followed by sorbitol could substitute sucrose for retrieval of plants. This is due to the fact that when mannitol is incorporated by the seeds *in vitro* reaching high concentration in the tissue, it can be metabolized into other carbohydrates and therefore promotes growth of the shoots *in vitro* more efficiently than sucrose or sorbitol as described by Leva *et al.* (1995).

Conclusion

From the foregoing results and discussion it can be concluded that the orchid seeds through *in vitro* culture can readily be regenerated in large scale. This facilitates the *in vitro* seed germination of rare, endangered and endemic orchid plants, which are in verge of genetic erosion thus allowing to obtain large number of plants in a relatively short period of time throughout the year under controlled conditions for *ex situ* conservation. This will help in future for large scale production of commercially important orchids to cater the domestic and international market.

Acknowledgement

Thanks are due to Department of Biotechnology, Ministry of Science and Technology, New Delhi, Government of India, for the financial assistance to carry out this work.

References

Acebedo, M.M., S. Lavee, J. Linan and A. Troncoso. 1997. *In vitro* germination of embryos for speeding up seedling development in olive breeding programmes. *Sci. Hort.*, 69: 207–215.

Arditti, J. 1966. The production of fungal growth regulating compounds by orchids. *Orch. Dig. Sci. Suppl.*, 30(3): 88–90.

Arditti, J. and Ernst, R. 1984. Physiology of germinating orchid seeds. In: *Orchid Biology: Reviews and Perspectives, III*. Cornell Univ. Press, New York, p. 177–222.

Bernard, N. 1909. L'evolution dans la symbiase, les orchidees et leurs champignons commensaux. *Ann. Sci. Nat. Bot. Ser.*, 9: 1–196.

Burgeff, H. 1959. Mycorrhiza of orchids. In: *The Orchids*, (Ed.) C.L. Withner. The Ronald Press Co., New York, p. 361–395.

Campbell, E.O. 1964. The fungal association in a colony of *Gastrodia sesamoides. R. Br. Trans. Roy. Soc.*, New Zealand, 2(18): 237–246.

Downie, D.G. 1957. *Corticium solani*, an orchid endophyte. *Nature*, 179: 160.

Frank, A.B. 1892. Lehrbuch der Botanik Bd. I.W. Engelmann, Leipzig, pp. 264.

Garcia, J.L., R. Sarmiento, A. Troncoso and C. Mazuelos. 1994. Effects of nitrogen source and concentrations on N fractions in olive seedlings. *Acta Hortic.*, 376: 193–196.

Garcia, J.L., J. Troncoso, R. Sarmiento and A. Troncoso. 2002. Influence of carbon source and concentration on the *in vitro* development of olive zygotic embryos and explants raised from them. *Plt. Cell Tiss. Org. Cult.*, 69: 95–100.

Knudson, L. 1922. Non-symbiotic germination of orchid seeds. *Bot. Gaz.*, 73: 1–25.

Knudson, L. 1929. Physiological investigations on orchid seed germination. *Proc. Int. Cong. Plant Sci.*, 2: 1183–1189.

Leva, A.R., R. Petruccelli, R. Muleo, R. Goretti and G. Bartolini. 1995. Influenza di fattori trofici, regolativi e condizioni del mezzo nutritivo sulla coltura *in vitro* di diverse cultivar di olivo. Atti del Convegno: L'Olivicoltura Mediterranea: stato e prospettive della coltura e della ricerca. Istituto Sperimentale per la Olivicoltura, Rende (Italy), p. 239–248.

Livingston, R.B. 1962. *Oncidium pusillum* blooming in flask. *Amer, Orch. Soc. Bull.*

Marchal, J., I. Sens and C. Teisson. 1992. Influence des sucre et de facteurs bioclimatiques sur la culture *in vitro* du bananier. *Fruit*, 47: 17–24.

Murashige, T. and Skoog, F. 1962. A revised medium for rapid growth and bioassays with tobacco tissue cultures. *Physiol. Plant.*, 15: 473–497.

Nadel, B.L., A. Altman and M. Ziv. 1989. Regulation of somatic embryogenesis in celery cell suspensions. I. Promoting effects of mannitol on somatic embryos, development. *Plt. Cell Tiss. Org. Cult.*, 18: 181–189.

Pua, E.C. and C. Chong. 1984. Requirement for sorbitol (D-glucitol) as carbon source for *in vitro* propagation of *Malus robusta* No. 5. *Can. J. Bot.*, 62: 1545–1549.

Rao, A.N. 1977. Tissue culture is the orchid industry. In: *Applied and Fundamental Aspects of Plant Cell Tissue Org. Culture*, (Eds.) Reinert, J. and Bajaj, Y.P.S.

Rosa, M.D. and Laneri, V. 1977. Modifications of nutrient solution for germination and growth *in vitro* of some cultivated orchids and for the vegetative propagation of *Cymbidium* cultivars. *Amer. Orch. Soc. Bull.*, 47: 813–820.

Roy, J. and Bannerjee, N. 2002. Rhizome and shoot development during *in vitro* propagation of *Geodorum densiflorum* (Lam.) Schltr. *Sci. Hort.*, 94: 181–192.

Sarmiento, R., J.L. Garcia, C. Mazuelos, J. Linan and A. Troncoso. 1994. Effect of the form and concentration of N on the growth and mineral composition of olive seedlings. *Acta Hortic.*, 356: 156–161.

Steinitz, B. 1999. Sugar alcohols display nonosmotic roles in regulating morphogenesis and metabolism in plants that do not produce polyols as primary photosynthetic products. *J. Plant Physiol.*, 155: 1–8.

Suttle, G.R.L. 1996. Commercial laboratory production. In: *Plant Tissue Culture: Concepts and Laboratory Exercises*, (Eds.) R.N. Trigiano, R.N. and Gray, D.J. CRC Press, Inc., New York, USA, p. 331–339.

Tanaka, H. 1957. Orchids flowered in flask. *Na Pua Okika O Hawaii. Nei.*, 7(3): 100–102.

Thompson, M.R. and T.A. Thorpe. 1987. Metabolic and non-metabolic role of carbohydrates. In: *Cell and Tissue Culture in Forestry*, (Eds.) Bonga, J.M. and Durzan, D.J. Martinus Nijhoff, Dordrecht, 1: 325–368.

Thorpe, T.A. 1982. Carbohydrate utilization and metabolism. In: *Tissue Culture in Forestry*, (Eds.) Bonga, J.M. and Durzan, D.J. Martinus Nijhoff/Dr W Junk, The Hague, p. 325–368.

Vacin, E. and Went, F. 1949. Some pH changes in nutrient solutions. *Bot. Gaz.*, 110: 605–613.

Vu, J.C.V., R.P. Neidz and G. Yelonosky. 1993. Glycerol stimulation of chlorophyll synthesis, embryogenesis, and carboxylation and sucrose metabolism enzymes in nucellar callus of 'Hamlin' sweet orange. *Plt. Cell Tiss. Org. Cult.*, 33: 75–80.

Chapter 19

Influence of INM on Availability and Uptake of Macronutrients to Rice (*Oryza sativa* L.) at Different Stages of Crop Growth

K. Hema and G. Swarajya Lakshmi

Department of Soil Science and Agricultural Chemistry, Agricultural College, Bapatla – 522 101

ABSTRACT

A pot culture experiment was conducted during *rabi* 2002, at net house, Agricultural College, Bapatla on clayey soil with rice (*Oryza sativa* L.) as test crop Var. Vijetha (MTU 1001) to study the effect of organic manures and biofertilizer alone and in combination with inorganic nitrogen on growth, yield and uptake of macro nutrients by rice and available nutrient status of the soil. The treatments consisted of control (without manure or fertilizer) 50, 75 and 100 per cent RDFN through inorganic fertilizer, application of entire N through organics (PM, GM and FYM) and biofertilizer (*Azolla*) alone and also 50 and 75 per cent substitution of inorganic N through organics (PM, GM and FYM) and biofertilizer (*Azolla*). In total there were sixteen treatments laid out in a completely randomized design replicated thrice. The highest grain and straw yields and uptake of macronutrients (N, P and K) were obtained by the application of 75 per cent RDFN through inorganic fertilizer along with 2.0 t PM ha^{-1} at all the stages of crop growth. The available nutrient status (N, P and K) was highest in application of 8.0 t PM ha^{-1} at all the stages of crop growth.

Introduction

Indiscriminate use of chemical fertilizers also results in the deficiency of nutrients other than the applied and decline in organic carbon (Singh *et al.*, 1999).

However, the present fertilizer consumption ratio ($N : P_2O_5 : K_2O$) is 8.9 : 2.8 : 1.0 as against the optimal ratio of 4 : 2 : 1. This leads to much greater drain on native soil fertility and soil will not support high production levels in future. It has been demonstrated that inorganic N fertilizers, generally are not efficiently utilized by rice and are prone to heavy losses as N gases (Dedatta and Buresh, 1989).

The current decline in organic matter levels in many potential areas of productivity is a matter of great concern for Indian agriculture. To compensate this, there is a need to develop integrated nutrient management (INM) system. Hence, proper blending of chemical fertilizers which locally available not only improves soil health but also helps to maximize the sustainable production.

Materials and Methods

A pot culture experiment was conducted during *rabi* 2002 at nethouse of Agricultural college, Bapatla. It is located at 80° 3' East longitude, 15° 84' North latitude at an elevation of 5.4 m above mean sea level. The soil needed for the experiment was collected from top 30 cm depth from the Kondubotla vari palem village near Bapatla. The experimental soil characterstics are presented in Table 19.1. In total there were sixteen treatments laid out in completely randomized design (CRD) and replicated thrice. The experiment was conducted with rice variety MTU 1001 (Vijetha). Different types of organic manures and biofertilizer were used in the experiment. The nutrient composition of the manures and biofertilizer is presented in the Table 19.2. 12 kg capacity cement pots were chosen, cleaned and created submerged conditions and measures were taken to prevent leaching of nutrients.

Treatmental details of the experiment were Azolla, Farmyard manure (FYM), Daincha (GM) and Poultry manure (PM) either alone or in combination with inorganic fertilizer.

Table 19.1: Physico-chemical Characteristics of the Experimental Soil

Sl.No.	*Characteristics*	*Value*
1.	Soil reaction (pH) (1 : 2.5 soil water suspension)	8.2
2.	EC (dS m^{-1}) (1 : 2.5 soil water suspension)	0.42
3.	Organic carbon (%)	0.432
4.	CEC (c.mol) (p^+) (kg^{-1})	39.79
5.	Available nitrogen (kg ha^{-1})	282.50
6.	Available phosphorus (P_2O_5 kg ha^{-1})	39.20
7.	Available potassium (K_2O kg ha^{-1})	504.00
8.	DTPA extractable iron (ppm)	2.079
9.	DTPA extractable manganese (ppm)	5.22
10.	DTPA extractable copper (ppm)	2.995
11.	DTPA extractable zinc (ppm)	0.858
	Particle size analysis	
	Sand (%)	26.0
	Silt (%)	12.0
	Clay (%)	62.0
12.	Texture	Clay

Recommended dose of fertilizer 120–60–30 kg NPKha^{-1} applied in the form of urea, single super phosphate and muriate of potash to supply N, P and K respectively. The entire dose of P, K and half of the N were applied basally and mixed thoroughly with soil to all treatments except control. The balance of the nitrogen was applied at active tillering and panicle initiation stages in two equal splits. Thirty days old seedlings were transplanted. Farmyard manure, poultry manure and green manure (Daincha) were applied fifteen days before transplanting for decomposition, where as *Azolla* was applied fifteen days after transplantation. Three hills were maintained in each pot and all the recommended measures were followed meticulously.

Table 19.2: Nutrient Composition of Different Manures Used in the Experiment

Source	*N (%)*	*P_2O_5 (%)*	*K_2O (%)*	*Fe (ppm)*	*Zn (ppm)*	*Cu (ppm)*	*Mn (ppm)*
FYM	1.22	0.62	0.90	308.8	62.2	83.9	291
Poultry manure	3.79	3.09	2.12	920.0	231.1	306.0	447.0
Daincha	2.01	0.22	2.03	405.0	64.7	68.0	329.0
Azolla	2.38	0.24	1.90	93.82	12.90	28.8	133.0

T_1: Control
T_2: Recommended Dose of Fertilizer Nitrogen (RDFN)
T_3: 75 per cent RDFN
T_4: 50 per cent RDFN
T_5: 6 t *Azolla* ha^{-1}
T_6: 75 per cent RDFN + 1.5 t *Azolla* ha^{-1}
T_7 50 per cent RDFN + 3.0 t *Azolla* ha^{-1}
T_8: 20 t FYM ha^{-1}
T_9: 75 per cent RDFN + 5.0 t FYM ha^{-1}
T_{10}: 50 per cent RDFN + 10.0 t FYM ha^{-1}
T_{11}: 20 t Daincha ha^{-1}
T_{12}: 75 per cent RDFN + 5.0 t Daincha ha^{-1}
T_{13}: 50 per cent RDFN + 10.0 t Daincha ha^{-1}
T_{14}: 8 t Poultry manure ha^{-1}
T_{15}: 75 per cent RDFN + 2.0 t Poultry manure ha^{-1}
T_{16}: 50 per cent RDFN + 4.0 t Poultry manure ha^{-1}

Plant samples were collected at active tillering, panicle initiation stages and also at harvest for computing dry matter production. The samples were dried at 60°C in an hot air oven to a constant weight and powdered with the help of plant sample grinder and were analysed for N, P and K. The plant N content was estimated by microkjeldhal method as outlined by Piper (1966). Wet digestion method as described by Piper (1966) was followed for determining phosphorus and potassium. Phosphorus was determined by vandomolybdate yellow colour method and potassium by Flame photometry.

Results and Discussion

Grain and Straw Yield

The results showed that manures and biofertilizer recorded significantly higher grain and straw yields over control and recommended dose of fertilizer (Table 19.3). In all the treatments the treatment

T_{15} (2.0 t PM + 75 per cent RDFN) recorded the highest grain and straw yields (33.10 g pot^{-1} and 60.6 g pot^{-1} respectively). While T_1 recorded the lowest grain and straw yields followed by T_5 (6 t *Azolla*). Among individual sources T_2 (100 per cent RDFN) recorded the highest grain and straw yields followed by poultry manure, which might be due to immediate availability of nutrients during early stages of crop growth. The low yield by organics and biofertilizer might be due to the low supply of N through organics in the initial stages and because of wider C : N ratio, slow mineralisation might have occurred. In fact nitrogen, phosphorus and potassium are the most common elements limiting rice growth and yield. Adequate supply of these nutrients to crop help in synthesis of carbohydrates, which in turn are required for the formation of protoplasm, thus resulting in higher cell division, cell enlargement, favouring improvement of growth characters and ultimately recording the highest grain and straw yields. However, among the combinations, 2.0 t PM along with 75 per cent RDFN resulted in the highest grain and straw yields. This might be due to the immediate availability of N through inorganic nitrogen and later by the mineralization of N through poultry manure resulting in steady supply of nutrients throughout the crop growth period. These results were in accordance with Singh *et al.* (1997).

Table 19.3: Grain and Straw Yields (g pot^{-1}) at Different Stages of the Crop Growth as Influenced by INM

Treatments	*Yield (g pot^{-1})*	
	Grain	*Straw*
T_1 Control	13.51	24.77
T_2 Recommended Dose of Fertilizer Nitrogen (RDFN) 120-60-30 kg NPK ha^{-1}	28.28	51.42
T_3 75% RDFN	26.39	48.35
T_4 50% RDFN	24.86	45.32
T_5 6 t *Azolla* ha^{-1}	23.22	43.49
T_6 75% RDFN + 1.5 t *Azolla* ha^{-1}	26.16	48.18
T_7 50% RDFN + 3.0 t *Azolla* ha^{-1}	24.00	44.23
T_8 20 t FYM ha^{-1}	25.37	46.76
T_9 75% RDFN + 5.0 t FYM ha^{-1}	29.04	53.18
T_{10} 50% RDFN + 10.0 t FYM ha^{-1}	26.86	49.16
T_{11} 20 t Daincha ha^{-1}	26.28	48.43
T_{12} 75% RDFN + 5.0 t Daincha ha^{-1}	31.36	57.10
T_{13} 50% RDFN + 10.0 t Daincha ha^{-1}	28.16	51.71
T_{14} 8 t Poultry manure ha^{-1}	27.66	56.50
T_{15} 75% RDFN + 5.0 t Poultry manure ha^{-1}	33.10	60.60
T_{16} 50% RDFN + 10.0 t Poultry manure ha^{-1}	31.74	57.40
S.E.m±	1.27	1.65
CD (P = 0.05)	3.53	4.5

Application of 50 per cent and 75 per cent N through inorganic fertilizer and rest of the N supplied through organic sources, recorded higher grain and straw yields compared to application of inorganic fertilizer alone. Similar results were reported by Bhandari *et al.* (1992) and Omarhattab *et al.*

(1998). A positive effect of organic and inorganic sources of nitrogen alone and in combination was also mentioned by Hemalatha *et al.* (2000) and Jagadeswari and Kumara Swamy (2000).

Available N, P and K

The available N, P_2O_5 and K_2O in soil was significantly influenced by the types and levels of manures and biofertilizer either alone or in conjunction with inorganic fertilizer N (Table 19.4). Significantly higher values were recorded at early stages of crop growth *i.e.,* active tillering as compared to the rest of the stages of crop growth. During later stages the available nutrients might have taken up by the crop.

Table 19.4: Effect of Integrated Nutrient Management (INM) on Soil Available Macronutrients N, P_2O_5 and K_2O (kg ha^{-1}) at Different Stages of Crop Growth

Treatments	Active Tillering Stage			Panicle Initiation Stage			Harvest		
	N	P_2O_5	K_2O	N	P_2O_5	K_2O	N	P_2O_5	K_2O
T_1	203.9	22.49	312.13	191.62	20.32	279.13	175.08	18.73	204.13
T_2	224.54	26.01	400.8	218.94	32.49	315.8	184.81	36.56	240.8
T_3	219.51	24.73	380.8	215.98	30.53	295.8	183.74	37.76	220.8
T_4	206.46	23.05	375.86	211.02	29.26	324.2	179.55	36.56	215.88
T_5	252.45	29.45	389.63	226.5	34.85	314.63	189.89	40.1	239.53
T_6	236.39	26.78	367.5	212.94	32.49	292.5	183.42	38.43	217.36
T_7	248.06	27.95	386.5	220.2	33.18	312.5	186.09	39.25	236.43
T_8	290.66	32.91	410	258.4	37.74	325	210.32	42.28	250.4
T_9	274.33	30.06	385	226.42	35.16	300	190.32	39.92	225
T_{10}	285.83	31.98	401.63	233.82	36.66	316.63	194.85	41.43	241.23
T_{11}	320.83	36.57	434.56	272.41	39.68	335	232.21	45.72	274.23
T_{12}	298.33	34.14	392.5	232.36	38.16	307.5	205.83	43.28	232.5
T_{13}	310.46	35.16	420	253.36	39.1	315.8	220.56	45.28	260.5
T_{14}	356.33	40.85	476.63	278.95	45.16	391.3	247.98	50.56	316.4
T_{15}	335.66	36.5	394.3	247.28	39.2	309.3	209.21	46.74	234.6
T_{16}	342.5	38.6	464.53	259.64	42.36	379.53	225.32	49.24	302.63
S.E.m±	16.12	1.53	20.83	8.22	1.47	12.56	8.02	0.95	10.82
CD (P = 0.05)	44.64	4.25	57.74	22.78	4.08	34.83	22.32	2.63	30.01

Among all the treatments T_{14} (8.0 t PM) recorded the highest availability of N (356.33 kg ha^{-1}, 278.95 kg ha^{-1} and 247.98 kg ha^{-1} at active tillering, panicle initiation stages and at harvest respectively), P_2O_5 (40.85 kg ha^{-1}, 45.16 kg ha^{-1} and 50.56 kg ha^{-1} at active tillering, panicle initiation stages and at harvest respectively) and K_2O (476.63 kg ha^{-1}, 391.63 kg ha^{-1} and 316.40 kg ha^{-1} at active tillering, panicle initiation stages and at harvest respectively) at all the stages of crop growth and was on par with T_{15} (2.0 t PM + 75 per cent RDFN) and T_{16} (4.0 t PM + 50 per cent RDFN) at active tillering stage. Similar trend was observed in GM and FYM treated pots. This might be due to the slow release of N, P and K by manures on mineralisation during the crop growth period. Similar results were observed in rice-rice system (Tiwari *et al.,* 1980) and rice-wheat system (Gill and Meelu, 1982).

The lowest available N, P_2O_5 and K_2O were recorded by treatment T_1 (Control) followed by T_4 (50 per cent RDFN), T_3 (75 per cent RDFN) and T_2 (100 per cent RDFN) which might be attributed to the maximum utilization of soil available nutrients by the crop.

Among the green manure combinations T_{11} (20 t GM), among the FYM combinations T_8 (20 + FYM) and among *Azolla* combinations T_5 (6.0 t *Azolla*) recorded the highest available N, P_2O_5 and K_2O. It might be due to slow release of nutrients by the manures under submerged conditions as observed by Sri Ramachandra Sekharan *et al.* (1996).

The build up of the higher available nitrogen in poultry manure treated pot might be due to the fact that PM is rich source of nutrients, which was reflected in high contents of these elements. Apart from this, addition of poultry manure might have also helped in elongation of root system because it contains growth promoting hormones that produce better root growth than fertilizer and thereby helping the crop to exploit more volume of soil for its nutrition (Srivastava and Khanna, 1974).

The build up of available P_2O_5 might be due to the organic acids, which were released during the microbial decomposition of organic matter which helped in the solubility of native insoluble phosphates in sub-merged conditions, thus increasing the available phosphorus content in soil. Besides this, appreciable quantities of carbon dioxide, released during the decomposition of organic matter might be formed into carbonic acid, which might have enhanced the solubility of phosphates thereby resulting in higher availability of phosphate in pots treated with organic manures. Similar results reported by Bhandari *et al.*, 1992, Bharadwaj and Omanwar (1994) and Sharma *et al.* (2001) and it is also possible that organic anions and hydroxy acids liberated during organic matter decay might have immobilized Fe and Althrough complex formulation or chelation thereby preventing phosphate ions from reacting with Fe and Al, and thus resulting in increased P availability to crop. Greater availability of nutrients to plants in the presence of organic manures might be due to release of organic acids which have solubilizing effect on fixed forms of nutrients in soil.

The build up of available K_2O in soil compared to control might be due to the direct addition of potassium to available pool of the soil, besides the reduction in potassium fixation and release of potassium due to interaction of organic matter with clay. Similar beneficial effects of organics on available K were reported in case of FYM by Bhardwaj *et al.* (1982) and Bharadwaj and Omanwar (1994). Similar results of Bindra and Thakur (1996), Omarhattab *et al.* (1998) and Sharma *et al.* (2001). The decrease in the available K_2O content at active tillering stage to harvest stage from the soil in all the treatments is attributed in uptake values Bhandari *et al.* (1992) reported the same.

Uptake of Macronutrients (N, P and K)

The results showed (Table 19.5) that the uptake of N, P and K were significantly influenced by the type and level of manure and fertilizer either alone or in conjunction with inorganic fertilizer N. Among all the treatments, the treatment T_{15} (2.0 t PM + 75 per cent RDFN) recorded the highest uptake of nitrogen, phosphorus and potassium at all the stages of the crop growth. It might be due to the steady and increased availability of nutrients from poultry manure due to rapid mineralisation which might have resulted in increased uptake of nutrients by plant. Increase in the uptake of nitrogen, phosphorus and potassium from soil amended with poultry manure was reported by Maskina and Randhawa (1983), Madhavi *et al.* (1996), Singh *et al.* (1997) and Dosani *et al.* (1999). The poultry manure combinations performed better than their corresponding combinations of GM, FYM and *Azolla*. It might be due to the steady increased availability of nutrients to plant.

Among the individual sources, the treatment T_2 (100 per cent RDFN) recorded the highest uptake of nitrogen, phosphorus and potassium which might be due to the immediate availability of N which

Table 19.5: Effect of Conjunctive Use of Organic and Inorganic Sources of Nitrogen on Macronutrient (N, P and K) Uptake (mg pot^{-1}) of Rice at Different Stages of Crop Growth

Treatments	*N Uptake (mg pot^{-1})*				*P Uptake (mg pot^{-1})*				*K Uptake (mg pot^{-1})*			
	Active Tillering Stage	*Panicle Initiation Stage*	*Harvest*		*Active Tillering Stage*	*Panicle Initiation Stage*	*Harvest*		*Active Tillering Stage*	*Panicle Initiation Stage*	*Harvest*	
			Grain	*Straw*			*Grain*	*Straw*			*Grain*	*Straw*
T_1 Control	65.37	135.88	99.61	171.73	29.30	37.30	94.53	42.68	98.88	138.50	54.93	262.55
T_2 Recommended Dose of Fertilizer Nitrogen (RDFN) 120-60-30 kg NPK ha^{-1}	140.32	337.60	249.91	508.48	111.9	150.16	311.03	175.40	260.32	356.14	207.38	791.86
T_3 75% RDFN	131.64	273.59	231.35	477.07	71.6	121.53	242.76	148.23	225.73	316.53	179.54	722.83
T_4 50% RDFN	123.10	249.55	208.82	424.71	70.5	110.56	208.78	100.10	162.65	280.98	154.41	626.92
T_5 6 t *Azolla* ha^{-1}	129.67	256.12	210.21	439.90	60.36	109.93	212.83	93.10	156.54	257.79	136.76	608.85
T_6 75% RDFN + 1.5 t *Azolla* ha^{-1}	166.37	287.96	252.63	449.68	131.83	138.10	320.00	155.10	268.06	340.25	200.56	783.72
T_7 50% RDFN + 3.0 t *Azolla* ha^{-1}	156.97	264.76	226.08	446.58	95.13	116.96	280.80	141.33	204.23	299.04	162.00	695.88
T_8 20 t FYM ha^{-1}	127.60	284.95	215.01	437.98	80.53	119.00	213.30	116.06	176.66	284.24	149.53	682.69
T_9 75% RDFN + 5.0 t FYM	184.21	334.37	287.29	513.18	155.66	170.60	379.56	193.13	301.76	434.64	232.32	916.46
T_{10} 50% RDFN + 10.0 t FYM ha^{-1}	165.74	303.68	262.86	467.02	133.30	153.00	330.33	186.33	257.79	353.55	179.06	806.22
T_{11} 20 t Daincha ha^{-1}	130.10	308.24	237.85	460.08	94.43	124.50	242.60	135.40	203.15	301.70	157.65	726.44
T_{12} 75% RDFN + 5.0 t Daincha ha^{-1}	199.05	378.14	357.08	607.16	176.03	219.06	439.00	255.40	350.65	492.10	282.24	1056.35
T_{13} 50% RDFN + 10.0 t Daincha ha^{-1}	182.67	350.14	332.05	585.50	134.06	187.20	373.53	214.73	320.13	430.31	234.66	818.73
T_{14} 8 t Poultry manure ha^{-1}	139.55	303.92	249.69	486.80	101.03	142.70	280.23	153.16	244.63	334.83	193.62	774.33
T_{15} 75% RDFN + 5.0 t Poultry manure ha^{-1}	238.41	412.91	460.30	711.03	231.10	297.16	517.56	431.60	416.74	591.25	308.90	1183.72
T_{16} 50% RDFN + 10.0 t Poultry manure ha^{-1}	208.03	397.72	439.17	688.80	191.63	260.23	457.03	386.16	354.16	525.94	275.08	1073.38
S.E.m±	5.79	19.14	12.02	26.24	7.27	8.72	9.35	8.75	11.89	21.14	12.05	32.08
CD (P = 0.05)	16.05	53.05	33.32	72.74	14.26	24.19	25.93	24.27	32.97	58.62	33.41	88.93

might have also enhanced the availability of P and K. Individual sources of manures (PM, FYM and GM) and biofertilizer (*Azolla*) recorded the low uptake which might be due to slow release of nutrients by the manures during the early stage of crop growth.

Among green manure combinations the treatment T_{12} (5.0 t PM + 75 per cent RDFN) recorded the highest nitrogen, phosphorus and potassium uptake. These are in conformity with findings of Sriramachandrasekharan *et al.* (1996). The combinations of FYM and *Azolla* recorded lower nitrogen, phosphorus and potassium uptake than GM combinations at all the stages of the crop growth. It might be due to higher availability of N in green manure (*Sesbania acculeata*) treated pots due to its high N content, coupled with succulent nature which might have gone rapid decomposition in submerged soil. The lower nitrogen, phosphorus and potassium uptake by *Azolla* treated pots might be due to higher lignin content which known to retard mineralization as reported by Sriramachandrasekhran *et al.* (1996). By increasing the dose of inorganic fertilizer N along with organic manures not only increased the uptake of N but also P and K which are inconcurrence with the results of Bhadari *et al.* (1992), Balasubramaniyan *et al.* (1993) and Deka Medhi *et al.* (1996).

However the treatment T_{15} (2.0 t PM + 75 per cent RDFN) appeared superior to T_{16} (4.0 t PM + 50 per cent RDFN) and T_{14} (8 t PM) which might be due to the addition of high mineral nitrogen along with low quantity of PM which reduce the C : N ratio and thus increased the rate of decomposition, resulting in faster availability of nutrients throughout the period of crop growth.

References

Balasubramaniyan, P., Palaniappan, S.P. and Francis, H.J. 1993. Effect of organic and inorganic N fertilization and population on nutrient uptake and yield of low land rice. *Madras Agril. J.*, 80(2): 1–3.

Bhandari, A.L., Anil Sood, Sharma, K.N. and Rana, D.S. 1992. Integrated Nutrient Management in a rice-wheat system. *J. Indian Soc. of Soil Sci.*, 40: 742–747.

Bharadwaj, V., Omanwar, P.K. and Bansal, S.K. 1982. Available potassium changes in soil under continuous rotational cropping of rice, wheat and cowpea. *J. Indian Soc. of Soil Sci.*, 30: 99–101.

Bharadwaj, V. and Omanwar, P.K. 1994. Long-term effect of continuous rotational cropping and fertilization on EC, PH, organic matter and available nutrients of soil. *J. Indian Soc. of Soil Sci.*, 42: 387–392.

Bindra, D. and Thakur, R.L. 1996. Influence of green manures along with fertilizers on nitrogen, Phosphorus and potassium contents in rice. *Oryza*, 33: 143–145.

Dedatta, S.K. and Buresh, R.J. 1989. *Advances in Soil Science*, 10: 143–169.

Dosani, A.A.K., Talashilkar, S.C. and Mehta, V.B. 1999. Effect of poultry manure applied in combination with fertilizers on the yield, quality and nutrient uptake of groundnut. *J. Indian Soc. of Soil Sci.*, 47(1): 166–169.

Gill, H.S. and Meelu, O.P. 1982. Studies on substitution of inorganic fertilizers with organic manures and their effect on soil fertility in rice-wheat rotation. *Fertilizer Res.*, 3: 303–314.

Hemalatha, M., Thirumurugan, V. and Balasubramaniyan, R. 2000. Effect of organic sources of nitrogen on productivity quality of rice (*Coryza sativa*) and soil fertility in single crop wet lands. *Indian J. Agron.*, 45 (3): 564–567.

Jagadeswari, V. and Kumaraswamy, K. 2000. Long-term effects of manure fertilizer schedules on the yield and nutrient uptake by rice crop in permanent manurial experiment. *J. Indian Soc. of Soil Sci.*, 48(14): 833– 836.

Madhavi, B.L., Suryanarayana, Reddy M. and Gopal Reddy, B. 1996. Effect of Poultry manure and fertilizers on yield and micronutrients uptake by maize. *J. Res., ANGRAU,* 24(2): 154–156.

Maskina, M.S. and Randhawa, N.S. 1983. Effect of organic manures and zinc levels on the availability of zinc, iron, manganese and copper to wet land rice. *Indian J. Agril. Sci.,* 53(1): 48–52.

Omarhattab, K., Natarajan, K. and Gopalswamy, A. 1998. Influence of different organic manures on yield and N use efficiency of rice. *J. Indian Soc. of Soil Sci.,* 40 (2): 239–242.

Piper, C.S. 1966. *Soil and Plant Analysis.* Hand Publishers, Bombay, p. 137–153.

Sharma, M.P., Bali, S.V. and Gupta, D.K. 2001. Soil fertility and productivity of rice (*Oryza sativa*)–wheat (*Triticum aestivum*) cropping system in an inceptisol as influenced by integrated nutrient management. *Indian J. Agril. Sci.,* 71(2): 82–86.

Singh, G.R., Parihar, S.S., Chaure, N.K. and Sharma, R.B. 1997. Integrated nutrient management in summer sesame. *Indian J. Agron.,* 42(4): 699–701.

Singh, N.P., Sachan, R.S., Panday, P.C. and Bishti. 1999. Effect of a decade long fertilizer and manure application on soil fertility and productivity of rice-wheat system in a Mollisol. *J. Indian Soc. of Soil Sci.,* 47: 72.

Sriramachanadrasekharan, M.V., Ramanathan, G. and Ravichandran. 1996. Effect of organic manures on the availability of nutrients in rice rhizosphere Soil. *Coryza,* 33: 126–131.

Srivastava, O.P. and Khanna, S.S. 1974. Organic manures as supplement nitrogen fertilizers. *News,* 19: 39–43.

Tandon, H.L.S. (Ed). 1993. *Methods of Analysis of Soils, Plants, Water and Fertilizers.* Fertilizers Development and Consultation Organisation, New Delhi, India, p. 144.

Tiwari, K.V., Tiwari, S.P. and Pathak, A.N. 1980. Studies on green manuring of rice in double cropping system in a partially reclaimed Saline Sodic Soil. *Indian J. Agron.,* 25: 136–145.

Chapter 20

Uptake of Nutrients by Maize and the Associated Weeds Under Integrated Weed Management

S.R. Ghodake, T.K. Tiwari and V.S. Pawar

Department of Agronomy, Post Graduate Institute, M.P.K.V., Rahuri – 413 722, District Ahmednagar, Maharashtra, India

ABSTRACT

A field experiment was conducted during *rabi* 1998–1999 at Post-graduate Institute Research Farm. Mahatma Phule Krishi Vidyapeeth, Rahuri, Distt. Ahmednagar (M.S.). Weed dry weight was significantly influenced by weed free treatment followed by alachlor 1 kg a.i. ha^{-1} with one hand weeding and metribuzin 1 kg a.i. ha^{-1} with one hand weeding. N, P and K uptake was significantly more in unweeded control while lowest uptake was registered in alachlor 1 kg a.i. ha^{-1} with one hand weeding.

N and P uptake by maize crop was highest in weed free treatment (141.98 and 27.24 Kg ha^{-1} respectively) followed by alachlor 1 kg a.i. ha^{-1} with one hand weeding. K uptake was highest in alachlor 1.5 kg a.i. ha^{-1} (126.62 kg ha^{-1}) followed by weed free treatment and alachlor 1 kg a.i. ha^{-1} with one hand weeding.

Keywords: *Maize, Weed control, Nutrient uptake, Alachlor, Metribuzin.*

Introduction

Weeds characterized by a fast growth rate in their seedling stage become a potential source of competition especially for soil nutrients. They interfere with efficiency of fertilizer utilization by crop plants, because of sizeable portion of the fertilizer added to the soil used by weeds. To minimize the nutrient drain by weeds and more effective utilization of fertilizers by crop plants require the control of weeds (Mani, 1975). Herbicides cause an appreciable decrease in nutrient depletion by weed growth as consequence of which considerable improvement in nutrient uptake of crop plants and enhancement in crop yields occur. Therefore, an attempt was made to develop suitable weed management practices for decreasing the nutrient removal by weeds and increasing the uptake by maize crop.

Materials and Methods

Field study was conducted on maize in medium black soil at Post-graduate Institute Research Farm, Mahatma Phule Krishi Vidyapeeth, Rahuri, Distt. Ahmednagar (M.S.), India during *rabi* 1998–1999. The soil of the experimental site was moderately alkaline (pH 8.1), with low N (263 kg ha^{-1}), medium P (18.20 kg ha^{-1}) and very high K (47.0 kg ha^{-1}) content. Nine treatments (Table 20.1) were tested in a randomized block design with four replications.

Table 20.1: Effect of Different Weed Management Treatments on Dry Weight of Weeds and Uptake of Nitrogen, Phosphorus and Potassium (kg ha^{-1}) by Weeds at Harvest of Maize Crop

Treatments	*Dose (kg a.i. ha^{-1})*	*Dry Weight of Weeds (q ha^{-1})*	*Nutrient Uptake by Weeds (kg ha^{-1})*		
			Nitrogen	*Phosphorus*	*Potassium*
Alachlor (PE)	1.5	3.54	2.89	1.74	5.31
Alachlor (PE) + 1 HW at 30 DAS	1.0	2.34	1.72	1.08	3.53
Alachlor (PE) + Glyphosate (Post em)	1.0	3.00	2.65	1.57	4.62
Metribuzin (PE)	1.0	3.94	3.05	2.12	5.68
Metribuzin (PE) + 1 HW at 30 DAS	1.5	2.53	1.78	1.24	3.76
Metribuzin (PE) + Glyphosate (Post em)	1.0	3.34	2.81	1.68	4.99
2 HW at 20, 40 DAS		3.61	2.93	1.80	5.28
Weed free		0.00	0.00	0.00	0.00
Unweeded control		12.35	19.37	5.62	17.99
CD (P = 0.05)		0.58	0.76	0.31	0.8

DAS: Days after sowing; HW: Hand weeding; PE: Pre-emergence; Post-em: Post emergence.

The 'African tall' variety was sown at a spacing of 75 cm × 30 cm and uniformly fertilized with 60 kg ha^{-1} each N, P and K as basal application. 60 kg N ha^{-1} was applied 30 days after sowing.

Alachlor and metribuzin were applied as pre-emergence spray on third day after sowing and glyphosate was applied as post-emergence spray on 30 day after sowing of maize. Three hand weedings were given on 30th, 45th and 60th day from sowing in weed free treatment. While two hand weedings were carried out on 20th and 40th day after sowing. One hand weeding was done in the plot with integrated weed management treatments on 30th day after sowing.

Species-wise weed population (per m^2), their dry weights and nutrient uptake by weeds and maize were recorded at harvest. The nitrogen content in maize plant and weed samples were estimated by Micro-Kjeldahl's method after destroying the organic matter by H_2O_2 digestion. Phosphorus was determined by vanadomolybdate phosphoric yellow colour colorimetric method in nitric acid system and potassium by flame photometer method (Chapman and Pratt, 1961). The uptake of NPK by crops and weeds were computed by multiplying per cent content and dry weight.

Results and Discussion

Various weed control practices significantly reduced the weed dry weight at harvest over unweeded control (Table 20.1). Pre-emergence application of alachlor 1 kg a.i. ha^{-1} + one hand weeding 30 DAS registered lowest weed dry weight which was statistically at par with pre-emergence application of metribuzin 1 kg a.i. ha^{-1} supplemented with one hand weeding at 30 DAS.

Nutrient Uptake by Weeds

The removal of nitrogen was maximum (19.37 kg ha^{-1}) in unweeded control, which was significantly more than those observed in rest of the treatments. Metribuzin 1.5 kg a.i. ha^{-1} was next in order of merit and was at par with those of alachlor 1.5 kg a.i. ha^{-1}, alachlor 1 kg a.i. ha^{-1} + glyphosate 1 kg a.i. ha^{-1}, matribuzin 1 kg a.i. ha^{-1} + glyphosate 1 kg a.i. ha^{-1} and two hand weeding treatments. Alachlor 1 kg a.i. ha^{-1} + one hand weeding recorded the lowest (1.72 kg ha^{-1}) uptake of nutrients by weeds and was at par with metribuzin 1 kg a. i. ha^{-1} + one hand weeding. Almost similar trend as that of nitrogen was observed in different treatments for phosphorus and potassium.

Nutrient Uptake by Maize Crop

The uptake of nitrogen and phosphorus was significantly higher (141.98 and 27.24 kg ha^{-1}) in weed free treatment over those by rest of the treatments followed by Alachlor 1 kg a.i. ha^{-1} + one hand weeding. But it was at par with that of Alachlor 1.5 kg a.i. ha^{-1}. Metribuzin treated plots recorded significantly lowest N and P uptake. This confirms the results reported by Sreenivas and Satyanarayan (1996).

Potassium uptake was highest in Alachlor 1.5 kg a.i. ha^{-1} (126.62 kg ha^{-1}) which was at par with those of weed free plot and Alachlor 1 kg a.i. ha^{-1} + one hand weeding.

Table 20.2: Effect of Different Weed Management Treatments on Uptake of Nitrogen, Phosphorus and Potassium (kg ha^{-1}) by Maize at Harvest

Treatment	*Dose (kg a.i. ha^{-1})*	*Nitrogen (kg ha^{-1})*			*Phosphorus (kg ha^{-1})*			*Potassium (kg ha^{-1})*		
		Grain	*Storer*	*Total*	*Grain*	*Storer*	*Total*	*Grain*	*Storer*	*Total*
Alachlor (PE)	1.5	57.44	65.57	123.01	8.85	16.25	25.10	11.55	115.07	126.62
Alachlor (PE) + 1 HW at 30 DAS	1.0	65.65	63.10	128.76	10.16	15.12	25.29	12.29	105.97	118.27
Alachlor (PE) + Glyphosate (Post em)	1.0	59.08	56.55	115.63	9.02	13.71	22.74	10.41	96.89	106.06
Metribuzin (PE)	1.0	19.50	23.05	42.55	3.09	6.28	9.37	4.04	43.89	47.94
Metribuzin (PE) + 1 HW at 30 DAS	1.5	32.29	29.37	61.66	5.13	7.86	13.00	6.65	60.03	66.69
Metribuzin (PE) + Glyphosate (Post Em)	1.0	26.03	22.01	48.04	4.14	6.07	12.40	5.40	42.14	47.54
2 HW at 20, 40 DAS	1.0	57.71	57.05	114.45	8.91	14.13	23.05	11.64	99.43	111.06
Weed free		72.56	69.42	141.98	11.13	16.10	27.24	14.29	111.88	126.17
Unweeded control		40.09	43.56	83.66	5.72	11.50	17.23	8.40	78.50	86.90
CD (P = 0.05)		4.10	6.38	8.01	0.06	1.64	2.92	1.58	8.82	8.89

However, there was drastic reduction in plant population, growth and yield of maize due to application of metribuzin and hence the total N, P and K uptake by maize crop was reduced in matribuzin treated plots.

Hence, it can be concluded that, weed free maintenance during first 60 days after sowing or pre-emergence application of Alachlor 1 kg a.i. ha^{-1} + one hand weeding 30 DAS is required to achieve potential crop yield.

References

Chapman, H.D. and Pratt, P.F. 1961. *Methods of Analysis of Soil, Plant and Water*. Div. of Agric. Sci. Univ. of California.

Mani, V.S. 1975. Nutrient drain by weed growth in crop fields. *Fertilizer News*, 20: 21–27.

Sreenivas, G. and Satyanarayan, V. 1996. Nutrient removal by weeds and maize. *Indian J. Agron.*, 41(1): 160–162.

Chapter 21

Effect of Different Levels of Gulkand on Chemical Composition and Organoleptic Quality of Ice Cream

J.N. Ahire, A.P. Chavan, S.P. Kalhapure and R.B. Walujkar
Mahatma Phule Krishi Vidyapeeth, Rahuri – 413 722, Maharashtra

ABSTRACT

Experiment were conducted to evaluate defect of different levels of Gulkand on chemical composition and organoleptic quality of ice cream revealed that the different levels of Gulkand had significant effect on total solids and solids not fat content in ice cream sample. The differences in flavour scores for various levels of Gulkand were significant ($P < 0.05$).

Keyword: *Ice cream, Chemical composition, Gulkand, Organoleptic quality.*

Introduction

Ice cream industry in India is comparatively of recent origin started in 1960's. The production of ice cream was estimated to 50,000 tonnes, having turn over of Rs. 600 crore (Anonymous, 1999). However, the production of ice cream has been showing upper trend ever since it introduced in Indian market different flavour ice cream is nowadays available and popularized in the market. Thus the ice cream industry in India has a bright future in a country as well as for export. The present investigation was therefore, undertaken to study the effect of incorporation of Gulkand in different levels on chemical and organoleptic quality of ice cream.

Materials and Methods

The milk and Gulkand used for preparation of ice cream was analysed for its composition as shown below by adopting the standard methods. It was observed from Table 21.1, that the values for different constituents of cow milk were well within the normal range. The chemical composition of cow milk was closely agreeable to that reported by Chavan (1985). The cream used for preparation of ice cream confined fat 39.00 per cent, total solids 44.81 per cent, solids not fat 5.81 per cent, with 0.11 per cent acidity (LA). The average composition of Gulkand used for trials had total soluble solids 68 per cent, protein 0.48 per cent, total sugar 66.67 per cent and acidity 0.08 per cent.

Table 21.1: Chemical Composition of Different Ingredients Used in Ice Cream Preparation (Mean of 5 Replications)

Constituents	*Mean*	
	Cow Milk	*Gulkand*
Fat %	3.90	–
Total solids %	12.15	–
Solids Not Fat %	9.25	–
Protein %	3.36	0.48
Acidity (% LA)	0.13	0.08
pH	6.70	–
Total soluble solids %	–	68.00
Total sugar	–	66.67

Results and Discussion

It could be seen from the Table 21.2, that the different levels of Gulkand had significant effect on the total solids content in ice cream sample. The ice cream prepared with addition of Gulkand 3 per cent (T_3) had maximum total solids (38.46), while minimum in T_0 (36.61 per cent). It was also noticed that as the per cent levels of Gulkand increased, the total solids content of ice cream increased. These differences may be due to the higher percentage of total solids content in Gulkand. The different levels of Gulkand had a non-significant effect on fat content in ice cream. This might be because Gulkand does not content fat and hence the results was non-significant. The different levels of Gulkand had non-significant effect on protein content of ice cream samples. The different Gulkand levels had a significant effect on the solids not fat content of ice cream.

Table 21.2: Effect of Different Levels of Gulkand on Chemical Composition of Ice Cream

Parameters	*Mean*				*SE±*	*CD at 5 % Levels*
	T_0	T_1	T_2	T_3		
Total Solids %	36.61	36.96	37.48	38.46	0.09	0.29
Fat %	10.66	10.66	10.66	10.67	0.008	NS
Protein %	4.10	4.10	4.10	4.20	0.06	NS
Solids Not Fat %	25.95	26.30	26.81	27.80	0.09	0.29

It was observed from Table 21.3, that the difference in flavour scores for various levels of Gulkand were significant ($P < 0.05$). This attributed to rose petals in Gulkand. The result of the current trials agree with the report of Kolpe (1995) that the ber pulp with 10 per cent kadaka ber pulp had significantly more rich flavour. The different levels of Gulkand had significant effect on texture of ice cream. The texture of milk product always depends upon total solids content. The different levels of Gulkand had a non-significant effect on colour of ice cream. These results agree with report of Kedar (1972) who stated that very little variation in colour of ice cream was observed between mango pulp and banana powder in ice cream mix. These results were moderate agreement with that results presented by Kolpe (1995) who reported that very little veriation in colour of ice cream was observed due to addition of different levels of Gulkand had a significant effect on overall acceptability of ice cream. The overall acceptability score was 71.01 indicating that the product was highly acceptable.

Table 21.3: Organoleptic Quality of Ice Cream Prepared by Using Gulkand

Parameters	*Mean*				*SE±*	*CD at 5 % Levels*
	T_0	T_1	T_2	T_3		
Flavour	34.00	38.00	39.80	42.80	0.22	0.68
Texture	20.70	20.60	22.40	23.70	0.04	0.14
Colour	4.25	4.39	4.35	4.51	0.06	NS
Acceptability	58.95	62.99	66.25	71.01	0.26	0.79

References

Anonymous. 1999. *Indian Agriculture.* Indian Economics Data Research Centre, New Delhi, p 481–483.

Chavan, K.D. 1985. Studies on chemical composition and processing properties of milk obtained from different crossbred. *M.Sc. (Agril.) Thesis,* MPKV, Rahuri, (MS), India.

Kedar, M.D. 1972. Effect of mixing of different levels of banana powder and mango pulp on over run and organoleptic quality of ice cream. *M.Sc. (Agril.) Thesis,* MPKV, Rahuri, (MS), India.

Kolpe, D.B. 1995. Utilization of ber pulp (*Zizyphus jujuba*) in preparation of ice cream. *M.Sc. (Agril.) Thesis,* MPKV, Rahuri, (MS), India.

Chapter 22

Seasonal Incidence of Diamondback Moth on Cabbage

A.P. Chavan, D.B. Pawar, D.B. Kadam and S.P. Kalhapure
Mahatma Phule Krishi Vidyapeeth, Rahuri – 413 722, Maharashtra, India

ABSTRACT

Studies on the seasonal incidence of *Plutella xylostella* Linn. with effect of different dates of transplanting in cabbage at MPKV, Rahuri during 2000–2001 revealed that the incidence of *P. xylostella* larvae was the highest in the month of March and April when the crop was transplanted in the month of January and February 2001, and lowest population in the month of April and May when the crop was planted in the month of March.

Keyword: *Cabbage, Diamondback moth, Plutella xylostella, Seasonal incidence.*

Introduction

Importance of vegetables as protective food in human diet is well known. Among vegetables crops, cruciferous (cole) crops are economically important throughout the world. One of the important constraints in the successful cultivation of these crops is the pest problem. Among lepidopterous pests recorded on cabbage crop, the diamondback moth, *Plutella xylostella* Linn. assumes greater significance and it is most serious and widely distributed throughout the world (Bonne Maison, 1965). Despite realizing the importance of pest management not much examples of very successful management of pests infesting different crops are available. In the present investigation an attempt was made to study the seasonal incidence of diamondback moth on cabbage.

Materials and Methods

The research work was conducted at the experimental field of All India Coordinated Vegetables Improvement Project, Mahatma Phule Krishi Vidyapeeth, Rahuri – 413 722 (India) during *rabi* and summer season 2000–2001 on 0.2 ha area. The experiment was planned using cabbage variety Golden Acre with 3 treatments in a Randomized Block Design. For studying the seasonal incidence of the *P. xylostella* L. the planting of cabbage was done every month starting from October to March spaced with 60 × 45 cm. The plants were raised by recommended agronomic practices without giving any insecticidal treatments. The larvae of the *P. xylostella* were counted on five randomly selected plants in each planting at weekly interval starting from 10 days after planting up to harvesting.

Results and Discussion

It could be seen from the Table 22.1, that the incidence of *P. xylostella* larvae was the highest in the month of March and April when the crop was transplanted in the month of January and February 2001. The lowest population of *P. xylostella* larvae was recorded in the month of April and may when the crop was planted in the month of March. The results are in conformity with the work carried out by Chen and Su (1986) who reported that the population peak in the field occurred from February to March. However, their further studies revealed that the peak in population were in the month of June to July and October to December which was not observed in the present study. Similarly Ong and Soon (1989) reported that the population peaks of diamondback moth occurred in February, April and November. The population peaks were noticed in the month of March which was also noticed in the present work.

Table 22.1: Incidence of Diamondback Moth from October 2000–March 2001 (No. of Larvae/Plant)

Date	Mean	Mean	Mean	Mean	Mean	Mean
	Oct.					
14–10–2000	1.70					
21–10–2000	2.40					
28–10–2000	2.50					
3–10–2000	2.23					
10–11–2000	5.36					
17–11–2000	5.50					
24–11–2000	5.76					
1–12–2000	5.00	Nov.				
8–12–2000	8.50	2.20				
15–12–2000	12.13	1.20				
22–12–2000		0.86				
29–12–2000		2.00				
5–1–2001		2.66				
12–1–2001		10.33				
19–1–2001		9.33				
26–1–2001		16.66				

Contd...

Table 22.1–Contd...

Date	*Mean*	*Mean*	*Mean*	*Mean*	*Mean*	*Mean*
2–2–2001		11.46	Dec.			
28–12–2000			3.43			
4–1–2001			5.93			
11–1–2001			5.26			
18–1–2001			4.46	Jan.		
25–1–2001			4.73	4.00		
1–2–2001			6.53	4.20	Feb.	
8–2–2001			2.46	4.40	4.06	
15–2–2001			5.20	5.16	4.30	
22–2–2001			4.66	5.70	5.23	
1–3–2001			4.40	5.93	5.86	
8–3–2001				10.13	6.53	Mar.
15–3–2001				15.13	6.66	–
22–3–2001				21.13	19.93	0.66
29–3–2001					23.73	1.93
5–4–2001					23.86	2.66
12–4–2001						0.86
19–4–2001						0.53
26–4–2001						0.60
3–5–2001						0.20
10–5–2001						0.066
17–5–2001						0.133
SE ±	0.09	2.08	0.96	0.80	0.86	0.24
CD at 5 % level	0.27	6.17	2.40	2.34	2.50	0.64

Khaire *et al.* (1987) reported that the incidence of diamondback moth peaked in the first fortnight of February. In the present studies, the incidence of *P. xylostella* was noticed in the month of October to May and peak was in the month of march. Similarly, Kadam and Patel (1956) reported that the incidence of *P. xylostella* was noticed from September to March.

It could be seen from the Table 22.2, that the yield of marketable cabbage heads was maximum in the crop planted during November, while yield decreased from the month of December and the lowest yield was recorded in the crop planted during March. Therefore, it could be seen that as the incidence of diamondback moth increased and the yield level decreased.

Table 22.2: Yield of Cabbage in Quintal/ha

Month	*Oct.*	*Nov.*	*Dec.*	*Jan.*	*Feb.*	*Mar.*
Yield q/ha.	128.78	133.5	66.99	86.60	47.84	31.89

Khaire *et al.* (1987) revealed in their study that the maximum yield of cabbage was obtained when transplanting took place during November and December. Shashidar *et al.* (1994) reported that highest average yield/plot were recorded in the crop planted in the first week of October.

References

Bonne Maison, L. 1965. Insect pests of crucifers and their control. *Ann. Rev. Ent.*, 10: 233–256.

Chen, C.N. and Su, W.Y. 1986. Ecology and control thresholds of the diamondback moth on crucifers in Taiwan. Proceedings of the first International workshop. Asian Vegetables Research and Development Center, Shanhua, Taiwan, p. 415–421.

Kadam, M.Y. and Patel, G.A. 1956. Pests of cruciferous, chillies, onions, cucurbits and bhendi: Crop pests and how to fight them. Govt. of Bombay Publication, p. 91–100.

Khaire, V.A., Lawande, K.E. and Ajri, D.S. 1987. Population dynamics of insect pests of cabbage. *Curr. Res. Rept.*, MPKV, 3(1): 27–31.

Ong, K.H. and Soon, S.C. 1989. Seasonal abundance and distribution pattern among host of diamondblack moth, *Plutella xylostella* L. *Singapore J. of Primary Industries*, 17(1): 29–40.

Shashidar Viraktamath, Shekarappa and Reddy, B.S. 1994. Effect of date of planting on the extent of damage by the diamondback moth, *Plutella xylostella* on cabbage. *Karnataka J. Agric. Sci.*, 7(2): 238–239.

Chapter 23

Genetic Diversity for Yield and its Attributing Traits in Rice (*Oryza sativa* L.)

K.K. Sarkar, K.S. Bhutia and S.K. Roy

Department of Plant Breeding, Bidhan Chandra Krishi Viswavidyalaya, Mohanpur, Nadia – 741 252, West Bengal

ABSTRACT

Genetic divergence between 41 genotypes of rice (*Oryza sativa* L.) from diverse sources was assessed using Mahalanobis D^2 statistics. The genotypes were grouped into five clusters. Cluster I consisted of highest number of genotypes (33) followed by cluster III with four genotypes, cluster II with two genotypes and cluster IV and V with a single genotype each respectively. The desirable yield and its attributing traits were mainly distributed in clusters III and V respectively. Owing to their highest inter-cluster divergence and possession of desirable traits, the genotypes from clusters III (Satabdi, Pusa Basmati, IR 69739-5-1-2 and IET 13546) and V (Addey) may be used in a hybridization programme to exploit heterosis and develop transgressive segregants.

Introduction

In any crop improvement programme through selection, genetic diversity in the crop is a prerequisite factor. This genetic diversity is due to the heritable potential variability already present in the crop due to prolonged natural and artificial selection (Roy *et al.*, 1993). An efficient crop breeding programme requires this genetic diversity in the genotypes involved, to produce high yielding varieties (Murthy and Arunachalam, 1996). In this context, the present investigation was carried out to determine the nature and magnitude of genetic diversity for yield and its attributing traits in a set of 41 rice genotypes.

Materials and Methods

The experimental material consisted of 41 genotypes of rice, collected from Regional Research Sub-Station (B.C.K.V.) Chakdah, Nadia, and were grown in the same research station in a randomized block design consisting of three replications. The seedlings were transplanted at a spacing of 20 × 15 cm between and within rows respectively. The required recommended agronomic practices were followed to obtain a good harvest. The observations were recorded from five randomly chosen plants from each replication for the following 13 characters:

1. Tillers/plant
2. Panicles/plant
3. Panicle length (cm)
4. Grains/panicle
5. Filled grains/panicle
6. 1000-grain wt (g)
7. Rough rice length (mm)
8. Rough rice breadth (mm)
9. Rough rice L/B ratio
10. Brown rice length (mm)
11. Brown rice breadth (mm)
12. Brown rice l/b ratio and
13. Yield/plant (g).

Analysis of variance, followed by multivariate analysis using D^2 statistics according to Mahalanobis (1936) was carried out with the data. The genotypes were grouped into different clusters using Tochers method (Rao, 1952).

Results and Discussion

Varietal differences were significant for yield and its attributing traits. On the basis of D^2 values the 41 genotypes were grouped into five clusters (Table 23.1). Cluster I consisted of the largest number of genotypes (33), followed by cluster III with four genotypes, cluster II with two genotypes and clusters IV and V were monogenotypic respectively. Maximum inter-cluster divergence was observed between the stocks Satabdi (IET 4786), Pusa Basmati, IR-69739-5-1-2 and IET 13546 falling in cluster III and Addey in cluster V (85.139), while minimum inter-cluster divergence was observed between clusters I and II respectively (Table 23.2). The intra-cluster divergence varied from 0.00 to 28.467, the maximum being in cluster III and minimum in clusters IV and V which were monogenotypic.

Table 23.1: Distribution of 41 Rice Genotypes in Different Clusters

Clusters No.	*Total No. of Genotypes*	*Name of Genotypes*
I	33	Vandana, Rupsail, IET 826, Sitasail, PNR 381, Jhapaka, Narendra, IRHON 1, Bidhan, RAU 1344, IR 50, Rasi, Kataribhog, Indrasail, Kakdwip SL-7, BJ 1, HR 12, Annada, Madhuri, IET 4094, PNR 579, IR 30, C9 IR 64, BG Dasheria, CRU-9, IR 71605-2-1-5-3, RAU 1345, Mala, Luit, Kakdwip SL 10, TN 1, Kankri
II	2	BT-111, Kakdwip selection II
III	4	Satabdi (IET 4786), Pusa Basmati, IR 697390-5-1-2, IET 13546
IV	1	Tulsibhog
V	1	Addey

Table 23.2: Average Intra and Inter-cluster Distance [sqrt. (D^2) values]

Cluster	*I*	*II*	*III*	*IV*	*V*
I	20.574				
II	29.277	15.422			
III	40.962	44.299	28.467		
IV	37.740	54.623	66.969	0.000	
V	53.118	53.162	85.139	42.882	0.000

Cluster mean of various traits (Table 23.3) revealed that cluster II was superior for 1000 grain weight, rough rice breadth, brown rice breadth and yield/plant. Cluster III exhibited highest value for tillers/plant, panicle/plant, rough rice length, rough rice L/B ratio, brown kernel length and brown kernel L/B ratio, whereas on the other hand cluster V was superior for panicle length, grains/ panicle and filled grains/panicle. Similar observations were made by Bansal *et al.* (1999), when he found in a stock of 34 genotypes that the clusters differed for yield/plant, tillers/plant and grains/panicle. The mean for the majority of yield attributing traits were highest in the clusters III and V and they had the highest inter-cluster divergence. So combination of genotypes from these two clusters may provide a large number of transgressive segregants, which would give an opportunity to select high yielding genotypes (Mokate *et al.,* 1998). The genotypes in cluster III (Satabdi, Pusa Basmati, IR 69739-5-1-2) have exhibited, appreciably high yield / plant and if they are intercrossed with the genotype in cluster V *i.e.,* Addey, there is enough scope for further yield improvement through exploitation of heterosis or by production of transgressive segregants, as cluster V is superior for important yield attributing traits like filled grains/panicle, besides panicle length and grains/ panicle. There is a similar scope in yield improvement on combination of stocks between cluster II (BT -111, Kakdwip selection II) and cluster III which have appreciable inter-cluster distance and cluster II showing highest yield/plant.

Table 23.3: Cluster Means for Yield and its Attributing Traits in Rice

Clusters	*Tillers/ Plant*	*Panicles/ Plant*	*Panicle Length (cm)*	*Grains/ Panicle*	*Filled Grains/ Panicle*	*1000-Grain wt (g)*	*Rough Rice Length (mm)*	*Rough Rice Breadth (mm)*	*Rough Rice L/B Ratio*	*Brown Rice Length (mm)*	*Brown Rice Breadth (mm)*	*Brown Rice L/B Ratio*	*Yield/ Plant (g)*
I	11.81	7.96	23.12	123.42	89.13	21.99	8.6	2.65	3.27	6.67	2.20	3.06	14.60
II	10.50	7.67	21.67	120.67	106.67	28.57	8.62	3.05	2.83	6.48	2.72	2.39	21.24
III	14.00	7.99	25.32	114.33	83.25	27.05	10.64	2.49	4.30	8.30	2.02	4.15	17.34
IV	8.67	5.00	26.27	145.00	103.67	13.21	6.93	2.30	3.01	5.87	1.77	3.32	6.84
V	9.67	7.00	26.63	165.67	115.33	18.17	5.63	3.03	1.86	4.53	2.50	1.81	14.69
Population Mean	11.83	7.85	23.41	123.96	90.41	22.50	8.69	2.65	3.31	6.75	2.21	3.11	15.00

References

Bansal, R.G., Saini, R.G., Rani, N.S. and Kaur, A. 1999. Genetic divergence in quality rice. *Oryza,* 36(1): 20–23.

Mahalanobis, P.C. 1936. On the generalized distance in statistics. *Proc. Nat. Inst. Sci.,* India, 2: 49–55.

Mokate, A.S., Mehetre, S.S. Bendale, V.W. and Birari, S.P. 1998. Genetic divergence in rice. *Advances in Plant Science,* 11(2): 189–192.

Murthy, B.R. and Arunachalam, V. 1996. The nature of genetic divergence in relation to breeding system in crop plants. *Indian J. Genet.,* 26: 188–198.

Rao, C.R. 1952. *Advance Statistical Methods in Biometrical Research.* John Wiley and Sons, New York, USA.

Roy, A. and Panwar, D.V.S. 1993. Genetic divergence in some rice (*Oryza sativa* L.) genotypes. *Ann. Agric. Res.,* 14(3): 276–281.

Chapter 24

Role of *Azospirillum* for Enhancing the Efficacy of Inorganic Nitrogen in Relation to Growth and Yield of Wheat (*Triticum aestivum* L.)

Gurkirpal Singh, K. Jatinder Singh, Sarbjit Singh Sooch and Sohan Singh Walia
Punjab Agricultural University, Ludhiana – 141 004, India

ABSTRACT

The effect of seed inoculation by biofertilizer *viz. Azospirillum* for judicious use of inorganic nitrogen (N) in wheat was studied for two crop seasons at Punjab Agricultural University, Ludhiana, India. Results revealed that application of 100 per cent N + 100 per cent P, which is recommended practice, yielded 48.5 q/ha wheat grains which increased to maximum value of 49.4 q/ha under application of 100 per cent N + 100 per cent P + Slurry + *Azospirillum*. However, the treatment 50 per cent N + 100 per cent P + Slurry + *Azospirillum*, with a saving of 50 per cent of nitrogen as compared to recommended dose of inorganic N, produced 47.8 q/ha wheat grains which was statistically at par with the grain yield of 48.5 q/ha recorded under recommended practice. Similarly, during second year of study, the treatment 50 per cent N + 100 per cent P + Slurry + *Azospirillum* produced 44.3 q/ha of wheat grain which was at par with grain yield of 46.1 q/ha recorded under the recommended practice thereby saving 50 per cent of nitrogen over the recommended dose of inorganic nitrogen. This saving of inorganic nitrogen could have been possible due to enhanced nitrogen supply as a result of biological fixation of atmospheric nitrogen by nitrogen fixing bacteria "*Azospirillum*" in the presence of biogas slurry.

Keywords: *Wheat, Slurry, Biofertilizer, Azospirillum.*

Introduction

The total dependence of agricultural production system on use of inorganic fertilizers is not only adversely affecting the natural resources reserve but also adds to the cost of production. Excessive and unplanned application of these fertilizers, besides polluting under groundwater, is also producing certain undesirable chemicals in the soil thereby playing havoc to soil health and causing various diseases in plants and animals. The natural balance of overall food chain has been distorted due to these changes. To check the ill effects of excessive use of inorganic fertilizers, there is a need to find alternative nutrient supply systems. Healthy evidence of favourable effects of eco-friendly bio-fertilizers *viz. Azospirillum* and Phosphobacteria in providing nutrients to several crops and in supplementing the inorganic fertilizers have been reported (Balasubramanian *et al.*, 2000 and Vendan *et al.*, 2000).

Azospirillum is non-symbiotic nitrogen fixing bacteria which–when used as an inoculant–substantially contribute to the nitrogen content of soil. These bacteria besides producing vitamins and antibacterial, antifungal and growth promoting substances, also help in better seed germination and proliferation of root system thereby playing an important role in developing and establishing a balanced and effective integrated nutrient management system. Gupta and Pandher (1996) have reported beneficial effects of these bacteria on crops like rice, maize and wheat. Capability of *Azospirillum* in increasing the crop yield in different types of soils and climatic regions have also been reported (Subba Rao *et al.*, 1980 and Ganguly and Manna, 1990). In another important study Kumar *et al.* (1998) have cited that *Azospirillum* could compensate the nitrogenous requirement in field crops like pearlmillet, sorghum, maize, wheat, jute and mustard. However, the information on effect of *Azospirillum* on yield and yield attributing characters of wheat for climatic conditions of states like Punjab is lacking. Therefore, the present study was undertaken to evaluate the effect of *Azospirillum* inoculation for judicious use of inorganic nitrogenous fertilizer in wheat.

Location and Climate

Ludhiana, India–the site of experiment–representing the Indo-Gangetic alluvial plains is situated at 30°56′ latitude and 75°52′ E longitude and at an altitude of 247 metre above mean sea level.

Ludhiana is characterized by sub-tropical and semi-arid type of climate with quite hot summers and very cold winters showing considerable difference between mean maximum and mean minimum atmospheric temperatures coupled with a noticeable fluctuation in routine daily temperature during whole of the year. The average annual rainfall is 705 mm with a greater portion of about 80 per cent of the total being received as summer monsoon during the months of July to September. The rains during winter are scanty but few showers of cyclonic rains are received during December and January or late spring season.

Materials and Methods

A field experiment was conducted to evaluate the influence and efficiency of biofertilizer *viz., Azospirillum* along with varying levels of inorganic nitrogen fertilizer on wheat (PBW-343) during *rabi* seasons of 1998–1999 and 1999–2000 at Punjab Agricultural University, Ludhiana, India. The experimental soil was loamy sand in texture with a pH value of 8.1 (1 : 2, Soil : Water). It was low in organic carbon (0.30 per cent) and available N (140 kg/ha) but medium in P (15.4 kg/ha) and K (117 kg/ha). The experiment with five treatments replicated thrice was laid out in a randomized block design. The treatments included T_1; 100 per cent recommended N + 100 per cent recommended P, T_2; 100 per cent recommended N + 100 per cent recommended P + Slurry, T_3; 100 per cent recommended N + 100 per cent recommended P + Slurry + *Azospirillum*, T_4; 75 per cent recommended N + 100 per cent recommended P + Slurry + *Azospirillum* and T_5; 50 per cent recommended N + 100 per cent recommended P + Slurry + *Azospirillum*.

Biogas slurry 50 per cent by weight of the seed weight was taken and zinc sulphate (21 per cent Zn), 2 per cent of the seed weight was added to it. Coated the wheat seed separately with gum equal to 1 per cent of the seed weight and allowed the seed to dry for half an hour under shade. Mixed the biogas slurry containing zinc sulphate with gum coated seed and allowed the seed to dry for 24 hours under shade. Inoculated this slurry coated seed with *Azospirillum* taken 2 per cent by weight of the seed weight and dried the same for half an hour under shade. Crop was sown using 100 kg seed per hectare on 14th and 16th November during 1998–1999 and 1999–2002 respectively. 120 kg nitrogen and 60 kg P_2O_5 per hectare was used as recommended dose of inorganic fertilizer. The sources of N and P were urea (46 per cent N) and diammonium phosphate (18 per cent N and 46 per cent P_2O_5) respectively. Whole of phosphorus and half dose of nitrogen was applied at sowing time and remaining half of nitrogen was applied after first irrigation 21 days after sowing (DAS). The periodical data on biomass accumulation and plant height were recorded 60, 90, 120 days after sowing and at the time of harvesting of crop. The data on yield attributing characters *viz.*, effective tillers, grains per ear and 1000-grain weight and grain yield were recorded and statistically analyzed as per the method described by Cochran and Cox (1963).

Results and Discussion

Grain Yield

Application of 100 per cent N + 100 per cent P (T_1), which is the recommended practice, yielded 48.5 q/ha wheat grain during 1998–1999 (Table 24.1). The grain yield further increased to 49.4 q/ha under the treatments T_2: 100 per cent N + 100 per cent P + Slurry and T_3: 100 per cent N + 100 per cent P + Slurry + *Azospirillum*. In treatment T_4: 75 per cent N + 100 per cent P + Slurry + *Azospirillum*, with a saving of 25 per cent of nitrogen and in treatment, T_5: 50 per cent N + 100 per cent P with a saving of 50 per cent of nitrogen as compared to recommended dose of inorganic fertilizer, the grain yield was recorded to be 48.0 q/ha and 47.8 q/ha respectively both being statistically at par with the maximum grain yield of 49.4 q/ha produced under treatment T_2 and T_3 and the one 48.5 q/ha produced under T_1.

Table 24.1: Influence of Graded Levels of Inorganic Fertilizer (N) and Biofertilizer (*Azospirillum*) on Wheat Grain Yield and Harvest Index

Treatments	%age of Recommended Dose		Grain Yield (q/ha)		Harvest Index (%)	
	N	P	1998–1999	1999–2000	1998–1999	1999–2000
T_1; Seed	100	100	48.5	46.1	34.5	32.8
T_2; Seed + Slurry	100	100	49.4	46.8	34.5	32.9
T_3; Seed + Slurry + *Azospirillum*	100	100	49.4	46.9	33.9	33.1
T_4; Seed + Slurry + *Azospirillum*	75	100	48.0	45.7	34.3	32.7
T_5; Seed + Slurry + *Azospirillum*	50	100	47.8	44.3	34.3	32.8
C.D. (0.05)			N.S.	N.S.		

During the second year of study *i.e.*, 1999–2000, the treatment T_3 again recorded maximum grain yield of 46.9 q/ha followed by 46.8 q/ha under T_2. Grain yields of 45.7 q/ha and 44.3 q/ha were recorded under treatments T_4 and T_5 respectively which were at par with the maximum grain yield

(46.9 q/ha) produced under treatment T_3 and the one 46.1 q/ha produced under treatment T_1. In view of the results obtained, treatment T_5 resulting in saving of 50 per cent N as compared to the recommended practice can be rated as optimum treatment under present study. Such a positive role of *Azospirillum* inoculation in wheat crop has also been reported by Panwar *et al.*, (2000). Harvest index value was the highest under treatments T_1 and T_2 followed by the value under T_4 and T_5 during first year but highest under treatment T_3 followed by that under T_2 and then that under T_I and T_5 during second year (Table 24.1).

Yield Attributing Characters

As listed in Table 24.2, maximum number of effective tillers (119) were recorded under treatment T_3; 100 per cent N + 100 per cent P + Slurry + *Azospirillum* followed by those (117) produced under treatment T_2; 100 per cent N + 100 per cent P + Slurry. Effective tillers produced under treatments T_4 and T_5 were at par with those produced under T_3. Further, treatment T_3 produced highest number of grains per ear *i.e.*, 82.7 followed by 82.6 produced under treatment T_4 and then 82.5 produced under treatment T_5. All the five treatments were found to be statistically at par in producing number of "grains per ear". Similar parity was observed for 1000-grain weight (gms). This may be due to the fact that inoculation with *Azospirillum* helped in increasing allocation of nitrogen and biomass to the seeds, resulting in heavier seeds, with a lower amount of nitrogen remaining in the straw at physiological maturity of wheat, ultimately influencing the grain yield positively (Didonet *et al.*, 2000). Similar results were obtained during the second year of experimentation for these three yield attributing characters. Keeping in view results of yield attributing characters discussed above treatment T_5: 50 per cent N + 100 per cent P + Slurry + *Azospirillum* again came out to be the optimum treatment which ultimately contributed to wheat grain yield and resulted in saving of 50 per cent nitrogen as compared to recommended practice.

Table 24.2: Yield Attributing Characters of Wheat as Influenced by Graded Levels of Inorganic Fertilizer (N) and Biofertilizer (*Azospirillum*)

Treatments	%age of Recommended Dose		Effective Tillers (per m. Row Length)		Grains/Ear		1000-Grain Weight (gms)	
	N	P	1998–1999	1999–2000	1998–1999	1999–2000	1998–1999	1999–2000
T_1; Seed	100	100	114	111	81.6	74.5	38.3	36.3
T_2; Seed + Slurry	100	100	117	113	81.7	74.3	38.6	36.5
T_3; Seed + Slurry + *Azospirillum*	100	100	119	116	82.7	75.8	38.8	36.6
T_4; Seed + Slurry + *Azospirillum*	75	100	113	112	82.6	74.8	38.4	36.4
T_5; Seed + Slurry + *Azospirillum*	50	100	113	108	82.5	74.6	38.0	36.3
C.D. (0.05)			N.S.	N.S.	N.S.	N.S.	N.S.	N.S.

Periodic Plant Height

The data regarding periodic plant height (Table 24.3) indicates the trend of plant height gained by crop under different treatments. The plant height under treatment T_3 was maximum followed by that, with non-significant difference, under other treatments. Similar results were obtained during second year of study.

Table 24.3: Effect of Graded Levels of Inorganic Fertilizer (N) and Biofertilizer (*Azospirillum*) on the Periodic Plant Height (cm) of Wheat

Treatments	%age of Recommended Dose		60 DAS		90 DAS		120 DAS		At Harvest	
	N	P	1998–1999	1999–2000	1998–1999	1999–2000	1998–1999	1999–2000	1998–1999	1999–2000
T_1; Seed	100	100	41.3	41.2	79.5	78.3	103.2	103.5	105.7	105.6
T_2; Seed + Slurry	100	100	41.7	41.4	79.6	78.6	104.1	104.3	106.3	106.2
T_3; Seed + Slurry + *Azospirillum*	100	100	41.9	41.7	79.9	78.9	105.9	104.7	107.3	106.9
T_4; Seed + Slurry + *Azospirillum*	75	100	41.1	41.3	79.4	78.0	104.5	103.2	106.4	106.0
T_5; Seed + Slurry + *Azospirillum*	50	100	40.2	39.9	77.1	75.0	104.0	102.7	105.2	104.3

DAS: Days After Sowing.

Periodic Biomass Accumulation

Biomass production by crop is one of the important growth parameters to be considered while assessing the effect of different treatments and competitive ability of the crop. The data mentioned in Table 24.4, shows that with advancement in age, there was gradual increase in biomass accumulation of crop in all the treatments. The relevant values of biomass monitored 60, 90, 120 days after sowing and at harvest time were highest under treatment T_3 closely followed by those recorded under other treatments. Similar trends were also observed during second year of study. Such a spectacular influence of this bio-fertilizer on growth parameters, yield contributing characters and ultimately on grain yield of wheat was due to favourable change in microbial population in rhizosphere, production of plant growth hormones, supply of additional nitrogen.

Table 24.4: Effect of Graded Levels of Inorganic Fertilizer (N) and Biofertilizer (*Azospirillum*) on the Periodic Biomass Accumulation (q/ha) of Wheat

Treatments	%age of Recommended Dose		60 DAS		90 DAS		120 DAS		At Harvest	
	N	P	1998–1999	1999–2000	1998–1999	1999–2000	1998–1999	1999–2000	1998–1999	1999–2000
T_1; Seed	100	100	16.1	15.6	75.6	75.0	130.4	129.2	152.6	154.4
T_2; Seed + Slurry	100	100	16.8	16.5	76.6	77.1	133.1	132.1	157.1	160.4
T_3; Seed + Slurry + *Azospirillum*	100	100	17.0	17.1	78.1	77.6	133.9	132.8	165.1	167.1
T_4; Seed + Slurry + *Azospirillum*	75	100	14.4	14.0	70.8	70.3	115.5	114.6	155.1	161.6
T_5; Seed + Slurry + *Azospirillum*	50	100	14.1	13.5	62.8	61.5	112.6	110.8	147.3	149.3

DAS: Days After Sowing.

Conclusion

The addition of biofertilizer *viz.*, *Azospirillum* along with 50 per cent recommended N and 100 per cent recommended P produced statistically similar wheat grains as compared to that produced through application of 100 per cent recommended N and 100 per cent recommended P alone and thus, resulted in saving of 50 per cent inorganic nitrogen.

References

Balasubramanian, A., Salvaraju, R., Duraisamy, V.K. and Subbian, P. 2000. Combined effect of inorganic fertilizer with *Azospirillum* and Phospobacteria on rainfed sorghum. *Madras Agril. J.*, 1–3: 151–152.

Cochran, W.G. and Cox, G.M. 1963. *Experimental Designs*. Asia Publishing House, Calicut–New Delhi.

Didonet, A.D., Lima, O. dos S., Caudaten, A.A. and Rodrigues, S. 2000. Translocation of nitrogen and biomass to seed in wheat inoculated with *Azospirillum*. *Pesquisa Agropecuaria Brasileira,* Brazil, 35(2): 401–411.

Ganguly, T.K. and Manna, M.C. 1999. *Azospirillum brausilense*: An associative organism in agriculture: A review. *Agricultural Reviews,* Karnal, 20(3–4): 220– 228.

Gupta, R.P. and Pandher, M.S. 1996. Biofertilizers in Agriculture. *J. Res.,* Punjab Agricultural University. Ludhiana, 33(1–4): 209–224.

Kumar Chunchun, Agarwal, M.M., Gupta, B.R. and Kumar, C. 1998. *Azospirillum* and its potential as biofertilizer. *Fertilizer News*, 43(11): 47, 50.

Panwar, J.D.S., Singh, Ompal and Singh, O. 2000. Response of *Azospirillum* and *Bacillus* on growth and yield of wheat under field conditions. *Indian J. Plant Physiol.*, 5(1): 108–110.

Subba Rao, N.S., Tilak, K.V.B.R., Lakshami Kumari, M. and Singh, C.S. 1980. *Azospirillum*: A new bacterial fertilizers. *Indian Farming*, 30: 3.

Vendan, R.T., Karunanithi, K. and Subramanian, M. 2000. Response of rice fallow cotton to bioinoculants in combination with inorganic fertilizers. *J. Ecobiol.,* 12(2): 99–102.

Chapter 25

Studies on the Efficacy of Five Botanical Extracts as Pupicidal against *Trogoderma granarium* (Everts)

S.C. Dwivedi and Nidhi Bala Shekhawat

Ecotoxicology Laboratory, Department of Zoology, University of Rajasthan, Jaipur – 302004, India

ABSTRACT

A laboratory trial was designed to find the potential of five seed extracts, namely *Trigonella foenum graecum* (Methi), *Nigella sativa* (Kalongi), *Eruca sativa* (Taramira), *Carica papaya* (Papeeta) and *Cucumis sativus* (Khira) in two different solvents (Acetone and Pet ether).

The per cent pupal mortality and per cent corrected pupal mortality was calculated of the khapra beetle, *Trogoderma granarium.* Each extract was applied at 25, 50, 75 and 100 per cent conc. along with control. Acetone extract of *Trigonella foenum graecum* exhibited maximum per cent corrected pupal mortality at 100 per cent conc.

Keywords: *Trogoderma granarium, Pupicidal action, Seed extract.*

Introduction

Synthetic pesticides opened a new era for the control of stored product pests and much reliance was placed on the use of these chemicals. But lately, their continuous and non-judicious use have resulted in serious problems like contamination of environment and food disruption of non-target organisms, pest resistance etc. These adverse effects of insecticides again drew attention of scientists on the safer alternatives *i.e.*, the insecticides of plant origin. In the recent past the use of indigenous plant material has acquired an important position in the modern approach of pest control. Ahmed and Ahmed (1992), Kachare *et al.* (1994), Mishra (1996), Srivastava and Bhatt (1993) and Suryakala *et al.* (1995) are comeparatively safe due to their rapid biodegradable nature. Many workers (El. Nadi *et al.*, 2001; Hassan, 2001; Jood *et al.*, 1996 and Mostafa, 1993) have reported use of regional plants for botanical pesticides to prevent infestation of *Trogoderma granarium*. However very little literature is available on the pupicidal activity of plant extracts against *Trogoderma granarium*.

Keeping this in view the present investigation has been tarageted to determine the efficacy of seed extracts of *Trigonella foenum graecum* (Linn.), *Nigella sativa* (Linn.), *Eruca sativa* (Mill), *Carica papaya* (Linn.) and *Cucumis sativus* (Linn.) against pupae of *Trogoderma granarium*.

Material and Methods

The culture of *Trogoderma granarium* was reared at 35 ± 2° C and 60 ± 10 per cent R.H.

All seed extracts were prepared by soxhlet extraction method using acetone and pet ether as solvents and then the extract was filtered which was kept in refrigerator as stock solution (100 per cent conc.).

Newly transformed 20 pupae were separated from sub-culture and were treated by dipping method in different concentration of extract for 2 seconds and then pupae were transferred into vials for observations. The experiment was replicated thrice for each dose level and a control was also run using solvent. After 120 hr. total emergence and mortality was counted and per cent mortality and per cent corrected mortality was calculated by Abotts formula (1925):

$$(i)\ \%\ \text{Mortality} = \frac{\text{Mortality}}{\text{No. of insects}} \times 100$$

$$(ii)\ \%\ \text{Corrected Mortality} = \frac{\%\ \text{MT} - \%\ \text{MC}}{100 - \%\ \text{MC}} \times 100$$

where,

MC: Mortality in control; MT: Mortality in treated.

Results and Discussion

All the five seed extracts in the present study exhibited effective pupicidal activity at four conc. tested when compared with control. The extracts of *Trigonella foenum graecum* in both the solvents (acetene and pet ether) was most effective at 100 per cent conc. showing 93.2 per cent and 75.72 per cent corrected mortality respectively. At the same dose level, *Nigella sativa, Carica papaya, Eruca sativa* and *Cucumis sativus* acetone extract recorded 62.88 per cent, 61.08 per cent, 59.22 per cent and 57.54 per cent corrected mortality respectively. Whereas their pet ether extract exhibited 74.68 per cent, 42.85 per cent, 65.21 per cent and 60 per cent corrected mortality respectively.

At 75 per cent dose level acetone extract of *Trigonella foenum graecum* showed 86.66 per cent corrected mortality, however its pet ether extract exhibited. 72.72 per cent corrected mortality which is at par with *Nigella sativa* pet ether extract (70.06 per cent). At the same dose level acetone extract of *Carica papaya, Nigella sativa, Eruca sativa* and *Cucumis sativus* recorded 69.41 per cent, 59.22 per cent

55.55 per cent and 51.45 per cent corrected mortality whereas their pet ether extract expressed 39.67 per cent, 70.06 per cent, 44.86 per cent and 41.28 per cent corrected mortality respectively.

Table 25.1: Pupicidal Action of Seed Extracts Against *Trogoderma granarium*

Sl.No.	Name of the Plant	Family	Conc.	% Mortality		% Corrected Mortality	
				Acetone	Pet ether	Acetone	Pet ether
1.	*Trigonella foenum graecum*	Papilionaceae	25	81.66	56.65	63.32	21.18
			50	88.33	78.3	76.66	60.34
			75	93.33	85	86.66	72.72
			100	96.6	86.65	93.2	75.72
			Control	50	45	–	–
2.	*Nigella sativa*	Ranunculaceae	25	76.65	55	48.11	37.93
			50	78.3	71.65	51.77	60.89
			75	81.65	78.3	59.22	70.06
			100	83.3	81.65	62.88	74.68
			Control	55	27.5	–	–
3.	*Eruca sativa*	Brassicaceae	25	80	50	55.55	13.04
			50	78.3	68.3	51.77	44.86
			75	80	68.3	55.55	44.86
			100	81.65	80	59.22	6521
			Control	55	42.5	–	–
4.	*Carica papaya*	Carricaceae	25	58.33	70	30.55	22.85
			50	75	66.65	58.33	36.47
			75	81.65	68.33	69.41	39.67
			100	76.65	70	61.08	42.85
			Control	40	47.5	–	–
5.	*Cucumis sativus*	Cucurbitaceae	25	68.3	53.3	42	25.28
			50	68.3	65	42	44
			75	73.3	63.3	51.45	41.28
			100	76.65	75	57.54	60
			Control	45	37.5	–	–

At 50 per cent dose conc. acetone extract of *Trigonella foenum graecum* exhibited 76.76 per cent corrected mortality and its pet ether extract showed 60.34 per cent corrected mortality which was again at par with pet ether extract of *Nigella sativa* (60.89 per cent). At the same dose level acetone extract of *Carica papaya, Eruca sativa, Nigella sativa* and *Cucumis sativus* recorded 58.35 per cent, 51.77 per cent, 51.77 per cent and 42 per cent corrected mortality respectively whereas, pet ether extract of *Carica papaya, Eruca sativa* and *Cucumis sativus* exhibited 58.33 per cent, 51.77 per cent and 42 per cent corrected mortality respectively.

At 25 per cent dose level acetone extract of *Trigonella foenum graecum* recorded 63.32 per cent corrected mortality. Whereas its pet ether extract recorded poor 21.18 per cent corrected mortality. At the same dose conc. 55.55 per cent, 48.11 per cent, 42 and 30.55 per cent corrected pupal mortality was observed in acetone extract of *Eruca sativa, Nigella sativa, Cucumis sativus* and *Carica papaya* respectively. However their pet ether extract gave poor performance exhibiting 13.04 per cent, 37.93 per cent, 25.28 per cent and 22.85 per cent corrected mortality respectively.

The pupicidal action can be attributed to the chemicals which enter into the puparium and disrupt the normal metabolic activities of the developing stage. These results fall in line with the previous observation made (Dwivedi *et al.*, 2000) who had recorded 92.66 per cent corrected pupal mortality in acetone extract of *Acacia nilotica* at 100 per cent dose level. Thus, it is evident from the results that these extracts assert significant mortality and can be safely encorporated in integrated pest management programmes.

Acknowledgement

The authors express their sincere thanks to Dr. A.L. Bhatia, Professor and Head, Department of Zoology, for providing necessary facilities.

References

Abbott, W.S. 1925. A method of computing the effectiveness of insecticide. *J. Econ. Ent.*, 18(2): 265–267.

Ahmed, S.M. and Ahmed, A. 1992. Efficacy of some indigenous plants as pulse protectants against *Callosobruchus chinensis* L. infestation. *Int. Pest Control*, 34(2): 54–56.

Dwivedi, S.C. and Bajaj, M. 2000. Efficacy of *Acacia nilotica* leaf extract as pupicidal against *Trogoderma granarium* (Everts). *J. Exp. Zool., India*, 3(2): 153–155.

El. Nadi, A.H., Ahag, E.A. Zaitoon, A.A. and At Paghairi, M.A. 2001. Toxicity of three plant extracts to *T. granarium* Everts (Coleoptera: Dermestidael. *Pakistan J. Bot. Sci.*, V4(12): 1503–1505.

Hassan, T.A. 2001. Effect of three plant oils (sesame, sunflower, castor), against stored grain insects, (*T. granarium* and *S. granarius*). *Uni. of Aden. J. Nat. Appl. Sci.*, 5(1): 103–110; 12 ref.

Jood, S., Kapoor, A.C. and Singh, R. 1996. Evaluation of some plant products against *Trogoderma granarium* Everts in sorghum and their effect on nutritional composition and organoleptic characteristics. *J. Stored Prod. Res.*, 32(4): 345–352.

Kachare, B.V., Khaire, V.M. and Mote, U.N. 1994. Efficacy of different vegatable oils as seed treatment in increasing storage ability of pigeonpea seed against pulse beetle, *Callosobruchus chinensis* Linn. *Ind. J. Ent.*, 56(1): 58–62.

Mishra, N.C. 1996. Management of *Callosobruchus chinensis* on black gram during storage using botanicals. *Insect Env.*, 1(4): 12–13.

Mostafa T.S. 1993. Effects of certain plant powders on mortality, disrupting growth and metamorphosis of *Trogoderma granarium* Everts fourth instar larvae. *Bull. Ent. Soc., Egypt*, 20: 67–75.

Srivastava, A. and Bhatt, R.S. 1993. Effect of *Eucalyputs globulus* leaf extract on the mortality of the rice with, *Corcyra cephalmica. J. Adv. Zool.*, 14(2): 113–114.

Suryakala, S.S., Thakur and Kishan, B.R. 1995. Ovicidal activity of plant extracts on *Spodoptera litura* and *Dysdercus koenigii. Ind. J. Ent.*, 57(3): 192–197.

Chapter 26

Bioconversion of *Parthenium hysterophorus* as an Organic Manure for Chilli (*Capsicum annuum* L.)

B. Vijayakumari and R. Hiranmai Yadav***
**Reader in Botany, ** Ph.D. Scholar,*
Department of Botany, Avinashilingam Deemed University, Coimbatore – 641 043

ABSTRACT

A pot culture experiment was conducted at Avinashilingam Deemed University, Coimbatore to assess the impact of fresh, composted and vermicomposted parthenium, on chilli. The germination per cent of chilli was observed on 14, 21 and 28 DAS (Days After Sowing). The biometric attributes like root length, shoot length, fresh weight, dry weight and vigour index were observed on 30, 60 and 90 DAS. The germination per cent was more in vermicomposted parthenium treatment on 21st and 28th DAS compared to control. The increased root length of chilli on 30, 60, 90 DAS and shoot length on 30 and 60 DAS was due to composted parthenium application and on 90 DAS, fresh parthenium application registered longer shoot. The fresh and dry weight on 30 and 90 DAS were increased in composted parthenium and on 60 DAS in fresh parthenium. The vigour index was more in composted parthenium applied pots on 30 and 60 DAS and in fresh parthenium on 90 DAS.

Keywords: *Compost, Vermicompost, Parthenium, Chilli.*

Introduction

Chilli is an indispensable condiment of every home. Besides, imparting pungency and red colour to the dishes, it is a rich source of vitamins–A, C, E and P and has certain medicinal properties. Chilli stimulates saliva and gastric juices and helps digestion. It is used as a counter irritant in prickly heat powders, cosmetics, skin ointments and pain balms (Anand Singh, 2001).

Carrot grass more popularly known as 'Parthenium' (*Parthenium hysterophorus*) is one of the most noxious weed found increasingly in agricultural and plantation fields. Eco-friendly use of parthenium should be considered to reduce this menace. Considering the positive effect and applying profitable alternate use can solve the problem of spread (Sukanya, 2002).

In the present study an attempt was made to compost and vermicompost the parthenium plants and applying it to chilli as an organic manure to find its impact on the crop.

Materials and Methods

An experiment was conducted at Avinashilingam Deemed University, Coimbatore to study the effect of fresh, composted and vermicomposted parthenium on biometric attributes of chilli.

The treatment details:

T_0: Control–Red soil
T_1: Fresh parthenium @ 35.0 g/pot
T_2: Composted parthenium @ 26.25 g/pot
T_3: Composted parthenium @ 35.0 g/pot
T_4: Composted parthenium @ 43.75 g/pot
T_5: Vermicomposted parthenium @ 26.25 g/pot
T_6: Vermicomposted parthenium @ 35.0 g/pot
T_7: Vermicomposted parthenium @ 43.75 g/pot
T_8: NPK @ 60 : 30 : 30.

The pots of 7 kg capacity were used and the design adopted for the study was completely randomised design. The pots were filled with red soil and sand in 2 : 1 ratio. The manures were mixed into the soil and chilli seeds were sown after moisturing the soil.

The germination per cent was calculated 14, 21 and 28 DAS using the formula of Abdul Baki and Anderson (1973).

Biometric characters like root length, shoot length, fresh weight, dry weight and vigour index were observed 30, 60 and 90 DAS.

Statistical Analysis

The data obtained from the various biometrical observations were subjected to the statistical analysis (Panse and Sukhatme, 1978).

Results and Discussion

Germination Per cent

The germination per cent was observed on 14, 21 and 28 DAS (Table 26.1).

The highest germination per cent of 78.33 per cent was in T_5 compared to lower germination per cent of 40.00 per cent (21 DAS) and 50.00 per cent (28 DAS) in control.

Biometric Parameters

The biometric parameters observed were significant (Table 26.1).

Table 26.1: Impact of Fresh, Composted and Vermicomposted Parthenium on the Biometric Characters of Chilli (*Capsicum annuum* L.) on 30, 60 and 90 DAS

	Germination (%)			*Root Length (cm)*			*Shoot Length (cm)*			*Fresh Weight (g)*			*Dry Weight (g)*			*Number of Leaves*			*Vigour Index*		
	14 DAS	*21 DAS*	*28 DAs*	*30 DAS*	*60 DAS*	*90 DAS*	*30 DAS*	*60 DAS*	*90 DAS*	*30 DAS*	*60 DAS*	*90 DAS*	*30 DAS*	*60 DAS*	*90 DAS*	*30 DAS*	*60 DAS*	*90 DAS*	*30 DAS*	*60 DAS*	*90 DAS*
T_0	31.67	40.00	50.00	5.90	9.97	24.20	5.17	20.40	40.20	0.26	1.93	12.33	0.03	0.39	2.47	5.67	19.67	82.67	545.50	1495.60	3185.50
T_1	35.00	73.33	73.33	6.93	10.60	26.80	6.57	24.80	63.53	0.47	4.19	24.67	0.04	1.15	4.93	7.67	27.00	108.00	1085.00	2640.83	6664.83
T_2	46.67	56.67	63.33	7.10	6.97	29.67	6.87	25.53	57.00	0.38	2.26	22.33	0.04	0.45	4.47	7.00	21.00	87.00	886.33	2054.00	5480.00
T_3	53.33	58.33	63.33	5.77	14.30	26.07	6.03	26.33	63.00	0.28	3.24	29.33	0.04	0.65	5.80	5.67	21.33	126.66	754.67	2589.50	5647.67
T_4	63.33	75.00	75.00	7.23	12.50	27.93	7.70	27.77	56.33	0.58	3.66	23.67	0.05	0.73	4.75	7.33	22.67	79.67	1118.50	3006.50	6336.67
T_5	55.00	78.33	78.33	6.40	9.03	25.90	5.70	20.00	50.83	0.25	2.40	16.00	0.03	0.48	3.88	7.00	18.66	72.00	891.67	2131.33	5619.00
T_6	61.67	70.00	76.67	5.17	11.37	21.30	5.87	22.33	38.73	0.28	2.13	14.33	0.04	0.43	4.08	6.33	20.67	80.00	845.17	2586.17	4606.10
T_7	31.67	51.67	51.67	5.83	11.00	18.43	6.23	17.80	42.17	0.25	2.38	13.33	0.04	0.48	2.95	6.66	17.00	58.00	502.50	1400.50	3033.17
T_8	41.67	55.00	63.33	4.37	7.02	20.17	7.73	18.20	48.20	0.40	2.85	16.33	0.04	0.19	5.24	9.00	13.00	50.67	766.83	1625.00	4314.83
S.E.d	NS	9.40	8.39	0.40	2.11	3.07	0.49	3.05	3.52	0.07	0.37	2.88	0.01	0.08	0.80	0.67	3.41	10.41	121.64	445.01	738.27
C.D. (0.05)	NS	19.74	17.62	0.83	4.44	6.46	1.03	6.40	7.39	0.14	0.78	6.04	0.03	0.16	1.68	1.40	7.16	21.87	255.56	934.95	1551.07

DAS: Days After Sowing.

Root Length

The root length was more in T_4 (7.23 cm) on 30 DAS, T_3 (14.30 cm) on 60 DAS and T_2 (29.67 cm) on 90 DAS. The least values were observed in T_8 (4.37 cm) on 30 DAS, T_2 (6.97 cm) on 60 DAS and T_7 (18.43 cm) on 90 DAS.

Shoot Length

The shoot length was more in T_4 (7.70 and 27.77 cm) on 30 and 60 DAS. On 90 DAS, it was in T_1 (63.53 cm) immediately followed by T_3 (63.00 cm). The shoot length was minimum in control on 30 DAS (5.17 cm) in T_7 on 60 DAS (17.80 cm) and T_6 on 90 DAS (38.73 cm).

Fresh and Dry Weight

The fresh and dry weights were more in T_4 (0.58 and 0.05 g) on 30 DAS, T_1 (4.19 and 1.15 g) on 60 DAS and in T_3 (29.33 and 5.80 g) on 90 DAS. On 30 DAS, the lesser fresh weight of 0.25 g was in T_7 and dry weight of 0.03 g in T_0 and T_5. On 60 and 90 DAS, least values for fresh weight (1.93 and 12.33 g) was in control. The minimum dry weight was in T_8 (0.19 g) on 60 DAS and in control (2.47 g) on 90 DAS.

Vigour Index

The vigour index was more in T_4 on 30 (1118.50) and 60 (3006.50) DAS. On 90 DAS, T_1 (6664.83) registered higher vigour index. The vigour index was least in control on 30 DAS (545.50). On 60 (1400.50) and 90 (3033.17) DAS it was least in T_7.

Govindan *et al.* (1995) has also reported that application of 100 per cent vermicompost resulted in maximum vegetative growth of bhindi.

Mayalagu (1983) observed increased growth of groundnut due to the composted coir pith application.

Any organic waste application aids in moisture conservation which is utilized for better root penetration and crop growth.

References

Abdul Baki, A.A. and J.D. Anderson 1973. Vigour determination in soybean seed by multiple criteria. *Crop Sci.,* 13: 360–363.

Govindan, M., Muralidharan, P. and S. Sasikumaran. 1995. Influence of vermicompost in the field performance of bhindi (*Abelmoschus esculentus* L.) in a laterite soil. *J. Tropic. Agric.,* 33: 173–174.

Mayalagu, K. 1983. Influence of different soil amendments on the physical properties of a heavy black soil and yield of groundnut, TMV 7 in the Periyar-Vaigai command area. *Madras Agric. J.* 70: 304–308.

Panse, V.G. and P.V. Sukhatme 1978. *Statistical Methods for Agricultural Workers.* ICAR, New Delhi, pp. 327.

Singh, Anand 2001. Chillies: An important cash crop. *Kisan World,* 28: 47–48.

Sukanya, T.S. 2002. Parthenium: The most noxious weed. *Kisan World,* 29: 47–49.

Chapter 27

Effects of Brewery Effluent on Photosynthetic Pigments, Starch, Nitrate Reductase Activity and Protein Content of *Vigna mungo*

A. Pragasam, R. Praveena and J. Prasena
Department of Botany, Kanchi Mamunivar Centre for Post Graduate Studies, Pondicherry – 605 008, India

ABSTRACT

The photosynthetic pigments, starch, nitrate reductase activity and protein content of leaves of *Vigna mungo*, treated with different concentrations (10, 25, 50, 75 and 100 per cent) of brewery effluent and mere water, were estimated. All the parameters studied were found to be more than control in the plants treated with 10, 25, 50 and 75 per cent concentrations of the effluent and less than control in those treated with 100 per cent concentration. Maximum response was recorded in the plants treated with 50 per cent concentration. The present work suggests that the diluted effluent can be used for better growth of the test crop.

Keywords: *Brewery effluent, Vigna mungo, Photosynthetic pigments, Starch, NRA, Protein.*

Introduction

A number of industries such as dairies, distilleries, breweries, tanneries, refineries, pharmaceutical factories, textile mills, sugar mills, paper mills, card board industries, automobile industries and sewage release an enormous quantity of effluents into the ecosystem. Though the effluents of many industries are known for the presence of toxic substances those of some industries contain the essential elements required for better growth of plants. Because of the presence of such elements, the effluents offer a possibility to make use of them for irrigation after proper treatment and dilution. A number of studies revealed the growth promoting effects of the effluents (Ajmal and Khan, 1984: brewery effluent; Gautam and Bishnoi, 1990: dairy effluent; Misra and Behera, 1991: paper factory effluent; Medhappan, 1993: tannery effluent; Vijayarengan and Lakshmanachary, 1993: textile mill effluent; Kumar and Bhargava, 1998: sugar mill effluent; Pragasam and Kannabiran, 2001: distillery effluent).

Perusal of literature revealed that there are several works with reference to the effluents of various industries on the growth of plants, but work on the effects of brewery effluent on plants is meagre. So the present work has been taken up to analyse the effects of brewery effluent on pigments, starch, nitrate reductase activity and protein contents of *Vigna mungo* (L.) Hepper cv. Vamban.

Materials and Methods

The South India Corporation Agencies (SICA) breweries limited, Pondicherry is the first in India to be awarded the ISO 9002 and 14001 certification. It has the production capacity of 2.5 million cases of beer per annum. For every litre of beer produced about 10 litres of effluent is released. An average of 650 kilolitres of effluent is released on every working day.

The effluent was collected from the main outlet of the brewery and analysed for its physico-chemical properties using standard methods as recommended by American Public Health Association (APHA, 1975). Different concentrations of the effluent (10, 25, 50, 75 and 100 per cent) were prepared by mixing the raw effluent with water (VV). Certified seeds of *Vigna mungo* were obtained from Perunthalaivar Kamaraj Krishi Vigyan Kendra, Pondicherry. Twenty seeds in each pot were sown and treated with 500 ml of each concentration of the effluent daily. A control set treated with water was also maintained. Three sets were maintained for each concentration. The leaves of 15 days old seedlings were used for analysis. The parameters such as photosynthetic pigments, starch, nitrate reductase activity and protein content were estimated. They were extracted and estimated following the standard procedures proposed by Hiscox and Israeltam, 1979 (Pigment extraction); Harborne, 1984 (Pigments estimation); Mc Cready *et al.,* 1950 (Starch); Jaworski, 1971 (Nitrate Reductase Activity) and Lowry *et al.,* 1951 (Protein).

Results and Discussion

The effluent was found to be very light brown in colour and had no objectionable smell. The temperature of the effluent was 30°C, pH 8.77, total suspended solids 24 mg/l, biochemical oxygen demand 20 mg/l, chemical oxygen demand 44 mg/l, oil and grease 3 mg/l, nitrogen 12 mg/l, phosphorus 1.8 mg/l, Chloride 12 mg/l and sulphate 20 mg/l. All the parameters observed were found to be within tolerance limits. The pigment contents were more in the plants treated with 10, 25, 50 and 75 per cent concentrations of the effluent. The contents were less than control in the plants treated with 100 per cent concentration and maximum in those treated with 50 per cent concentration of the effluent (Figure 27.1). Starch, nitrate reductase activity (NRA) and protein contents also showed similar results (Figures 27.2 and 27.3).

An increase of 27.273 per cent total chlorophyll, 36.165 per cent Carotenoid, 36.446 per cent starch, 56.976 per cent NRA and 46.391 per cent protein and a decrease of 17.531 per cent total chlorophyll, 4.576 per cent Carotenoid, 13.323 per cent starch, 9.431 per cent NRA and 7.293 per cent protein were observed in the plants treated with 50 and 100 per cent concentrations of the effluent respectively over control.

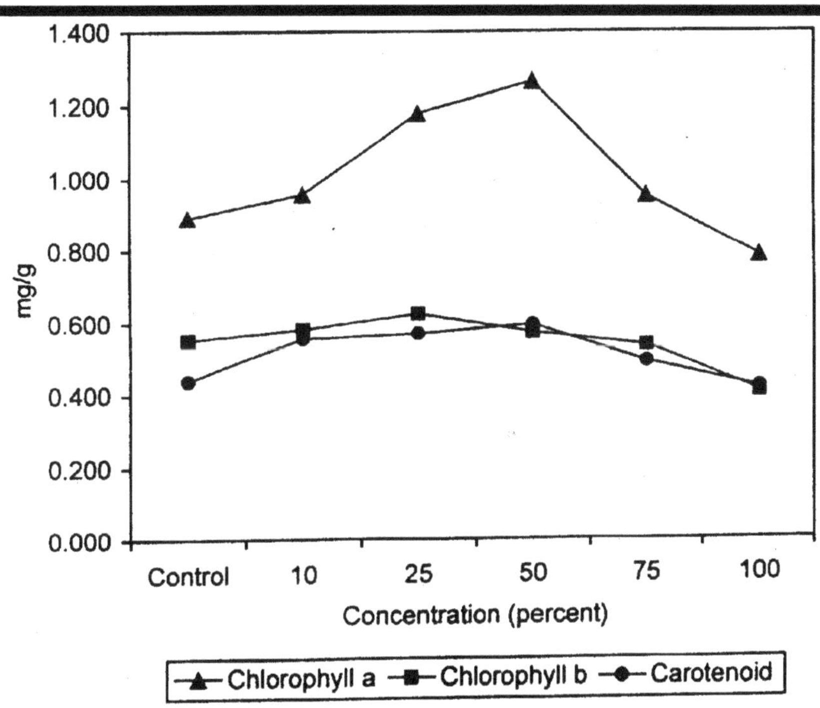

Figure 27.1: Effects of SICA Brewery Effluent on the Photosynthetic Pigments of Leaves of *Vigna mungo* (mg/g fresh weight)

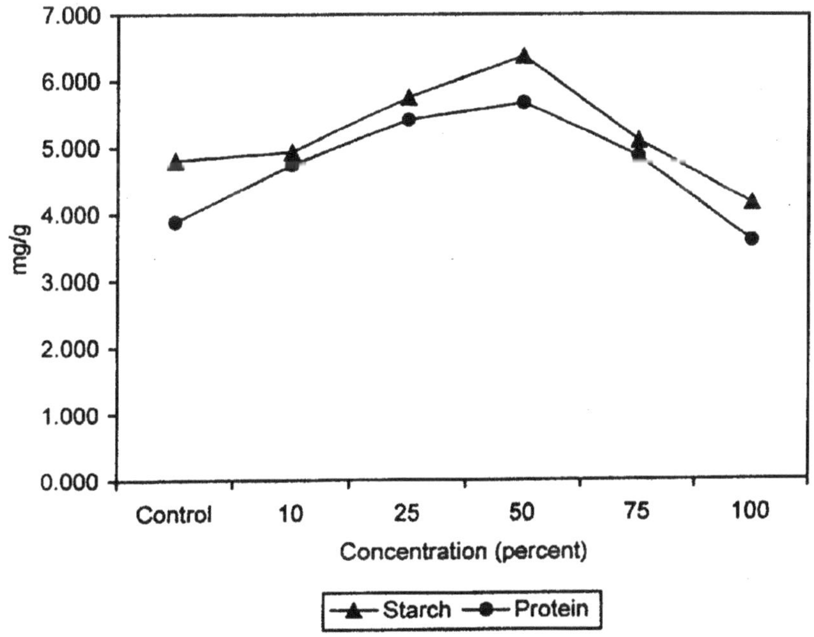

Figure 27.2: Effects of SICA Brewery Effluent on the Starch and Protein Contents of Leaves of *Vigna mungo* (mg/g fresh weight)

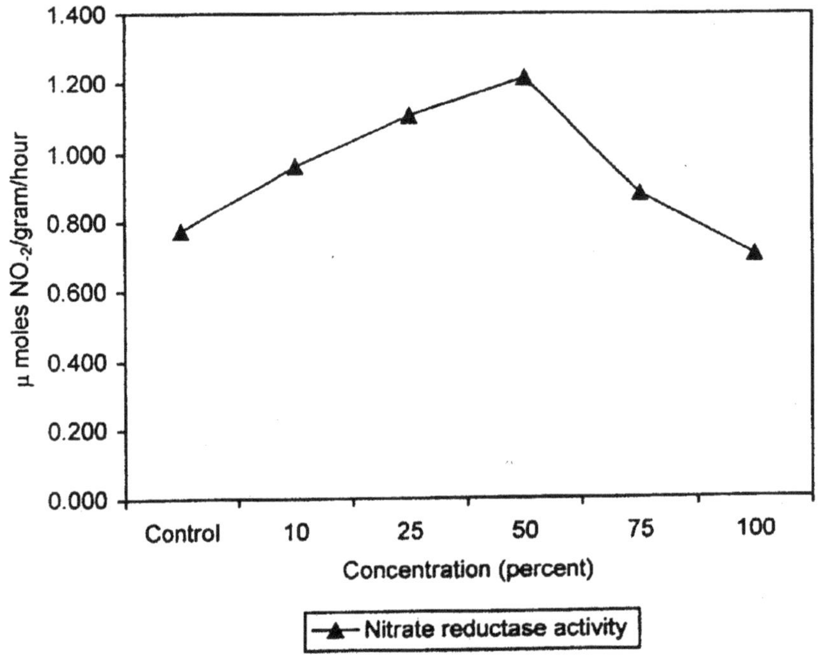

Figure 27.3: Effects of SICA Brewery Effluent on the Nitrate Reductase Activity of *Vigna mungo* (µ moles NO^{-2} formed /gram fresh weight/hour)

The works of Sahai and Srivastava (1986) on *Cajanus cajan*, Pragasam and Kannabiran (2001) on *Vigna mungo* with distillery effluent; Misra and Behera (1991) on rice seedlings with paper factory effluent; Madhappan (1993) on *Phaseolus mungo* and *Phaseolus aureus* with tannery effluent and Vijayarengan and Lakshmanchary (1993) on *Vigna radiata* with textile mill effluent showed similar increase and decrease in pigment content of the plants treated with lower and higher concentrations of the effluent respectively. The increased pigment contents in the plants treated with lower concentrations of the effluent must be due to the presence of optimum quantities of the nutritive elements and favourable osmoticum of that concentration of the effluent. Singh (1982) has suggested the osmotic imbalance prevailing in the concentrated effluent might be responsible for the reduction in the uptake of the mineral elements which resulted in the reduction of pigments.

Swaminathan and Vaitheeswaran (1991) have reported more starch content in *Arachis hypogaea* due to dyeing factory effluent. Manonmani *et al.* (1992) have reported similar results in *Arachis hypogaea* and *Zea mays* due to photofilm industry effluent. This must be due to the presence of more photosynthetic pigments and increased rate of photosynthesis.

Sahai *et al.* (1985), Muthukumar and Arokiasamy (1994) have also reported more nitrate reductase activity in the plants treated with lower concentrations of distillery effluent. It must be due to the presence of more absorbable nitrogen. Sahulka (1978) has reported reduced rate of NRA in the plants treated with higher concentrations of chloride. According to Pandey and Srivastava (1993) it was due to the restricted transport of nitrate to the metabolic pool.

Experiments of Kadiaglu and Algur (1990) with distillery effluent on *Helianthus annuus* and *Pisum sativum* have revealed that nitrate nitrogen present in optimum quantities in the diluted effluent was responsible for increased protein content in the plants. In the plants treated with cent per cent concentration of the effluent, the absorption and metabolism of nitrogen might have been reduced leading to the reduction in total physiological activities and resulted in the reduction of protein content.

Thus the present work suggests the use of diluted brewery effluent up to 75 per cent, especially 50 per cent, concentration for better growth of the test crop.

References

Ajmal, M. and A.U. Khan 1984. Effect of brewery effluent on soil and crop plants. *Environ. Pollut.*, 33: 341–352.

APHA, AWWA and WPCF 1975. *Standard Methods for the Examination of Water and Wastewater*, 14th ed. American Public Health Association, Inc., New York.

Gautam, D.D. and S. Bishnoi 1990. Studies on the effects of Urmul dairy effluent on soil characteristics and growth of wheat plant. *Ad. Plant Sci.*, 3: 236–244.

Harborne, J.B. 1984. *Phytochemical Methods: A Guide to Modern Techniques of Plant Analysis*, 2nd ed. Chapman and Hole, London, p. 215–217.

Hiscox, J.D. and G.M. Israelstam 1979. A method for the extraction of chlorophyll from leaf tissue without maceration. *Can. J. Bot.*, 57: 1332–1334.

Jaworski, E.G. 1971. Nitrate reductase in intact plant tissue. *Biochem. Biophys. Res. Commun.*, 43: 1274–1279.

Kadioglu, A. and O.F. Algur 1990. The effect of vinasse on the growth of *Helianthus annuus* and *Pisum sativum*. Part I: The effects on some enzymes and chlorophyll and protein contents. *Environ. Pollut.* 67: 223–232.

Kumar, Rajesh and A.K. Bhargava 1998. Effects of Sugar mill effluent on the vegetative growth and yield of *Triticum aestivum* cv. U.P. 2003. *Ad. Plant Sci.*, 11(2): 221–227.

Lowry, O.H., Rosenbrough, N.J., Farr, A.L. and R.J. Randall 1951. Protein measurement with the folin-phenol reagent. *J. Biol. Chem.*, 193: 265–275.

Madhappan, K. 1993. Impact of tannery effluent on seed germination, morphological characters and pigment concentration of *Phaseolus mungo* L. and *Phaseolus aureus* L. *Poll. Res.*, 12: 159–163.

Manonmani, K., Murugeswaran, R. and K. Swaminathan 1992. Effect of photofilm factory effluent on seed germination and seedling development of some crop plants. *J. Ecobiol.*, 4: 99–105.

Mc Cready, R.M., Guggole, J., Silviera, V. and H.S. Owners 1950. Determination of starch and amylase in vegetables: Application to peas. *Anal. Chem.*, 29: 1156–1158.

Misra, R.N. and B.K. Behera 1991. The effect of paper factory effluent on growth, pigments, carbohydrates and protein of rice seedlings. *Environ. Pollut.*, 72: 159–167.

Muthukumar, B. and D.I. Arokiasamy 1994. Effect of distillery waste on the nitrate reduction system in *Vigna mungo* (Black gram). *J. Ecotoxicol. Environ. Moni.*, 4: 27–33.

Pandey, M. and H.S. Srivastava 1993. Inhibition of nitrate reductase activity and nitrate accumulation by mercury in maize leaf segments. *Indian J. Environ. Hlth.*, 35: 110–114.

Pragasam, A. and B. Kannabiran 2001. Effect of distillery effluent on the morphology, biochemistry and yield of *Vigna mungo. Ad. Plant Sci.*, 14(11): 547–552.

Sahai, R., Jabeen, S. and P.K. Saxena 1985. Nitrate reductase activity of *Vigna mungo* (*L.*) Hepper as affected by distillery waste. *J. Environ. Biol.*, 6: 205–209.

Sahai, R. and N. Srivastava 1986. Effect of distillery waste on the seed germination, seedling growth and pigment content of *Cajanus cajan. J. Indian Bot. Soc.*, 65: 208–211.

Sahulka, J. 1978. The effect of chloride on nitrate reductase level as anaerobic nitrate reduction and on nitrate content in excised *Pisum sativum* L. roots. *Biol. Plant* (PRAHA), 20: 201–209.

Singh, D.K. 1982. Effect of distillery effluent on some germination parameters of certain legumes. *Sci. and Environ.*, 4: 61–67.

Swaminathan, K. and P. Vaidheeswaran 1991. Effect of dyeing factory effluent on seed germination and seedling development of groundnut (*Arachis hypogaea*). *J. Environ. Biol.*, 12: 353–358.

Vijayarengan, P. and A.S. Lakshmanachary 1993. Effect of textile mill effluent on growth and development of green gram seedlings. *Ad. Plant Sci.*, 6: 359–365.

Chapter 28

Influence of New Molecules Against Sucking Pest Complex of Brinjal

B.M. Mhaske, A.P. Chavan, D.B. Kadam

Department of Entomology, Mahatma Phule Krishi Vidyapeeth, Rahuri – 413 722, Distt. Ahmednagar, Maharashtra, India

ABSTRACT

The experiment were conducted during summer 2002 on brinjal crop with an object to study the influence of new molecules against sucking pests. The results revealed that the higher doses of imidacloprid 18 and 22.5 g a.i. per ha and thiamethoxam 25 and 50 g a.i. per ha were found effective for the control of jassids up to the 14th day and thrips and whitefly up to 10th day after spray.

Keywords: *New molecules, Brinjal pests, Imidacloprid, Thiamethoxam.*

Introduction

The brinjal (*Solanum melongena L.*) is an important vegetable crop and it is grown all over India and extensively cultivated in all the South East Asian countries. In Maharashtra, the crop is grown during all the seasons *viz., kharif, rabi* and summer, usually under irrigated condition. Brinjal is suffered heavily due to sucking pests *viz.,* jassids, thrips whitefly, aphids and mites. Besides direct damage, sucking pest acts as a vector of virus borne diseases. Losses due to the sucking pests vary from 10–15 per cent depending on the intensity of infestation. Thus sucking pests affects the vigour of the plant and ultimately yield. A number of pesticides have been advocated for the control of brinjal pest complex. However, they are partially effective or become ineffective in due course of time due to several reasons. A number of new molecules are put forth in the market. During the experiments, an attempt was made to study the efficacy of new molecules against sucking pest complex of brinjal.

Materials and Methods

Experimental Layout

The field experiments were conducted during summer 2002 on the Instructional Farm of Post Graduate Institute, Department of Entomology, Mahatma Phule Krishi Vidyapeeth, Rahuri, Distt. Ahmednagar, Maharashtra, India. The experiments were laid out in a randomized block design with twelve treatments and three replication on brinjal variety Mahyco 10 (Kalpataru) in gross plot size 4.5 × 3.0 mt. and net plot size was 3.60 × 1.80 mt. The treatment for sucking pest complex were imidacloprid 17.8 SL @ 13.5, 18 and 22.5 g a.i. per ha., thiamethoxam 25 WG 12.5, 25 and 50 g a.i. per ha., azadirachtin 1 per cent EC 10, 12.5 and 15 g a.i. per ha., profenofos 50 EC 500 g a.i. per ha., triazophos 40 EC 800 g a.i. per ha. and untreated control.

Method of Sprays

The spraying was done with due care and thorough coverage of all the plant parts which are recognized the main feeding sites of the pests. Spraying were done in morning hours and care was taken to wash the pump with water while switching on from one insecticide to another. The insecticidal spray were applied at an interval of 15 days, initiating the first 35 days after transplanting as soon as pests appear. In all three sprays were given to the experiments.

Observation

The observation for sucking pests such as jassids, thrips, aphids, whitefly and mites were recorded just before first spray (pre treatment count) and 3rd, 7th, 10th and 141th days after each spray (post treatment count). The counts were taken on the five randomly selected plants in each treatment plot, on each these plant six leaves, two from top, two from middle and two from bottom portion were observed to note the pest count. Since, there was a negligible infestation of aphids and mites, only number of jassids, thrips and adults of whitefly were recorded. Average number of pest individuals per six leaves of a plant for each plot was calculated. The insect numbers were transformed is to $\sqrt{n+1}$ value for statistical analysis, where, 'n' is number of insects.

Results and Disucssion

Influence of New Molecules Against Sucking Pest Complex of Brinjal

Jassids (*Amrasca biguttula biguttla* Ish.)

The total effect of all the three applications on jassids nymph population on brinjal at 3, 7, 10 and 14th day after application was studied (Table 28.1). AT 3rd day after application, the treatment with azadirachtin 1 EC @ 15 g a.i. per ha recorded the least number of jassids nymphs. However, all the other treatment were at par with it. The treatment with thiamethoxan 25 WG @ 50 g a.i. per ha emerged as a significantly superior treatment in checking jassids at 7th day after application, but treatment with imidacloprid 17.8 SL @ 22.5 and 18 g a.i. per ha and azadirachtin 1 EC @ 15 g a.i. per ha were at par with it. At 10th day and 14th days after application, imidacloprid 17.8 SL @ 22.5 g a.i. per ha significantly reduced the jassid population except thiamethoxam 25 WG @ 50 and 25 g a.i. per ha and imidacloprid 17.8 SL @ 18 g a.i. per ha.

Thrips (*Scirtothrips dorsalis* H.)

The data given in Table 28.1 revealed that, at 3rd and 14th day after application, imidacloprid 17.8 SL @ 22.5 g a.i. per ha recorded lesser number of thrips and all other treatments were at par with it. At

Table 28.1: Influence of New Molecules Against Sucking Pests of Brinjal (Average of Three Sprays)

Treatment	*g a.i. per ha*	*Average Number of Jassid Nymphs Six Leaves per Plant*					*Average Number of Thrips Six Leaves per Plant*					*Average Number of Whitefly Adults Six Leaves per Plant*				
		Pre Count	*3 DAS*	*7 DAS*	*10 DAS*	*14 DAS*	*Pre Count*	*3 DAS*	*7 DAS*	*10 DAS*	*14 DAS*	*Pre Count*	*3 DAS*	*7 DAS*	*10 DAS*	*14 DAS*
Imidacloprid 17.8 SL	13.5	2.00	0.43	0.36	0.75	1.60	8.47	2.07	2.87	4.02	5.03	3.53	3.73	3.38	3.76	3.80
			(1.20)	(1.17)	(1.32)	(1.61)		(1.75)	(1.97)	(2.24)	(2.46)		(2.17)	(2.09)	(2.18)	(2.19)
Imidacloprid 17.8 SL	18.0	2.60	0.30	0.28	0.54	1.43	8.87	1.97	2.45	2.71	4.63	3.40	3.13	2.29	3.60	3.70
			(1.14)	(1.13)	(1.24)	(1.56)		(1.72)	(1.86)	(1.93)	(2.37)		(2.03)	(1.81)	(2.14)	(2.17)
Imidacloprid 17.8 SL	22.5	2.60	0.20	0.12	0.33	1.00	9.67	1.80	1.92	2.44	4.03	3.47	2.10	1.78	2.38	2.67
			(1.10)	(1.06)	(1.15)	(1.41)		(1.67)	(1.71)	(1.85)	(2.24)		(1.76)	(1.67)	(1.84)	(1.92)
Thiamethoxam 25 WG	12.5	2.60	0.70	0.56	0.73	1.80	9.07	2.40	3.56	4.51	4.83	3.13	3.30	3.76	4.00	4.00
			(1.30)	(1.25)	(1.32)	(1.67)		(1.84)	(2.14)	(2.35)	(2.41)		(2.07)	(2.18)	(2.24)	(2.24)
Thiamethoxam 25 WG	25.0	2.87	0.30	0.56	0.44	1.28	8.73	2.17	2.87	3.09	2.90	3.47	2.93	3.04	3.25	4.00
			(1.14)	(1.25)	(1.20)	(1.51)		(1.78)	(1.97)	(2.02)	(1.97)		(1.98)	(2.01)	(2.06)	(2.24)
Thiamethoxam 25 WG	50.0	2.73	0.20	0.09	0.42	1.16	9.67	2.10	2.29	2.27	2.80	3.67	2.27	2.38	2.82	2.77
			(1.10)	(1.04)	(1.19)	(1.47)		(1.76)	(1.81)	(1.81)	(1.95)		(1.81)	(1.84)	(1.95)	(1.94)
Azadirachtin 1% EC	10.0	2.67	0.50	0.47	0.81	2.20	9.27	4.53	3.05	4.64	4.93	2.80	3.47	2.89	4.04	5.17
			(1.22)	(1.21)	(1.35)	(1.79)		(2.35)	(2.01)	(2.37)	(2.44)		(2.11)	(1.97)	(2.24)	(2.48)
Azadirachtin 1% EC	12.5	2.67	0.43	0.38	0.80	1.60	9.20	3.60	3.09	4.89	4.90	3.07	3.17	2.89	4.84	5.50
			(1.20)	(1.17)	(1.34)	(1.61)		(2.14)	(2.02)	(2.43)	(2.43)		(2.04)	(1.97)	(2.42)	(2.55)
Azadirachtin 1% EC	15.0	2.67	0.17	0.27	0.84	1.64	9.40	3.50	2.87	4.38	4.53	3.40	2.73	3.02	3.49	4.23
			(1.08)	(1.13)	(1.36)	(1.62)		(2.12)	(1.97)	(2.32)	(2.35)		(1.93)	(2.00)	(2.12)	(2.29)
Profenofos 50 EC	500.0	2.73	0.50	0.67	0.82	2.20	9.50	3.03	6.60	4.87	4.60	3.20	3.27	4.00	4.44	4.10
			(1.22)	(1.29)	(1.35)	(1.79)		(2.01)	(2.76)	(2.72)	(2.37)		(2.07)	(2.24)	(2.33)	(2.26)
Triazophos 40 EC	800.0	2.67	0.40	0.37	0.86	1.93	9.40	2.30	2.35	4.43	4.63	3.80	2.40	3.33	3.03	3.30
			(1.18)	(1.17)	(1.36)	(1.71)		(1.82)	(1.83)	(2.33)	(2.37)		(1.84)	(2.08)	(2.01)	(2.07)
Untreated control		2.67	1.96	1.60	1.00	2.70	9.30	8.30	7.20	6.90	8.35	3.00	7.23	6.60	7.24	7.95
			(1.72)	(1.61)	(1.41)	(1.92)		(3.05)	(2.86)	(2.81)	(3.06)		(2.87)	(2.76)	(2.87)	(2.99)
S.E.		N.S.	0.04	0.05	0.04	0.06	N.S.	0.16	0.18	0.16	0.11	N.S.	0.12	0.15	0.12	0.13
C.D. at 5% level		N.S.	0.15	0.12	0.12	0.16	N.S.	0.47	0.53	0.47	0.32	N.S.	0.35	0.45	0.36	0.39

*: Figures in parantheses are $\sqrt{n+1}$ transformation.

DAS: Days after spray.

7th day after application again imidacloprid 17.8 SL @ 22.5 g a.i. per ha was superior over others in reducing thrips population. However, all other treatment except profenofos 50 EC (@ 500 g a.i. per ha were at par with it. At 10th day after application, treatment with imidacloprid 17.8 SL @ 22.5 g a.i. per ha again found significantly superior in checking thrips population. However, other lower doses of imidacloprid *i.e.*, 13.5 and 18 g a.i. per ha, thiamethoxam 25 WG @ 25 and 50 g a.i. per ha and azadirachtin 1 EC @ 15 g a.i. per ha were at par with it.

Whitefly (*Bimisia tabaci* G.)

The data given in Table 28.1 revealed that at 3rd day after spray, the treatment with imidacloprid 17.8 SL @ 22.5 g a.i. per ha are showing minimum number of whitefly adults but rest of the treatments were equally effective and were at par with it. The efficacy trend remained more or less same at 7th day as that of 3rd day after spray. At 10th day after application, imidacloprid 17.8 SL @ 22.5 g a.i. per ha maintained its superiority over other treatments in reducing whitefly population. Similar results were observed at 14th day after application of spray.

It can be concluded that the new molecules of insecticides *i.e.*, imidacloprid 17.8 SL 18 and 22.5 g a.i. per ha and thiamethoxam 25 WG @ 25 and 50 g a.i. per ha were found to control jassids up to 14th day after application and thrips and whitefly up to 10th day after application significantly.

Elbert *et al.* (1991) reported that imidacloprid showed outstanding insecticidal activity against pests. Kumar (1998) reported good control of jassids and aphids with foliar spray of imidacloprid on cotton. Katyar (1999) reported the effectiveness of thiamethoxam against sucking pests of summer okra. Krishnakumar *et al.* (2001) tested imidacloprid and thiamethoxam against leaf hopper on okra and same insecticide were tested by Khutwad (2001) on sucking pests of green gram. Dhawan and Sinwat (2002) reported that thiamethoxam was significantly better and imidacloprid was at par with it in controlling jassids on upland cotton, but no literature was found to compare the present results of foliar sprays of these molecules against sucking pests of brinjal.

References

Dhawan, A.K. and G.S. Simwat. 2002. Field evaluation of thiamethoxam for the control of cotton jassids *A. biguttula biguttula* on upland cotton. *Pestology*, 26(1): 15–19.

Elbert, A.B., Becker, J. Hartwig and C. Erdelen. 1991. Imidacloprid: A new systemic insecticide. *Pflanzenschutz–Nachrichten, Bayer*, 44(2): 113–136.

Katyare, J.P. 1999. Efficacy of newer insecticide as seed dressers and foliar sprays on the infestation of sucking pests of summer okra. *M.Sc. (Agril.) Thesis*, M.P.K.V., Rahuri.

Khutwad, D.S. 2001. Efficacy of newer insecticides as seed dresser and foliar sprays on sucking pests of green gram (*Vigna radiata L.*). *M.Sc. (Agril.) Thesis*, M.P.K.V., Rahuri.

Krishnakumar, N.K., P.N. Krishnamoorthy and S.G. Eswar Reddy. 2001. Imidacloprid and thiamethoxam for the control of okra leaf hopper *A. biguttula biguttula* pest management. *Hort. Ecosystem*, 7(2): 117–123.

Kumar, K. 1998. Studies on bio-efficacy and determination of residues of imidacloprid applied against sucking pests of cotton. *Ph. D. Thesis*, T.N.A.U. Coimbatore, India.

Chapter 29

Growth and Development of Weeds in Sodic Soil

J.S. Tripathi, R.D. Vaishya, S.S. Singh and A.H. Khan

Department of Agronomy, Narendra Deva University of Agriculture and Technology, Narendra Nagar (Kumarganj), Faizabad – 224 229, U.P., India

The nature of soils affect the growth and reproductive behaviour of weeds considerably. Studies on salt toxicity and germination of facultative halophytes have been made by Choudhary and Siddiqui (1986). Choudhary and Siddiqui (1987) studied growth and reproductive capacity of *Eclipta alba* relation to soil salinity. The present study was undertaken to find out the effect of sodic soil on growth and reproductive behaviour of *Chloris barbata, Cyperus iria, Echinichloa colonum, Echinocloa crusgalli, Phalaris minor, Vicia sativa, Chenopodium album* and *Anagallis arvensis* weeds in *kharif* and *rabi* season.

Experiment was conducted in Complete Randomized Design (CRD) during *kharif* and *rabi* season. Normal soil (pH 8.0 and Ec 0.21 dS/m) and sodic soil (pH 9.5 and Ec 0.92 dS/m) were placed in separate cement colars of one meter diameter. One hundred seeds of each weed species were sown in each cement colar under both the conditions. After germination, seedlings were counted and germination percentage was calculated and thereafter uniform plant population was maintained in each colar. Plant height was measured at the maturity of crop. After maturity, one plant from each of the cement colar was taken to record the grain number per plant for calculation of reproductive capacity.

Height of *Echinochloa colonum, Echinocloa crusgalli, Cypeus iria, Chloris barbata* under normal soil condition was found greater than under sodic soil.

Germination percentage of *Cyperus iria, Echinocloa crusgalli, Echinochloa colonum* and *Chloris barbata* were found 57.00, 35.00, 31.00 and 80.00 respectively under normal soil condition, which was greater than under sodic soil.

Reproductive capacity of *Echinochloa colonum, Echinocloa crusgalli, Cyperus iria* and *Chloris barbata* under normal soil were recorded greater than reproductive capacity under sodic soil.

Height of *Phalaris minor, Vicia sativa, Chenopodium album* and *Anagallis arvensis* under normal soil, was greater than height recorded under sodic soil.

Reproductive capacity of *Phalaris minor Vicia sativa, Chenopodium album* and *Anagallis arvensis* were recorded 150.00, 48.21, 472.66 and 181.22 respectively under normal soil which were greater than under sodic soil condition. Germination of these weeds under normal soil was better than under sodic soil (Table 29.1).

Table 29.1: Growth and Reproductive Behaviour of Weeds Under Different Soil Habitat

	Normal Soil			*Sodic Soil*		
	Plant Height (cm)	*Germination Percentage*	*Reproductive Capacity*	*Plant Height (cm)*	*Germination Percentage*	*Reproductive Capacity*
***Kharif* weeds**						
Echinichloa colonum	39.92±3.75	57.00±2.08	697.84	34.19±2.70	33.00±1.75	214.50
Echinocloa crusgalli	45.88±3.21	35.00±3.21	861.49	30.33±1.63	20.00±2.64	160.42
Cyperus iria	51.57±2.88	31.00±2.30	620.00	24.28±1.75	17.00±1.73	255.00
Chloris barbata	41.22±1.53	80.00±2.08	167.30	34.55±2.70	43.00±2.64	49.80
***Rabi* weeds**						
Phalaris minor	31.35±2.33	75.00±2.88	150.00	25.78±2.08	40.00±2.88	40.00
Vicia sativa	22.23±3.75	91.20±2.88	48.21	18.44±2.70	67.20±2.70	17.67
Chenopodium album	17.30±2.70	20.00±2.88	472.66	10.45±2.08	10.00±1.15	136.00
Anagallis arvensis	15.22±2.08	53.30±1.73	181.22	9.08±1.73	35.00±2.70	84.00

Thus growth and reproductive behaviour of *kharif* and *rabi* weeds were better under normal soil than under sodic soil habitat.

References

Choudhary, G.N. and K.A. Siddiqui 1986. Salt toxicity and seed germiantion of a facultative halophyte. *Indian J. Ecol.,* 13: 325–327.

Choudhary, G.N. and K.A. Siddiqui 1987. Growth and reproductive behaviours of *Eclipta alba* in relation to salinity. *Indian J. Ecol.,* 14: 203–206.

Chapter 30

Groundwater Potential of Bist Doab Tract

Sarbjit Singh Sooch*, Baljeet S. Kapoor**, and N.S. Grewal***

**Assistant Professor, Department of Civil Engineering, Punjab Agricultural University, Ludhiana, Punjab*
***Principal, Chandigarh College of Engineering and Technology, Chandigarh*
****Professor-cum-Head (Retd.), Department of Civil Engineering, Punjab Agricultural University, Ludhiana, Punjab*

ABSTRACT

A study has been carried out to estimate the seasonal groundwater potential for North-east Punjab comprising Hoshiarpur, Jalandhar, Nawanshehar, Kapurthala and Ropar districts. This part of State is known as Bist Doab Tract. The information regarding meteorological parameters, daily canal water releases, number of tube-wells, water levels, cropping pattern and L-sections of canal branches, distributaries and minors in the study area were obtained from various State government agencies for the period June, 1989 to June, 2001. The information was used for assessing the surface and groundwater resources of Bist Doab Tract.

Recharge from rainfall, return flow from canal and tube-well irrigated area was estimated by using norms given by Groundwater Resource Estimation Committee in 1997. Water balance equation was used for computation of seasonal groundwater potential.

Average seasonal recharge from rainfall, canal network and return flow from irrigated area in the study was found to be 147.8, 19.3 and 115.6 mm during *kharif* season and 27.0, 13.3 and 66.3 during *rabi* season. The groundwater potential 1,40,473 ha-m is available in *kharif* however there is no additional groundwater potential available in *rabi* season. In the study area on an average groundwater potential is depleting 29,935.6 ha-m annually.

Keywords: *Groundwater, Water Balance, Net Recharge.*

Introduction

Punjab is an agricultural State of India where surface/subsurface water resources are being fully exploited for irrigation and are still inadequate to meet the total needs. Since surface water resource of Punjab is inadequate to meet the total needs, therefore, the stress has been on greater exploitation of groundwater resource to meet the irrigation needs. The total number of tubewells in the state has increased from 1.92 lakh in 1970–1971 to 9.50 lakh in 2000–2001 (Agriculture in Punjab, 2000–2001). This has resulted in decline of water table in about three-fourth area of the State except the saline groundwater tract lying in Southwestern part of the State. The area under water table depth less than 5 metre (which is optimum for pumping by centrifugal pumps) has decreased from 36.9 per cent in June 1973 to 25 per cent in June 1997; while the area under water table depth of 5 to 10 metre has increased from 41.4 per cent to 47 per cent during the corresponding period (Dhingra *et al.*, 1997). It is felt that if the present rate of groundwater exploitation is allowed to continue, it will result in decrease in well yield due to increased suction lift/total head that in turn would increase the energy consumption. Therefore, there is an urgent need of quantification of region water resources. This can be done, by seasonal water balance study. The study was carried out for Bist Doab through estimation of recharge of groundwater from various sources, pumpage and resulting net change in storage.

Study Area

The study area is located in the Northeast part of Punjab State and is known as Bist Doab Tract. It lies between latitudes of 30° 57′ 25″ N and 32° 8′ 15″ N and longitudes of 75° 01′10″ E and 76° 48′15″ E. The total area of Bist Doab Tract is 8,93,600 hectare, comprising the districts of Hoshiarpur, Jalandhar, Nawanshehar, Kapurthala and Nurpur Bedi block of Ropar district (Figure 30.1). On two sides it is bounded by Beas and Sutlej river and in the third side it is bounded by the Katar Dhar range of Shivalik hills.

The annual rainfall varies from 630 mm to 1200 mm. Most of the rainfall occurs during the months of July, August and September.

The topography of most of the area in tract is plain except some portion of Hoshiarpur district and Nurpur Bedi block of Ropar district. The tract slopes from North to South with a gradient varying from 1 in 500 to 1 in 2000. The northern part of the tract called *Kandi* area, contains large number of hill torrents known as "choes".

The soils of the area are alluvial in origin, varying from sandy to clay loam but most of the area is having loam soils. Sub soils are invariably heavier than top soil.

The area has a network of canal system namely Bist Doab canal, Shah Nahar canal and *Kandi* Canal (Phase-I) and their branches, distributaries and minors. The irrigation is also being carried out using groundwater through tube-wells.

Methodology

In order to assess the groundwater potential of a basin, the seasonal groundwater recharge was estimated. Groundwater recharge consists of recharge from rainfall, seepage from canal distribution system and deep percolation from the irrigated fields. The seasonal groundwater recharge (June–October) was estimated using the following equation:

$$P_g = P_{erc} + Q_{per} + P_w + P_c \tag{1}$$

where,

P_g = Total groundwater recharge, mm

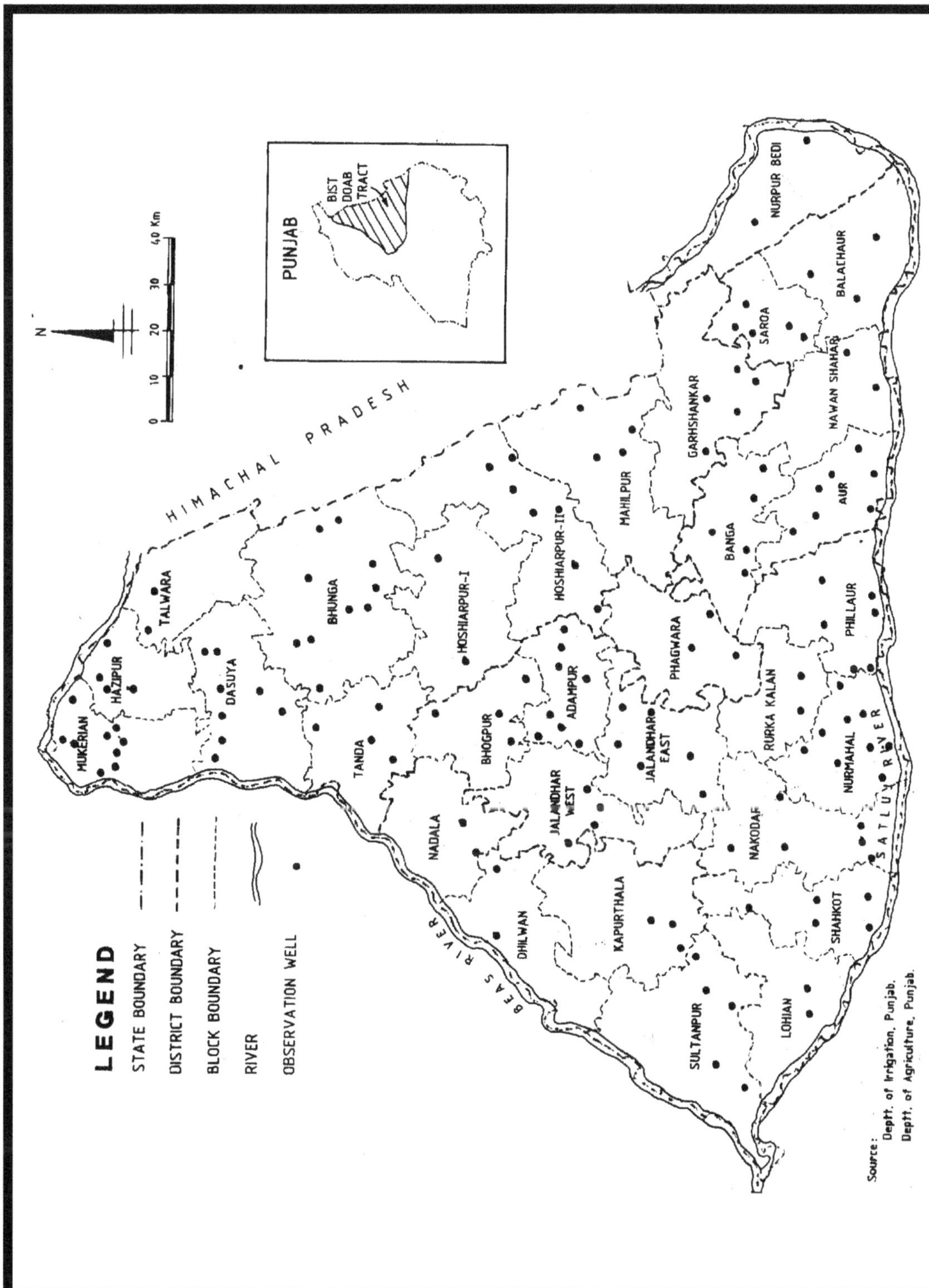

Figure 30.1: Location Map of Observation of Wells in Bist Doab Tract (Punjab)

P_{erc} = Recharge to groundwater from percolation of rainfall, mm

Q_{per} = Percolation of water through canal distribution system, mm

P_c = Percolation from canal irrigated areas, mm

P_w = Percolation of water from areas irrigated by wells, mm

The norms adopted for assessment of various components of Eq. (1) were as given by Ministry of Water Resources, Government of India (Anonymous, 1997) and Prihar *et al.* (1993).

Groundwater inflow/outflow from the study area to the neighbouring areas was calculated from water balance studies. The following form of water balance equation was used based on the available information:

$$(P + I) - (ET + E + Q_p + Q_g + Q_s \pm \Delta S_m + \Delta S_s) = \pm \Delta S_g \qquad (2)$$

where,

P = Total precipitation over the area, mm

I = Irrigation water applied over the area from all sources, mm

ET = Evapotranspiration, mm

E = Evaporation from uncultivated areas, mm

Q_p = Quantity of groundwater pumped (draft), mm

Q_g = Outflow of groundwater across the basin boundaries, mm

Q_s = Surface run-off from the area, mm

ΔS_m = Change in soil moisture storage, mm

ΔS_s = Change in surface storage in canal distribution system, mm

ΔS_g = Change in groundwater storage, mm

To estimate the total groundwater resource available for utilization for irrigation, municipal and industrial use, it is necessary to know the net change in groundwater storage. The net change in groundwater storage was calculated as:

Net change in groundwater storage = Gross recharge to groundwater
± Groundwater inflow/outflow
– Groundwater draft (3)

After computing the net change in groundwater storage, 85 per cent of this is taken as net recoverable recharge, which may be termed as additional utilizable resource available for irrigation.

The procedure used and norms adopted for estimating various components of equations (1) and (2) are given below:

An average value of 25 per cent of total rainfall has been assumed as groundwater recharge. It was also assumed that there is no recharge to groundwater if annual rainfall is less than 220 mm (Llyod *et al.*, 1966) or monthly rainfall is less than 50 mm or non monsoon rainfall is less than 10 per cent of normal annual rainfall. The seepage from unlined canal was taken as 25 ha-m/day/10^6 m_2 of the wetted area. For lined canal seepage losses were taken as 20 per cent of the value for unlined canals. Percolation from canal-irrigated areas was taken as 35 per cent of the water delivered at the outlet. In addition, to account for continuous deep percolation from paddy fields a value of 3 mm/day for 106 days of crop season is used. Percolation from area irrigated by wells was taken as 30 per cent of water delivered at the outlet. The quantity of groundwater pumped in different seasons during the year 1989–1990 to 2000–2001 was obtained based on the number of tubewells in the area and using draft, norms as given by Groundwater Resource Estimation Committee (Anonymous, 1997). The total draft values thus obtained were modified using multiplication factor based on variation in amount of rainfall from normal seasonal rainfall value. The multiplication factors varied from 0.6 to 1.4. The monthly crop evapotranspiration of each crop was calculated using Papadakis method (Mavi, 1994). Evaporation from shallow water table areas was obtained using potential evapotranspiration values calculated by Papadakis method. The monthly crop evapotranspiration was computed by multiplying potential evapotranspiration values with appropriate crop coefficient.

The surface runoff has been taken as 30 per cent of the monthly rainfall (Prihar *et al.*, 1993). Change in soil moisture storage was taken as 6 cm in *kharif* and it was assumed that 50 per cent of soil moisture stored in *kharif* is used during *rabi*. Change in surface storage in canal distribution was assumed to be zero.

Based on the observations of groundwater levels (June and October), the average seasonal rise or fall of water table was calculated for the study period. The corresponding, volume of water storage was computed by using an average value of 0.15 as specific yield.

Results and Discussion

The results presented in Tables 30.1 and 30.2 show that the average total groundwater recharge in the Bist Doab Tract during the period from 1989–1990 to 2000–2001 was 399.7 mm in *kharif* and 106.6 mm in *rabi* which is 40.92 per cent and 26.61 per cent of the total water inputs namely rainfall, irrigation from tubewells and canal network during corresponding *kharif* and *rabi* seasons. The recharge due to rainfall varies from 91.8 to 201.1 mm with average value of 147.8 mm in *kharif* season whereas in *rabi* season it varies from 10.3 to 74.4 mm with average 27.0 mm. The recharge due to canal seepage varies from 14.6 mm to 27.2 mm with average value of 19.3 mm in *kharif* season whereas in *rabi* season it varies from 8.4 mm to 16.9 mm with average value of 13.3 mm. The recharge due to canal-irrigated area varies from 10.5 mm to 17.2 mm with average value of 13.3 mm in *kharif* season whereas in *rabi* season it varies from 6.2 mm to 9.7 mm with average value of 8.0 mm. The recharge due to tubewell-irrigated area varies from 67.1 mm to 150.0 mm with average value of 102.3 mm in *kharif* season whereas in *rabi* season it varies from 53.7 mm to 61.7 mm with average value of 58.3 mm. In *kharif* season recharge due to Paddy cultivation varies from 109.6 mm to 127.8 mm with average value of 117.0 mm. Average seasonal recharge from rainfall, canal network and return flow from irrigated area in the study period was found to be 36.98, 4.83 and 58.19 per cent of total recharge during *kharif* and 25.33, 12.48, and 62.19 per cent of total recharge during *rabi*.

From the residual of water balance equation (2), it was observed that during *kharif* season there was subsurface in/out flow varies from – 215.0 mm to 40.8 mm with an average value of – 105.3 mm. During *rabi* season, the subsurface in/out flow varies from 447.5 mm to 543.2 mm with an average value of 491.7 mm. The average seasonal net change in groundwater storage was computed as 157.24 mm in *kharif* and – 190.7 mm in *rabi* season. The average annual net change in groundwater storage was observed as – 33.5 mm. The negative value of net change in groundwater storage shows that the groundwater resource is over exploited. Thus, there is further scope of groundwater development in Bist Doab Tract to control declining water table problem. In the above analysis, a number of assumptions have been made for estimating various parameters. Therefore, the result of seasonal water balance

Table 30.1: Net Change in Groundwater Storage of South West Punjab for *Kharif* Season (mm)

Year	*Rainfall Recharge*	*Canal Seepage*	*Percolation from Canal Irrigated Area*	*Percolation from Tube-well Irrigated Area*	*Recharge Due to Paddy*	*Gross Recharge*	*Tube-well Draft*	*Outflow*	*Evaporation from Shallow Water Table*	*Net Recharge*	*Computed Rise/Fall in Water Table*	*Observed Rise/Fall in Water Table*
(1)	*(2)*	*(3)*	*(4)*	*(5)*	*(6)*	*(7)*	*(8)*	*(9)*	*(10)***	*(11)**	*(12)*	*(13*
1989–1990	163.7	26.7	17.2	78.1	116.9	402.6	260.4	40.8	1.8	99.7	0.664	0.517
1989–1991	186.6	21.5	14.1	70.7	116.7	409.6	235.6	32.8	3.1	138.2	0.921	0.982
1991–1992	128.2	25.9	14.8	121.3	119.0	409.2	404.2	– 112.1	1.2	115.9	0.773	0.758
1992–1993	98.7	27.2	15.9	132.4	114.1	388.3	441.4	– 121.1	5.3	62.7	0.418	0.464
1993–1994	170.1	16.9	12.3	85.1	110.6	395.0	283.6	– 88.7	15.0	185.1	1.234	1.338
1994–1995	163.9	17.9	11.4	86.5	111.2	390.9	288.5	– 91.4	13.9	179.9	1.199	1.280
1995–1996	201.1	15.0	10.5	67.1	109.6	403.3	223.6	– 136.5	22.0	294.2	1.961	2.170
1996–1997	153.2	16.2	12.1	104.7	112.9	398.7	247.7	– 177.6	11.3	217.3	1.449	1.572
1997–1998	152.1	16.5	12.1	104.7	121.5	406.9	349.1	– 121.7	2.9	176.7	1.178	1.234
1998–1999	175.3	14.6	10.6	78.9	127.8	407.2	263.0	– 86.1	2.1	228.2	1.521	1.528
1999–2000	91.8	17.0	14.2	148.6	120.7	392.3	495.2	– 215.0	0.6	111.6	0.744	0.952
2000–2001	88.6	16.6	14.8	150.0	122.6	392.6	500.1	– 187.4	2.5	77.4	0.516	0.708
Average	147.8	19.3	13.3	102.3	117.0	399.7	341.0	– 105.3	6.8	157.2	1.048	1.125

*: Specific yield value taken as 0.15.

**: (10) = (2) + (3) + (4) + (5) + (6) – (8) – (9) – (10).

(2) to (11) values are in mm.

(12) to (13) values in m.

Table 30.2: Net Change in Groundwater Storage of South West Punjab for *Rabi* Season (mm)

Year	*Rainfall Recharge*	*Canal Seepage*	*Percolation from Canal Irrigated Area*	*Percolation from Tube-well Irrigated Area*	*Gross Recharge*	*Tube-well Draft*	*Outflow*	*Evaporation from Shallow Water Table*	*Net Recharge*	*Computed Rise/Fall in Water Table*	*Observed Rise/Fall in Water Table*
(1)	*(2)*	*(3)*	*(4)*	*(5)*	*(6)*	*(7)*	*(8)*	*(9)*	*(10)***	*(11)**	*(12)*
1989–1990	53.8	12.3	7.2	53.7	127.0	179.1	513.8	0	−207.06	−1.384	−1.183
1989–1991	33.4	13.7	7.6	54.8	109.5	182.8	431.7	0	−139.3	−0.929	−0.751
1991–1992	24.8	16.0	8.6	56.3	105.7	187.7	518.3	0	−224.9	−1.499	−1.312
1992–1993	10.3	14.6	9.0	56.3	90.2	187.8	484.7	0	−206.7	−1.378	−1.325
1993–1994	12.3	13.2	7.8	57.4	90.7	191.2	447.5	0	−165.6	−1.104	−1.280
1994–1995	31.3	16.9	9.2	58.3	115.7	194.4	479.0	0	−168.9	−1.126	−1.008
1995–1996	21.2	14.6	8.7	59.3	103.8	197.7	448.0	0	−146.5	−0.977	−1.136
1996–1997	12.5	16.1	8.8	59.8	97.2	199.5	482.5	0	−188.8	−1.259	−1.416
1997–1998	74.4	8.4	5.8	60.2	148.8	200.8	543.2	0	−193.5	−1.290	−1.103
1998–1999	15.5	8.7	6.2	60.4	90.8	201.5	531.5	0	−239.2	−1.595	−1.630
1999–2000	21.8	8.9	7.7	61.0	99.4	203.4	494.2	0	−191.3	−1.276	−1.314
2000–2001	13.2	16.5	9.7	61.7	101.1	205.6	522.6	0	−215.9	−1.439	−1.528
Average	27.0	13.3	8.0	58.3	106.6	194.3	491.7	0	−190.7	−1.271	−1.259

*: Specific yield value taken as 0.15.

**: (10) = (2) + (3) + (4) + (5) + (6) − (8) − (9).

(2) to (10) values are in mm.

(11) to (12) values in m.

studies were carried by comparing the observed and computed rise/fall of water level in both the seasons, respectively. Tables 30.1 and 30.2 shows that the computed seasonal rise/fall in water table compare favourably well with corresponding observed values.

The groundwater potential 1,40,473 ha-m is available in *kharif* however there is no additional groundwater potential available in *rabi* season. In the study area on an average groundwater potential is depleting 29,935.6 ha-m annually.

References

Anonymous. 1982. *Feasibility Report on Reclamation of Waterlogged Area by Tube-well Drainage in Faridkot Tract of Punjab.* Punjab State Tubewell Coproration Ltd., Chandigarh.

Anonymous. 1997. *Groundwater Resources Estimation Methodology.* Report of the Groundwater Resource Estimation Committee, Ministry of Water Resources, Government of India, New Delhi, pp. 107.

Doorenbos, J. and Pruitt W.O. 1977. *Guidelines for Predicting Crop Water Requirements.* Irrigation and Drainage Paper No. 24, FAO, Rome, pp. 144.

Llyod, J.W., Drennan, D.S.H. and Bennel, B.M.U. 1966. A groundwater recharge study in NE. *Jordan. Proc. Inst. Civil Engineers,* 35: 615–631.

Mavi, H.S. 1994. *Introduction to Agrometeorology.* Oxford and IBH Publishing Co. Pvt. Ltd., New Delhi, pp. 281.

Prihar, S.S., Khepar, S.D., Singh, R., Grewal S.S. and Sondhi S.K. 1993. *Water Resources of Punjab: A Critical Concern for the Future of its Agriculture.* Punjab Agril. Univ., Ludhiana, pp. 56.

Chapter 31

Comparison of Immunostimulant Activity of *Ocimum Sanctum* Linn. Leaf Extracts

M.S. Kondawar and S.B. Bhise***

**Appasaheb Birnale College of Pharmacy, South Shivajinagar, Sangli, M.S.*

***Government College of Pharmacy, Vidyanagar, Karad, M.S.*

ABSTRACT

The focus of the present study is to investigate the immunostimulant effect of *Ocimum sanctum* Linn., a sacred herb in India, also known as Holy Basil, popular as traditional medicine in variety of diseases and used as an ingredient in lots of Ayurvedic medicaments, The methanolic, aqueous and hydro-alcoholic extracts of *Ocimum sanctum* has been investigated for *in vitro* immunostimulant activity on human neutrophils by Nitrobluetetrazolium, Phagocytosis of killed *Candida albicans,* Candida cidal assay and Neutrophils locomotion and chemotaxis methods. The result of present study shows significant immunostimulant activity.

Keywords: *Ocimum sanctum, Immunostimulant, Nitrobluetetrazolium, Phagocytosis, Chemotaxis, Candida cidal assay, Candida albicans.*

Introduction

Ocimum sanctum Linn. (Lamiaceae) (syn: Tulsi), is a shrub reaching a height of 0.5 to 1.5 m, the leaves are 2–4 cm in length. There are several varieties of the plant, however commonly used one is with dark leaves. It is widely distributed in India, Burma, Sri Lanka, Malaya, Australia and the Islands of the Pacific. It is observed within the Himalayas up to the altitude of 6000 feet. The plant has great importance in the medicinal use, the root decoction is useful in malarial fever, leaves are useful in bronchitis as well as locally in ringworm infections, and the seeds are useful in the disorders of genito-urinary system (Kokate *et al.*, 2000). Fixed oil of the plant has been shown to produce anti-ulcer activity (Singh and Mujumdar, 1999) as well as anti-inflammatory activity (Singh *et al.*, 1996). The dried leaf powder has hypoglycemic effect (Rai *et al.*, 1997). A leaf extract of *Ocimum sanctum* has been shown to stimulate the release of ACTH from pituitary cells in vitro (Wagner *et al.*, 1994). Six phenolic compounds isolated from Tulsi including eugenol, rosmarinic acid, apigenin and the other flavonoids showed good to excellent antioxidant activity *in vitro* (Kelm *et al.*, 2000). *Ocimum sanctum* leaf has shown both prophylactic action and arrest the process of cataract formation in animal models (Sharma *et al.*, 1998).

Neutrophils play an important role in the host immune mechanism system. They are attracted by different stimuli. A foreign body, thermal and chemical burns and bacterial infections provoke an intense neutrophil response. The neutrophils are highly effective at killing certain bacteria and their ability to digest cellular debris and exogenous particulate matter provides an important step in healing process (Levinson and Jawetz, 1994). The clinical immunological laboratory tests (qualitative and quantitative analysis of immune responses) are used by the physicians in the diagnosis, treatment and assessment of cellular functions (Stites *et al.*, 1994). Hence the leaves of *Ocimum sanctum* were used for screening *in vitro* immunomodulatory activity on human neutrophils by the clinical immunological laboratory tests.

Materials and Methods

Materials

The chemicals required for the test of immunostimulant activity were obtained from various sources. *i.e.*, *Candida albicans* culture in Sabourad's broth was obtained and incubated at HI-TECH Healthcare and Diagnostic Center Private Limited, Belgaum. Fresh human serum was obtained from the donors. Ficoll hypaque was procured from Sigma. MEM, Nutrient agar, Nutrient broth and Tissue culture medium (RPMI) were purchased from Himedia. Nitrobluetetrazolium (NBT) was obtained from British Drug House and Neubauer heamocytometer and Boyden chamber (chemotaxis chamber) was obtained from HI-TECH laboratory. All other chemicals and stains required for the procedures *i.e.*, Dextran, Phosphate buffer, Glacial acetic acid, Sodium desoxycholate, Hank's solution, Casein, Gentian violet, Methylene blue, Haematoxylin stain and May-Grunwald-Giemsa stain were procured from HI-TECH laboratory, Belgaum.

Plant Extraction

For the present study the plant was procured from the reliable sources of Sangli region, and the identity of the drug was established by morphological study (Kirtikar and Basu, 1991) at the department of Botany, Willingdon College Sangli. Then after it was cut, shade dried at 37° to 40° and coarsely powdered through mesh 20. The powdered drug was again shade dried for 5 days and finely powdered through mesh 80. The powdered plant material was extracted separately with methanol (MK1) using Soxhlet's extractor, with water (MK2) and hydro-alcoholic mixture (MK3) using maceration at room temperature for 7 days. The extracts were concentrated to dryness on steam bath; these concentrated extracts were tested for the immunostimulant activity by different methods. All the extracts were used in the concentration of 20, 40, 60, 80 and 100 mg/1 00 ml.

Collection of Blood

All the tests were based on the collection of venous blood into sterile preservative free heparin containing test tubes. Blood for serum sample was collected into sterile glass tubes. About 20 ml of heparinized blood was used for the tests. In addition 3–5 ml of clotted blood was also used.

Separation Technique

Heparinized blood sample was diluted with equal volume of tissue culture medium (RPMI). 10 ml diluted blood was mixed with 6 ml dextran carefully to avoid air bubbles, and incubated at 37° for 25 min without being disturbed. The supernatant was removed in 3 to 4 centrifugal tubes and were diluted with an equal volume of RPMI. The mixtures were centrifuged at 500 g for 10 min the supernatant was discarded and the cells from two tubes were flicked carefully in one tube. These tubes were washed with RPMI and centrifuged at 500 g for 10 min; this procedure was repeated until all the cells were flicked into a single tube. The cells were washed with RPMI, centrifuged and the supernatant was discarded followed by dilution with RPMI up to 1 ml, finally the neutrophils were counted by using Neubauer heamocytometer.

All the extracts were screened for Immunomodulatory activity by *in vitro* methods on neutrophils of the blood withdrawn from normal healthy volunteers. Phosphate buffer salt solution (PBS) was used as a vehicle and extracts were used in the concentration range of 20 mg/ml to 100 mg/ml.

Delay in transport of material, especially in a warm environment may lead to adherence of phagocytic cells and reduction to yield. Room temperature is best for the transportation since Neutrophils perform less well after being kept at room temperature for any length of time.

Nitrobluetetrazolium Test (NBT)

E. coli endotoxin was used in the screening of respiratory burst by NBT test. Whole blood was mixed with NBT and endotoxin. The mixture was incubated for 30 min at 37°, counting fluid (0.2 per cent w/v of gentian violet in 2 per cent w/v of glacial acetic acid in distilled water) was added and mixed well. Sample was loaded in Neubauer chamber and counted for responded non-lymphoid cells as given in Table 31.1.

Table 31.1: Per cent of Responded Cells by NBT Stimulation

Concentration in (mg/100 ml)	*Response**			
	MK1	*MK2*	*MK3*	*Control*
20	35.4±1.14	45.2±0.45	41.6±0.89	
40	42.6±0.89	51.8±0.84	45.4±0.55	
60	57.8±0.84	64.6±0.55	47.0±1.00	9.2±0.45
80	60.4±0.55	68.8±0.84	48.8±0.84	
100	61.4±0.55	79.2±0.84	51.6±0.55	

*Mean of five experiments ± S.D.

Candida Cidal Assay

Candida albicans were harvested, washed and isolated by centrifugation. They were detected for dead blastospores (Dead cells should not exceed 5 per cent) and resuspended in RPMI. The neutrophils suspension of 7 × 10^6/ml was prepared in Hank's solution; the latter was used to setup different cultures. To a single culture of RPMI with normal serum and RPMI with patient serum, suspension of *Candida albicans* in RPMI were added and incubated for 30 min at 37° with shaking to resuspend cells after 15 min. Sodium desoxycholate was added to each tube and mixed vigorously for lysis of the neutrophils, followed by addition of methylene blue, and at least 200 blastospores were counted for both viability and death. The percentages of dead blastospores are reported in Table 31.2.

Table 31.2: Per cent of Dead *Candida albicans* Blastospores

Concentration in (mg/100 ml)	*Response**			
	MK1	*MK2*	*MK3*	*Control*
20	11.6±0.55	15.4±0.55	20.4±0.55	
40	15.2±0.84	17.8±0.45	23.8±0.84	
60	16.8±0.45	21.2±0.45	27.8±0.45	8.6±0.55
80	18.8±0.45	23.8±0.84	29.8±0.45	
100	20.6±0.55	26.4±0.55	31.6±0.55	

*Mean of five experiments ± S.D.

Phagocytosis of Killed *Candida albicans*

Candida albicans were grown on Sabourad's 2 per cent dextrose broth for 48 h to obtain organism in baker's yeast phase only. The yeasts were maintained at 5 × 10^6 cells/ml, heating to 100° for 30 min killed these yeasts. To a duplicate set of tubes Hank's solution, pooled normal human serum, heat killed *Candida albicans* and leukocyte suspension were added. The contents were mixed and incubated at 37° for 30 min followed by centrifugation at 200 g for 5 min. The supernatant was removed, the smear was prepared and it was stained with May-Grunwald-Giemsa. At least 200 cells were examined and counted for number of infested *Candida albicans* associated with each cell as reported in Table 31.3.

Table 31.3: Number of Infested *Candida albicans* Associated with Each Cell

Concentration in (mg/100 ml)	*Response**			
	MK1	*MK2*	*MK3*	*Control*
20	2.2±0.45	2.4±0.55	2.2±0.45	
40	2.6±0.55	2.8±0.45	3.0±0.00	
60	3.0±0.00	3.4±0.55	3.6±0.55	2.4±0.55
80	3.4±0.55	4.2±0.45	4.0±0.71	
100	4.0±0.00	4.8±0.45	4.6±0.55	

* Mean of five experiments ± S.D.

Neutrophil Locomotion and Chemotaxis

Cells were adjusted in Hank's solution approximately to 10^6 cells/ml and then lower compartment of the Chemotaxis chamber was filled with chemo-tactic reagent casein, pre-adjusted to the pH 7.2 in duplicate. One set was tested against the normal cells while other against the patient's cells. 3.6 ml of each solution was then added in 5 ml beaker, the upper compartment of the Chemotaxis chamber was filled with suspension of cells by ensuring that the fluid level in upper chamber equals to that of lower chamber. The upper compartment was placed in the lower compartment such that the level of fluid matches. The test was incubated at 37° in air for 90 min by protecting it against the disturbances. Finally the test sample was removed and inverted to drain the fluid out and fixed by immersing in 70 per cent ethanol. At last the test was stained by Haematoxylin stain, and it was counted for number of cells migrated (Thomson, 1987). The results were reported in mean per cent of cells migrated in Table 31.4.

Table 31.4: Per cent of Cells Migrated

Concentration in (mg/100 ml)	*Response**			
	MK1	*MK2*	*MK3*	*Control*
20	30.4±0.55	40.2±0.45	14.4±0.55	
40	32.8±0.84	42.2±0.84	16.0±1.00	
60	34.6±0.55	44.8±1.10	17.8±0.45	2.4±0.55
80	36.8±0.45	47.0±0.71	18.8±0.45	
100	39.6±0.55	49.6±0.55	20.4±0.55	

* Mean of five experiments ± S.D.

Results and Discussion

The immunostimulant activity of any drug is essentially dependent on its response towards external stimuli, *i.e.*, taking up the stain, migration of cells and response towards phagocytosis. The results of the present study reveal that the methanolic (MK1), aqueous (MK2) and hydro-alcoholic (MK3) extracts of the leaves of *Ocimum sanctum* Linn. possess potent immunostimulant activities that are observed by NBT test, candida cidal assay, phagocytosis of killed *Candida albicans* and chemotaxis assay.

In this study methanolic, aqueous and hydro-alcoholic extracts were used to compare their activities. The methanolic extract of the *Ocimum sanctum* leaves possesses immunostimulant activity (Godhwani *et al.*, 1988). Alcohol and water possesses good solvent property therefore the mixture of alcohol and water is used to extract the leaves of *Ocimum sanctum* Linn.

All the three extracts of *Ocimum sanctum* leaf were shown significant activity in stimulating non-lymphoid cells by NBT test. They stimulated the neutrophils to give phagocytic activity to the extent of 79.2±0.84 per cent, 61.4±0.55 per cent and 51.6±0.55 per cent at the concentration of 100 mg/100 ml of MK2, MK1 and MK3 respectively as compared to 9.2±0.45 per cent of neutrophils stimulated by negative control (Table 31.1).

In *Candida cidal* assay the mean per cent dead *Candida albicans* blastospores were observed maximum in hydro-alcoholic extract (MK3) at 100 mg/100 ml followed by aqueous extract (MK2), whereas poor activity was observed in methanolic extract (MK1) *i.e.*, 31.6±0.55, 26.4±0.55 and 20.6±0.55 respectively (Table 31.2). It is much more significant at all concentrations when compared to negative control.

All the three extracts have stimulated the phagocytosis of killed *Candida albicans.* 4–5 MPN when compared to control less than 3 MPN (Table 31.3). The aqueous and hydro-alcoholic extract shows significant activity.

In chemotaxis assay the results so obtained indicate that the per cent of cells migrated was maximum in case of MK2 followed by MK1 and MK3, *i.e.*, 49.6±0.55 per cent, 39.6±0.55 per cent and 20.4±0.55 per cent respectively at all the concentrations as compared to 2.4±0.55 per cent of cells migrated by negative control (Table 31.4). MK3 extract fails in showing potent activity by chemotaxis assay.

The dose dependency of the activity was tested by using five different doses. The negative control is used to check the activity of extracts over solvents. All the three extract shows dose dependant activity. In all the models that are used to test the immunostimulant activity, the aqueous extract was found to be more potent.

The increase in the per cent responded non-lymphoid cells indicate the immunostimulant activity. Leaves of *Ocimum sanctum* have been found to stimulate macrophages as evidenced by all above tests. Activated macrophages secret a number of cytokines such as GM-CSF and IL-l that in turn stimulates neutrophils (Reeves and Gedes, 1982), this gives host defense and the ability to fight diseases.

Acknowledgements

We thank Prof. D.D. Chougule (Principal, Appasaheb Birnale College of Pharmacy, Sangli) for providing necessary facilities. Thanks are also due to Dr. N.A. Rajmane (Ex-Reader, Department of Botany, Willingdon College, Sangli) for his advice and valuable suggestions. Dr. Kishor Bhat (Consulting Pathologist, HI-TECH Diagnostic Center, Belgaum) for his expertise in immunological techniques.

References

Godhwani, S. and Godhwani, J.L. *et al.* 1988. *Ocimum sanctum:* A preliminary study evaluating its immunoregulatory profile in *albino* rats. *J. Ethnopharmacology,* 24(2–3), 193–198.

Kelm, M.A. *et al.* 2000. Antioxidant and cyclooxygenase inhibitory phenolic compounds from *Ocimum sanctum* Linn. *Phytomedicine* 7: 7–13.

Kirtikar, K.R. and Basu, B.D. 1991. *Indian Medicinal Plant*. Vol. III. Periodical Experts Book Agency, Delhi, pp. 1965.

Kokate, C.K., Purohit, A.P. and Gokhale, S.B. 2000. *Pharmacognosy*. Nirali Prakashan, Pune, pp. 302.

Levinson, W.E. and Jawetz, E. 1994. *Medical Microbiology and Immunology*. Applenton and Lange Norwalk, CT, pp. 34.

Rai, V., Iyer, U. and Mani, U.V. 1997. Effect of Tulsi (*Ocimum sanctum*) leaf powder supplementation on blood sugar levels, serum lipids and tissue lipids in diabetic rats. *Plants food for Human Nutrition* 50: 9–16.

Reeves, D.S. and Gedes, A.M. 1982. *Recent Advances in Infection*. Churchill Livingstone, New York, pp. 89.

Sharma, P., Kulshreshtha, S. and Sharma, A.L., 1998. Anticataract activity of *Ocimum sanctum* on experimental cataract. *Indian J. Pharmacol.*, 30: 16–20.

Singh, S. and Mujumdar, D.K. 1999. Evaluation of Antiulcer activity of fixed oil of *Ocimum sanctum*. *J. Ethnopharmacology,* 65: 13–19.

Singh, S., Mujumdar, D.K. and Rehan, H.M. 1996. Evaluation of anti-inflammatory potential of fixed oil of *Ocimum sanctum* and its possible mechanism of action. *J. Ethnopharmacology,* 54: 19–26.

Stites, D.P., Terr, A.I. and Parslow, T.G. 1994. *Basic and Clinical Immunology*. Lange International, New Jersey, pp. 195.

Thomson, R.A. 1987. *Techniques in Clinical Immunology*. Blackwell Scientific Publication, pp. 279.

Wagner, H., Norr, H. and Winterhoff, H. 1994. Plant adaptogens. *Phytomedicine,* 1: 63–76.

Chapter 32

Combining Ability Studies for Yield and Yield Components in Sunflower (*Helianthus annuus* L.)

K. Venkata Siva Reddy and M.R. Manjare***

**Research Associate, **Senior Scientist (Breeding), Mahatma Phule Krishi Vidyapeeth, Department of Botany, Rahuri, Maharashtra*

ABSTRACT

Combining ability studies using line x tester (4 lines and 7 testers) mating design in sunflower revealed that non-additive genetic variance was important for circumference of head, total number of seeds per head, number of filled seeds per head, hundred seed weight and seed yield per plant. Non-additive gene action was found to be predominant for all the characters under study *i.e.*, circumference of head, total number of seeds per head, number of filled seeds per head, hundred seed weight and seed yield per plant. From the estimates of GCA it was observed that the females CMS 343A and CMS 852A and testers RHA 274 and Morden were found to be good combiners for most of the yield component traits *viz.*, circumference of head, total number of seeds per head, number of filled seeds per head, hundred seed weight, besides seed yield per plant. The crosses CMS 343A x Morden and CMS 89A x RHA 274 exhibited high SCA effects and *per se* performance for various yield components traits besides seed yield per plant.

Keywords: *Combining ability, Line x Tester, Non-additive, Sunflower.*

Introduction

Sunflower is one of the most important oil seed crop in the world and in India. Sunflower is good substitute for other oil seed crops because of its favourable characters like suitability to late sown conditions, short duration, drought and saline tolerance, photo-insensitivity, high adaptability, good market price and oil content. The discovery of Cytoplasmic male sterility (Leclercq, 1969) in France and fertility restoration (Kinman, 1970) in U.S.A had provided a break through for heterosis breeding in sunflower. As a result plant breeders have extensively used and exploited heterosis to improve seed and oil yields in sunflower. In the systematic breeding programme, it must be necessary to identify superior lines for hybridization and crosses to expand the variability reservoir for selecting superior genotypes. Combining ability analysis helps the breeders in selecting desirable parents and provides information on the importance of GCA and SCA for interpreting the genetic basis of important traits. The present study was under taken to obtain information on general and specific combining ability effects with regard to yield and yield component characters.

Materials and Methods

The experimental material for the present investigation consisted of 28 F1' s obtained by crossing four cytoplasmic male sterile lines (CMS-86A, CMS-89A, CMS-343A and CMS-852A) with seven fertility restorer lines (RHA-274, RHA-278, RHA-296, RHA-297, RHA-348, SS-56 and Morden) in a line x tester fashion at Post Graduate Institute farm, Department of Botany, Mahatma Phule Krishi Vidyapeeth, Rahuri. The 28 F1's along with their parents were grown in a randomized block design with 3 replications during *kharif* 2000. Recommended package of practices for the region were followed. Each entry was sown in a single row of 5 m length, with a spacing of 60 cm between rows and 30 cm between plants within a row. Observations were recorded on five random competitive selected plants in each treatment, replications wise for yield and its component characters *viz.*, circumference of head (cm), total number of seeds per head, number of filled seeds per head, hundred seed weight (g) and seed yield per plant (g). The mean data of five plants over three replications were used for statistical analysis. The analysis of variance for the design was carried out following Panse and Sukhatme (1978) and the combining ability as proposed by Kempthorne (1957).

Results and Discussion

The analysis of variance for combining ability revealed highly significant differences among the genotypes for majority of the traits indicating significant genetic differences existed in the material for yield and yield components (Table 32.1). The cytoplasmic male sterile lines differed significantly for circumference of head (cm), total number of seeds per head, number of filled seeds per head, hundred seed weight (g) and seed yield per plant (g). Fertility restorer lines differed significantly for circumference of head (cm), total number of seeds per head, number of filled seeds per head, hundred seed weight (g) and seed yield per plant (g). The line x testers showed significant differences for circumference of head (cm), total number of seeds per head, number of filled seeds per head, hundred seed weight (g) and seed yield per plant (g). The mean sum of squares due to crosses was highly significant for all the characters under study. It could be because of diverse nature of lines and testers and the significant interaction between lines and testers.

The relative importance of GCA and SCA variances was assessed by the ratio of σ^2 gca/σ^2 sca. The estimates of this ratio indicated the predominant role of non-additive effects over additive effects for all the characters under study. These results are in broad agreement with the results reported by Vanishree *et al.*, 1988; Govindaraju *et al.*, 1992; Alone *et al.*, 1996; Lande *et al.*, 1997; Gangappa *et al.*,

Table 32.1: Analysis of Variance for Combining Ability

Sl.No.	Source of Variation	DF	Mean Sum of Squares									
			Days to 50% Flowering	Days to Maturity	Plant Height at Maturity (cm)	Stem Girth (cm)	Circumference of Head (cm)	Total Number of Seeds per Head	Number of Filled Seeds per Head	Hundred Seed Weight (g)	Per cent Oil Content	Seed Yield per Plant (g)
1.	Replication	2	0.062	12.937	124.312	0.134**	81.796**	9704.00	5806.00	0.616**	9.898	36.459
2.	Female	3	63.218**	10.708	1187.75**	0.228**	85.281**	126869.3**	71940.00**	4.963**	84.468**	795.294*
3.	Male	6	17.052**	50.375**	2611.18**	0.196**	162.117**	167604.0**	131735.3**	3.0173**	50.776**	451.046
4.	Female x male	18	8.967**	18.572**	611.389**	0.101*	92.984**	90991.11**	60223.33**	3.0226**	23.551**	264.78**
5.	Error	54	2.689	5.483	65.592	0.023	12.508	16711.410	9883.408	0.119	5.361	25.153

**: Significant at 1 per cent level of significance; *: Significant at 5 per cent level of significance.

Table 32.2: Estimate of General Combining Ability Effects of 11 Parents for 10 Characters

Sl.No.	Parents	Days to 50% Flowering	Days to Maturity	Plant Height at Maturity (cm)	Stem Girth (cm)	Circumference of Head (cm)	Total Number of Seeds per Head	Number of Filled Seeds per Head	Hundred Seed Weight (g)	Per cent Oil Content	Seed Yield per Plant (g)
Females											
1.	86A	−0.06	−0.44	−10.78**	−0.14**	−0.93	9.50	7.45	0.09	1.35**	−2.42*
2.	89A	2.22**	0.61	5.86**	0.04	−0.36	32.40	12.78	−0.68**	−2.08**	−7.48**
3.	343A	−0.15	−0.77	0.60	0.10**	2.92**	68.78*	60.07**	0.46**	2.04**	6.08**
4.	852A	−2.01**	0.61	4.31*	−0.004	−1.61*	−110.69**	−80.31**	0.13	−1.30*	3.82**
	S.E. (g_i)±	0.35	0.51	1.76	0.03	0.77	28.20	21.69	0.07	0.50	1.09
	CD at 5%	0.71	1.02	3.54	0.06	1.54	56.58	43.51	0.15	1.01	2.19
	CD at 1%	0.95	1.36	4.72	0.08	2.06	75.37	57.96	0.20	1.35	2.92
Males											
1.	RHA–274	−1.64**	−1.77*	16.24**	0.16**	3.71**	171.89**	144.63**	0.65**	2.98**	−0.19
2.	RHA–278	−0.39	3.06**	−20.29**	−0.20**	−3.77**	−146.60**	−120.78**	−0.20*	−1.77*	−1.91
3.	RHA–296	2.10**	1.56*	3.11	−0.12**	−4.97**	−145.35**	−141.86**	−0.72**	−3.08**	−5.56**
4.	RHA–297	0.52	−2.52**	4.16	−0.004	−0.95	104.31**	97.96**	−0.30**	1.75*	−8.99**
5.	RHA–348	−0.72	−1.69*	3.54	0.009	0.13	−15.85	15.71	−0.25*	0.62	2.13
6.	SS–56	0.52	0.89	13.46**	0.04	0.67	15.06	−11.61	0.35**	−0.60	7.30**
7.	Morden	−0.39	0.47	−20.23**	0.12**	5.18**	16.56	15.96	0.47**	0.09	7.21**
	S.E. (g_i) ±	0.47	0.67	2.33	0.04	1.02	37.31	28.69	0.09	0.66	1.44
	CD at 5%	0.94	1.35	4.68	0.08	2.04	74.85	57.56	0.19	1.34	2.90
	CD at 1%	1.26	1.80	6.24	0.11	2.72	99.71	76.68	0.26	1.78	3.86

**: Significant at 1 per cent level of significance; *: Significant at 5 per cent level of significance.

1997b; Kandalkar, 1997; Limbore *et al.*, 1997; Shekar *et al.*, 1998; Ashok Kumar *et al.*, 1998; Hussain *et al.*, 1998; Singh *et al.*, 1999.

The estimates of GCA effects of the lines and testers for yield and yield components characters are presented in Table 32.2. The estimates of GCA effects indicated that the line CMS-343A and tester RHA-274 recorded significant GCA effect for maximum number of yield component traits besides seed yield per plant. Besides these CMS-852A line and testers RHA-297, SS-56 and Morden were found to be good combiners for one or two yield components traits *viz.*, circumference of head, number of filled seeds per head, hundred seed weight besides seed yield per plant.

The estimates of SCA effects of superior performing crosses for different traits are presented in Table 32.3. The cross CMS-343A x Morden recorded high SCA effects for the traits circumference of head, hundred seed weight and seed yield per plant. The cross CMS-89A x RHA-274 had high SCA effects for total number of seeds per head and number of filled seeds per head. In the case of crosses which exhibited significant SCA effects, the grading of parents based on their GCA effects indicated that for circumference of head, hundred seed weight and seed yield per plant, some of the crosses *i.e.*, CMS 343A x Morden, CMS-86A x RHA-278 and CMS-89A x RHA-296 involved both the parents with good combining ability, it may be attributed to the high frequency of additive alleles in the desirable direction. The cross CMS-89A x RHA-274 had high SCA effects for total number of seeds per head and filled seeds per head. Of the parental lines involved in the former cross combination, CMS-89A had significant negative GCA effects, where as RHA-274 had significant positive GCA effects. Inspite of GCA effects in opposite direction, the significant positive SCA effect of this combination is due to high frequency of dominant alleles in the desirable direction.

Table 32.3: Crosses Exhibiting High Mean Performance and Significant SCA Effects

Character	*Crosses*	
	Mean Performance	*SCA Effects*
Days to 50 per cent flowering	852A x RHA 348	89A x RHA-348
	852A x RHA 274	852A x RHA-348
	852A x RHA 278	343A x Morden
Days to maturity	852A x RHA 348	852A x RHA-348
	852A x RHA 274	89A x RHA-296
	343A x Morden	343A X Morden
Plant height at maturity	852A x SS 56	343A x Morden
	89A x RHA 274	86A x RHA-348
	852A x RHA 274	86A x 274
Stem girth	343A x Morden	343A x Morden
	89A x RHA 274	86A x 348
	852A x RHA 274	89A x 278
Circumference of head	343A x Morden	343A x Morden
	343A x RHA 274	343A x RHA-296
	89A x RHA 274	89A x RHA-278
Total number of seeds per head	89A x RHA 274	89A x RHA-274
	86A x RHA 274	86A x RHA-278
	343A x RHA 297	343A x 296

Contd...

Table 32.3–Contd...

Character	Crosses	
	Mean Performance	SCA Effects
Number of filled seeds per head	89A x RHA 274	89A x RHA-274
	86A x RHA 274	343A x RHA-296
	343A x RHA 297	852A x Morden
Hundred seed weight	343A x Morden	343A x Morden
	343A x SS 56	86A x RHA-278
	86A x RHA 274	86A x RHA-297
Seed yield per plant	343A x Morden	343A x Morden
	852A x Morden	86A x RHA-348
	343A x SS-56	343A x RHA-348
		343A x RHA-274

From the present study it could be concluded that among CMS lines 343A, 852A and among testers RHA-274, SS-56 and Morden possessed high GCA effects for different economic characters and can be utilized in the conventional breeding programme. The crosses CMS-343A x Morden, CMS-89A x RHA-274 and CMS-89A x RHA-348 with high SCA effects can be exploited for hybrid vigour. Non-additive gene action was found to be predominant for all the characters under study indicating usefulness by straight exploitation of heterosis breeding.

References

Alone, R.K., Patil, R.D., Mate, S.N. and Manjare, M.R. 1996. Combining ability in sunflower. *Indian J. Agric. Res.*, 30(3–4): 215–220.

Ashok Kumar, A., Ganesh, M. and Janila, P. 1998. Combining ability analysis for yield and yield contributing characters in sunflower. *Ann. Agric. Res.*, 19(4): 437–440.

Bajaj, R.K., Aujla, K.K. and Chahal, G.S. 1997. Combining ability studies in sunflower. *Crop improvement*, 24(1): 50–54.

Giriraj, K., Hiremath, S.R. and Senappa, K. 1987. Combining ability of converted male sterile lines of sunflower. *Indian J. Genet.*, 47(3): 315–317.

Gangappa, E., Channa Krishnaiah, K.M., Harini, H.S. and Ramesh, S. 1997b. Studies on combining ability in sunflower. *Helia*, 20(27): 73–84.

Govindaraju, T.A. and Sindagi, S.S. 1992. Combining ability for achene yield and its attributes in sunflower. *J. Oil Seeds Res.*, 9(2): 314–319.

Javed, N. and Aslam, M. 1995. Combining ability effects in sunflower F1 hybrids. *Helia*, 18(23): 41–45.

Kandalkar, B.S. 1997. Combining ability analysis in sunflower. *Bhartiya Krishi Anusandhan Patrika*, 12(2): 77–80.

Kempthorne, O. 1957. *An Introduction to Genetical Statistics*. John Wiley and Sons, Inc., New York, p. 217–568.

Khurana, S.R. and Bhatia, N. 1996. Combining ability studies in sunflower. *Indian J. Genet.*, 56: 533–539.

Kumar, A.A., Ganesh, M. and Janila, P. 1998. Combining ability analysis for yield and yield contributing characters. *Ann. Agric. Res.*, 19(4): 437–440.

Limbore, A.R., Weginwar, D.G., Gite, B.D. and Ghorade, R.B. 1997. Combining ability in sunflower. *J. Soils and Crops*, 7(1): 39–42.

Naware, M.S. 1999. Genetic analysis of some quantitative traits in sunflower. *M.Sc. (Agri.) Thesis*, M.P.K.V., Rahuri, M.S.

Shekar, G.C., Jayaramaiah, H., Virupakshappa, K. and Jagadesh, B.N. 1998. Combining ability of high oleic acid in sunflower. *Helia*, 21(28): 7–14.

Singh, D.P., Singh, S.B. and Raheja, R.K. 1999. Combining ability analysis for seed yield, oil and oil quality in sunflower. *J. Oil Seed Res.*, 16: 38–42.

Chapter 33

Economic Heterosis for Yield and its Component Traits in Sunflower (*Helianthus annuus* L.)

K. Venkata Siva Reddy and M.R. Manjare***

**Research Associate, **Senior Scientist (Breeding), Mahatma Phule Krishi Vidyapeeth, Department of Botany, Rahuri, Maharashtra*

ABSTRACT

An experiment was conducted to assess the magnitude of economic heterosis of 28 F 1 obtained by crossing four cytoplasmic male sterile lines with seven fertility restorer lines in a line x tester fashion. Economic heterosis over standard check KBSH-1 was studied using four cytoplasmic male sterile lines and seven fertility restorer lines for yield and its component characters *viz.*, circumference of head, total number of seeds per head, number of filled seeds per head, hundred seed weight and seed yield per plant. Significant economic heterosis in positive direction was expressed for all the characters under study. The crosses CMS-343A x Morden, CMS-852A x Morden and CMS-89A x RHA-274 showed superior performance than the check for the yield and its component characters studied.

Keywords: *Economic heterosis, Yield and yield components, Sunflower.*

Introduction

Sunflower is one of the most important oil seed crop grown in India because of its suitability for late sown conditions, short duration, drought and saline tolerance, photo-insensitivity, high adaptability and tolerant to common pests and diseases compared to other oil seed crops. Availability of cytoplasmic male sterility (CMS), fertility restoration sources and the highly cross pollinating nature of sunflower has made the exploitation of heterosis possible on the commercial scale. As a result, plant breeders have extensively used and exploited heterosis to improve yield and its components. This has resulted in the development and release of several hybrids suitable for different agro climatic situations in India as well as in the world. In India the first ever sunflower hybrid BSH-1 was released from Bangalore. Owing to the several advantages of hybrids over open pollinated varieties more than 70 per cent of the sunflower area is under hybrids. Hence in the present study, an attempt has been made to study the levels of economic heterosis in 28 cross combinations derived by crossing 4 CMS sources with 7 restorer lines in a line x tester design.

Materials and Methods

The experimental material was comprised of four cytoplasmic male sterile lines namely CMS-86A, CMS-89A, CMS-343A and CMS-852A and seven fertility restorer lines *viz.*, RHA-274, RHA-278, RHA-296, RHA-297, RHA-348, SS-56 and Morden. The seven fertility restorer lines were crossed with each of the CMS line in a line x tester manner to develop 28 hybrids. All the 40 genotypes (28 hybrids + 11 parents + 1 hybrid check KBSH-1) were grown in a randomized block design with three replications. The experiment was conducted at Post Graduate Institute Farm, Department of Botany, Mahatma Phule Krishi Vidyapeeth, Rahuri. The crossing was carried out during summer season, 2000 and the evaluation of the crosses along with their parents and standard check was done during *kharif* 2000. Each entry was sown in a single row of 5 m length, with a spacing of 60 cm between rows and 30 cm between plants with in a row. Standard package of practices were followed for raising the crop. Observations were recorded for circumference of head (cm), total number of seeds per head, number of filled seeds per head, hundred seed weight (g) and seed yield per plant (g) on five randomly selected competitive plants per replication in each genotype. The data were pooled and per cent heterosis over standard check was calculated for seed yield per plant and other related traits.

Results and Discussion

The values of economic heterosis over the check hybrid KBSH-1, for the characters under study are given in Table 33.1. Economic heterosis for circumference of head ranged between –31.28 to 38.16 per cent. Out of 28 cross combinations fourteen hybrids showed positive heterosis. Hybrids CMS-343A x Morden, CMS-343A x RHA-274 and CMS-89A x RHA-274 were found significantly superior over check. The maximum heterotic effect (38.16 per cent) was observed in the hybrid CMS-343A x Morden. Similar heterotic effect for this character was reported by Gill and Punia (1996); Gangappa *et al.* (1997a); Yenice and Arslan, (1997); Gill *et al.* (1998) and Naware (1999).

Maximum standard heterosis for total number of seeds per head was recorded by hybrid CMS-89A x RHA-274 (35.19 per cent). The economic heterosis was in the range of –35.33 to 35.19 per cent for total number of seeds per head. Eleven hybrids exhibited positive heterosis, of which 3 hybrids showed significant values *viz.*, CMS-89A x RHA-274, CMS-86A x RHA-274 and CMS-343A x RHA-297. Heterosis for this character has also been reported earlier by Rao (1989); Gill and Punia (1996); Lande *et al.* (1997); Limbore *et al.* (1998) and Naware (1999).

The range of useful heterosis was between –37.75 to 39.37 per cent for the character number of filled seeds per head. Twelve hybrids exhibited positive heterosis, of which the hybrids CMS-89A x RHA-274, CMS-86A x RHA-274 and CMS-343A x RHA-297 exhibited significant positive heterosis over the check KBSH-1. The maximum heterosis effect (39.17 per cent) was observed for the hybrid

CMS-89A x RHA-274. Similarly were obtained by Govindaraju *et al.* (1992); Madrap and Makne (1993); Gangappa *et al.* (1997a); Lande *et al.* (1998) and Naware (1999).

Table 33.1: Extent of Economic Heterosis (%) for Seed Yield per Plant and its Component Traits

Sl.No.	*Cross*	*Circumference of Head (cm)*	*Total Number of Seeds per Head*	*Number of Filled Seeds per Head*	*Hundred Seed Weight (g)*	*Seed Yield per Plant (g)*
1.	86A x RHA-274	– 0.07	22.93*	25.89**	9.28	– 30.06**
2.	86A x RHA-278	– 2.46	0.89	– 1.84	8.76	– 7.45
3.	86A x RHA-296	– 31.28**	– 26.67**	– 33.72**	– 36.63**	– 58.19**
4.	86A x RHA-297	– 5.28	2.68	10.97	– 0.57	– 42.81**
5.	86A x RHA-348	8.25	– 0.94	– 3.17	– 0.87	– 10.09
6.	86A x SS-56	4.63	– 0.39	– 2.10	– 21.82**	– 17.07
7.	86A x Morden	3.76	– 20.27*	– 14.70	– 18.11**	– 24.24**
8.	89A x RHA-274	9.05	35.19**	39.37**	– 11.55*	– 56.31**
9.	89A x RHA-278	2.75	– 15.50	– 13.37	– 18.34**	– 28.59**
10.	89A x RHA-296	– 4.92	– 9.07	– 20.57*	– 24.68**	– 32.15**
11.	89A x RHA-297	– 9.91	1.62	3.95	– 39.14**	– 62.82**
12.	89A x RHA-348	– 0.14	– 9.49	3.62	– 17.36**	– 25.15**
13.	89A x SS-56	– 4.70	– 18.71*	– 14.99	– 5.45	– 13.62
14.	89A x Morden	– 5.86	7.62	– 12.55	– 38.15**	– 46.02**
15.	343A x RHA-274	10.21	– 4.24	4.98	– 0.12	– 1.84
16.	343A x RHA-278	– 27.08**	– 35.33**	– 37.75**	– 31.01 **	– 56.60**
17.	343A x RHA-296	8.32	3.63	7.38	– 20.95**	– 5.91
18.	343A x RHA-297	1.66	19.60*	16.21	3.01	– 36.53**
19.	343A x RHA-348	3.62	6.48	13.11	– 15.90**	1.04
20.	343A x SS-56	1.37	15.67	2.88	9.52	2.67
21.	343A x Morden	38.16**	7.17	15.29	40.71**	34.90**
22.	852A x RHA-274	8.40	– 11.89	– 20.13*	7.66	– 1.49
23.	852A x RHA-278	– 10.71	– 14.83	– 14.55	– 13.98**	– 11.62
24.	852A x RHA-296	– 19.98**	– 32.23**	– 29.95**	– 9.11	– 38.74**
25.	852A x RHA-297	0.58	– 4.58	– 1.69	– 25.26**	– 21.75*
26.	852A x RHA-348	– 15.27*	– 16.98	– 20.57*	– 23.91**	– 33.78**
27.	852A x SS-56	– 0.14	– 7.15	– 4.91	1.86	1.55
28.	852A x Morden	4.34	– 4.60	5.06	7.71	8.13
	LSD at 5%	5.35	208.79	155.58	0.56	8.16
	LSD at 1%	7.11	276.98	206.39	0.74	10.83

For the character hundred seed weight the useful heterosis ranged from –39.14 to 40.71 per cent. Out of 28 hybrids, 21 hybrids exhibited positive heterosis. The hybrids CMS-343A x Morden, CMS-343A x SS-56 and CMS-86A x RHA-274 showed high and significant positive heterosis. The highest standard heterosis (40.71 per cent) was observed for the hybrid CMS-343A x Morden. Similar heterotic effects for this character was reported by Govindaraju *et al.* (1992); Sugoor *et al.* (1994); Kandhola *et al.* (1995a); Gill and Punia (1996); Yenice and Arslan (1997); Gangappa *et al.* (1997a) and Lande *et al.* (1998).

Economic heterosis for the character seed yield per plant ranged from –62.82 to 34.90 per cent. Out of 28 hybrids four hybrids exhibited positive heterosis. Three hybrids showed significant positive heterosis *viz.*, CMS-343A x Morden, CMS-852A x Morden and CMS-343A x SS-56. Maximum heterotic effect was observed for the hybrid CMS-343A x Morden (34.90). Similar results were reported by Madrap and Makne (1993); Sugoor *et al.* (1994); Kandhola *et al.* (1995a); Yenice and Arslan (1997); Gill *et al.* (1998); Kumar *et al.* (1999) and Shekar *et al.* (1999).

From the above results, it can be concluded that the circumference of head, number of filled seeds per head and hundred seed weight were the main components for getting good yield besides seed yield per plant. Therefore selection for these characters might result in the improvement of yield and the promising hybrids like CMS-343A x Morden and CMS-89A x RHA-274 may be further tested on larger plots over different locations and season before recommending them for commercial cultivation.

References

Gangappa, E., Channa Krishnaiah, K.M., Ramesh, S. and Harini, H.S. 1997a. Exploitation of heterosis in sunflower (*Helianthus annuus* L.). *Crop Res.*, 13(2): 339–348.

Gill, H.S. and Punia, M.S. 1996. Expression of heterosis in single, double and three way cross hybrids of sunflower. *Helia*, 19(25): 111–117.

Gill, H.S., Khurana, S.R., Yadav, T.P. and Sheoran, R.K. 1998. Expression of heterosis for different characters in sunflower over environments. *Haryana. Agric. Univ. J.*, 28(2–3): 95–100.

Govindaraju, T.A., Sindagi, S.S., Virupakshappa, K. and Ranganatha, A.R.G. 1992. Heterosis for achene yield and its components in sunflower. *Mysore J. Agric. Sci.*, 26: 236–238.

Kandhola, S.S., Behl, R.K. and Punia, M.S. 1995a. Heterosis in sunflower. *Ann. Bio.*, 11(1–2): 98–102.

Kumar, A.A., Ganesh, M., Kumar, S.S. and Reddy, A.V.V. 1999. Heterosis in sunflower. *Ann. Agric. Res.*, 20(4): 470–480.

Lande, S.S., Narkhede, M.N., Weginwar, D.G., Patel, M.C. and Golhar, S.R. 1998. Heterotic studies in sunflower. *Ann. Plant Physiol.*, 12(1): 15–18.

Limbore, A.R., Weginwar, D.B., Lande, S.S. Gite, B.D and Ghodke, K.M. 1998. Heterosis in sunflower. *Ann. Plant Physiol.*, 12(1): 38–42.

Madrap, I.A. and Makne, V.G. 1993. Heterosis in relation to combining ability effect and phenotypic stability in sunflower. *Indian J. Agric. Sci.*, 63(8): 484–488.

Naware, M.S. 1999. Genetic analysis of some quantitative traits in sunflower. *M.Sc. (Agri). Thesis*, M.P.K.V., Rahuri. M.S.

Shekar, G.C., Jayaramaiah, H., Virupakshappa, K. and Nehru, S.D. 1999. Standard heterosis of early maturing CMS lines and restorers in sunflower. *J. Oilseeds Res.*, 16: 32–35.

Sugoor, R.K., Giriraj, K. and Salimath, P.M. 1994. Influence of induced mutation on heterosis for seed yield and its attributes in sunflower. *J. Oilseeds Res.*, 11(2): 185–188.

Yenice, N. and Arslan, O. 1997. Heterosis reported for a synthetic variety obtained from selfed sunflower. *Turkish J. Agric. and Forestry*, 21(3): 307–309.

Chapter 34

Interaction Effect of *Rhizobium* and Pressmud Compost on Yield of Gram (*Cicer arietinum*)

A.M. Deshmukh

Department of Microbiology, Yashwantrao Chavan College of Science, Karad – 415 124, Maharashtra

ABSTRACT

Pressmud compost obtained from local sugar factory and soil samples around the Karad region were analyzed chemically. The pressmud compost was found rich in ash content, protein, carbohydrates and wax. Light soil was found to contain less amount of total nitrogen, organic matter and moisture than that of medium black soil and deep black soil.

The interaction effect of pressmud compost with locally isolated *Rhizobium* was tested on the growth of gram (*Cicer arietinum*) plant by performing pot experiments in laboratory. Three different soil samples *viz.*, Light soils, Medium black soils and Deep black soils and three different concentrations of pressmud compost *viz.*, 100 g/kg of soil, 200 g/kg of soil and 300 g/kg of soil were used for testing. It was found that pressmud compost in association with *Rhizobium* cause increase in nodule number, nodule biomass, plant biomass, root growth and shoot growth of gram in all soil types. The interaction effect of pressmud at the rate of 200 g/kg of soil on the growth of gram plant was found maximum in light soil.

Keywords: *Rhizobium, Pressmud compound compost, Gram (Cicer arietinum).*

Introduction

Pressmud is one of the main by products of sugar industry. Pressmud is used on agricultural field as manure after composting. The recycling and utilization of pressmud is of serious concern in view of the environmental hazards resulting from their non-utilization and rapid development of sugar industries in India.

Pressmud compost is a vital source of macro-micro nutrients, plant protecting steroids and variety of beneficial microorganisms increasing soil fertility and promoting plant growth.

Rhizobium is nitrogen fixing bacterium in association with roots of leguminous plants increasing the crop productivity. More the nodule number and nodule biomass, more the crop productivity.

The present investigation was aimed to ascertain the role of nitrogen fixing root nodulating bacterial inoculation with pressmud compost as coinculant in different types of soils.

Material and Methods

Pressmud Compost Analysis

The pressmud compost was obtained from Sahyadri Sugar Factory Limited, Yashwantnagar, Karad (Maharashtra). Chemical analysis of pressmud compost was performed as follows.

Total volatile solids by gravimetric method, Ash content by burning at 600°C for 30 minutes, organic carbon as per Walkley Black method–Piper (1986), wax by extraction in benzene, protein as per Lowry *et al.* (1951), reducing sugar by DNSA method as per Miller (1969), cellulose as per Updegraff (1969), hemi-cellulose as per Descatelets and Yu (1986) and lignin as described in methods of Wood Chemistry (1967).

Soil Analysis

Deep black soil, medium black soil and light soil samples were collected from Karad (Maharashtra) around the college premises. They were analyzed for the followings:

Moisture and pH as described by Piper (1966), organic carbon by rapid pitration method as described by Walkley and Black (Piper, 1966), organic matter by multiplying the values of organic carbon with conventional factor of 1.724 as described by Piper (1966), total phosphorous by sulphomolybdic method as described by Jackson (1973), total nitrogen by Kjeldahl's method as directed by Piper (1966) and total potassium by flame photometric method as directed by Jackson (1973).

Rhizobium Isolation

Healthy root nodules of gram (*Cicer arietinim*) samples were collected from the field and *Rhizobium* was isolated on Congo red yeast extract mannitol agar as decribed by Subbarao (1988). The isolated *Rhizobium* was confirmed by performing gram staining motility and biochemicals (as per Bergey's Manual of Systematic Bacteriology, 1988).

Pot Experiment

In order to study interaction effect of a *Rhizobium* with pressmud compost in a different soil types on seedling growth physiology and root nodule formation of Gram (*Cicer arietinum*) a pot experiment was designed as follows.

Experiment was conducted in triplicate of five sets using 15 earthen pots measuring 17 cm in height and having an internal diameter 20 cm. Pots were initially dry sterilized at 160°C for 60 min.

And each was packed uniformly with 3 kg of light soil previously steam sterilized at 15 psi for 2 hrs. Pressmud compost was mixed thoroughly at the rate of 100 g, 200 g and 300 g per kg of soil in three different posts.

Gram seeds of good quality and uniform in size were selected and surface sterilized by using 0.2 per cent mercuric chloride solution and then washed with sterile distilled water aseptically. A bunch of gram grain was soaked in a suspension of *Rhizobium* of density 10^8 cells/ml prepared in 10 per cent jaggery solution and then air dried for 1 hr in shed. Grains from this bunch were the seeded uniformly eight each in four pots (in triplicate) packed with soil. As a control uncoated sterile gram seeds were inoculated in one pot (set-I).

All above treatments consist of nitrogen application at the rate of 20.5 mg N/kg soil as a urea.

All the pots were kept under natural environment for a period 45 days and watere at an alternative day by using sterile distilled water. Thining of seedling were carried out during the course of incubation, suitably, if necessary to avoid crowding. At the end of 45 days of incubation five seedling from each pot were uprooted carefully and studied for nodule number, nodule biomass, plant biomass, root growth and shoot growth and the average values were noted.

Results and Discussion

Pressmud Compost Analysis

All the experiments were performed in triplicates and average values are taken. The results of the analysis are presented in Table 34.1. It is seen from the table that pressmud compost is rich in cellulose (14 per cent) and hemi cellulose (10 per cent) content. Right amount of these polysaccharides support the growth of various bacteria including nitrogen fixing and phosphate solubilizing. The presence of 3.9 per cent protein is a source of sustained release of aminoacids by continuous proteolysis by rhizobial microflora. Aminoacid absorption by roots of plants promote increased chlorophyll synthesis and increased plant growth rate (Sharma *et al.*, 1992, 1993, 1995 and Yadav *et al.*, 1995). Cellulolytic and hemi cellulolytic microorganisms degrade these polysaccharide releasing simpler sugar supporting the growth of *Rhizobium* and promoting root nodulation. It is seen from the table that wax is 4.2 per cent in pressmud compost. It has the dual potential to serve as a source of plant regulators such as tricontanol and phytosterols which have the potential for neutralizing the virulence of plant pathogens.

Table 34.1: Pressmud Compost Analysis

Constituents	%
Total Volatile Solids	57 per cent
Ash content	43 per cent
Organic carbon	27 per cent
Wax	4.2 per cent
Protein	3.9 per cent
Lignin	2.5 per cent
Cellulose	14 per cent
Hemi cellulose	10 per cent
Reducing sugar	0.4 per cent
Total carbohydrates	2 per cent

The presence of 43 per cent ash content in the pressmud compost increase the fertility value by complementing several physiological process in plant and rhizobial microflora. Thus pressmud compost chemicals support the plant growth rate and also promotes the rhizobial propogation.

Soil Analysis

Deep black soil (D.B.S.), medium black soil (M.B.S.) and light soil (L.S.) samples were collected from Karad soils around college premises. The data on the chemical analysis of the soil samples with reference to pH, moisture, organic carbon, organic matter, nitrogen, phosphorus, and potassium are given in Table 34.2.

Table 34.2: Chemical Composition of Deep Black, Medium Black and Light Soils

	Deep Black Soil	*Medium Black Soil*	*Light Soil*
Moisture (per cent)	16	9.2	5.7
pH	7.8	7.7	7.7
Organic matter (per cent)	1.81	1.12	0.52
Organic carbon (per cent)	1.05	0.65	0.35
Total phosphorous (mg/100 g)	57.65	57.43	48.65
Total nitrogen (per cent)	0.083	0.061	0.043
Total potassium (mg/100 g)	90.6	86.0	98.0

When the data on the chemical nature of the three soil types are compared with each other it becomes clear that deep black soil is rich in moisture, organic matter and nitrogen than that in medium black and light soils. Light soils are poorest in this regard. The pH in all soil types is not differing significantly.

Interation Effect of Pressmud Compost and *Rhizobium*

Results of interaction effect of *Rhizobium* with pressmud compost in different soil types, *viz.*, deep black, medium black and light soil, on seedling growth physiology and root nodule formation in gram (*Cicer arietinum*) are presented in Tables 34.3–34.5.

Table 34.3 represents interaction effect of *Rhizobium* with pressmud compost on the growth of gram (*Cicer arietinum*) in light soil. It is seen from the table that there is a increase in nodule number, nodule biomass, plant biomass, root growth and shoort growth of the gram plant by addition of *Rhizobium* and pressmud compost in the soil. It is also seen that nodule number, nodule biomass, plant biomass, root growth and shoort growth was maximum with treatment 5 followed by that with treatment 4 and treatment 3. But there is no marked difference between these values with treatment 4 and treatment 5, indicating that increase in concentration of pressmud compost beyond 200 g / kg of soil does not affect markedly the growth rate of the plant.

Table 34.4 represents interaction effect of *Rhizobium* with pressmud compost on the growth of gram (*Cicer arietinum*) in medium black soil. It is seen from the table that there is a increase in nodule number, nodule biomass, plant biomass, root growth and shoort growth of the gram plant by addition of *Rhizobium* and pressmud compost in the soil. It is also seen that nodule number, nodule biomass, plant biomass, root growth and shoort growth was maximum with treatment 5 followed by that with treatment 4 and treatment 3. But there is no marked difference between these values with treatment 4

Table 34.3: Interaction Effect of *Rhizobium* with Pressmud Compost in Light Soil on Seedling Growth Physiology and Root Nodule Formation in Gram (*Cicer arietinum*) After 45 Days

Treatment No.	*Nodule No./Plant*	*% Increase Over T2*	*Nodule Biomass mg/Plant*	*% Increase Over T2*	*Plant Biomass g/Plant*	*% Increase Over T2*	*Root Growth cm/Plant*	*% Increase Over T2*	*Shoot Growth cm/Plant*	*% Increase Over T2*
T1	–	–	–	–	3.01	–	11.6	–	17.3	–
T2	12	–	37	–	3.32	–	13.7	–	19.0	–
T3	15	25	44	18.9	3.69	11.1	14.9	8.8	20.2	6.3
T4	20	66.7	69	86.5	4.17	14.5	16.2	18.4	21.8	14.7
T5	21	75	71	91.9	4.34	30.7	16.8	22.6	22.4	17.9

Table 34.4: Interaction Effect of *Rhizobium* with Pressmud Compost in Medium Black Soil on Seedling Growth Physiology and Root Nodule Formation in Gram (*Cicer arietinum*) After 45 Days

Treatment No.	*Nodule No./Plant*	*% Increase Over T2*	*Nodule Biomass mg/Plant*	*% Increase Over T2*	*Plant Biomass g/Plant*	*% Increase Over T2*	*Root Growth cm/Plant*	*% Increase Over T2*	*Shoot Growth cm/Plant*	*% Increase Over T2*
T1	–	–	–	–	3.12	–	11.9	–	17.6	–
T2	13	–	37	–	3.37	–	13.9	–	19.3	–
T3	15	15.4	45	21.6	3.71	10.1	14.8	6.5	20.6	6.7
T4	21	61.5	69	86.4	4.25	26.1	16.0	15.1	22.3	15.5
T5	23	76.9	72	94.6	4.32	28.2	16.5	18.7	22.9	18.7

T1: Without *Rhizobium* and pressmud compost.

T2: With *Rhizobium* and without pressmud compost.

T3: With *Rhizobium* and pressmud compost at rate of 100 g/kg of soil.

T4: With *Rhizobium* and pressmud compost at rate of 200 g/kg of soil.

T5: With *Rhizobium* and pressmud compost at rate of 300 g/kg of soil.

Table 34.5: Interaction Effect of *Rhizobium* with Pressmud Compost in Deep Black Soil on Seedling Growth Physiology and Root Nodule Formation in Gram (*Cicer arietinum*) After 45 Days

Treatment No.	*Nodule No./Plant*	*% Increase Over T2*	*Nodule Biomass mg/Plant*	*% Increase Over T2*	*Plant Biomass g/Plant*	*% Increase Over T2*	*Root Growth cm/Plant*	*% Increase Over T2*	*Shoot Growth cm/Plant*	*% Increase Over T2*
T1	–	–	–	–	3.60	–	12.8	–	18.6	–
T2	15	–	39	–	3.82	–	14.2	–	19.9	–
T3	18	16.7	46	17.9	4.20	9.9	15.00	5.6	20.9	5.0
T4	22	46.7	70	79.5	4.52	18.3	15.9	12.0	22.4	12.6
T5	27	80.0	72	84.6	4.62	20.9	16.3	14.8	23.2	16.6

T1: Without *Rhizobium* and pressmud compost.

T2: With *Rhizobium* and without pressmud compost.

T3: With *Rhizobium* and pressmud compost at rate of 100 g/kg of soil.

T4: With *Rhizobium* and pressmud compost at rate of 200 g/kg of soil.

T5: With *Rhizobium* and pressmud compost at rate of 300 g/kg of soil.

Table 34.6: Interaction Effect of *Rhizobium* with Pressmud Compost (at the rate of 200 g/kg of soil) in Different Soil Types on Seedling Growth Physiology and Root Nodule Formation in Gram (*Cicer arietinum*) After 45 Days

Soil Type	*% Increase Over Control/Plant*				
	Nodule No.	*Nodule Biomass*	*Plant Biomass*	*Root Growth*	*Shoot Growth*
Light Soil	66.7	86.5	14.5	18.4	14.7
Medium Black Soil	61.5	86.4	26.1	15.1	15.5
Deep Black Soil	46.7	79.5	18.3	12.0	12.6

Control: *Rhizobium* without pressmud.

and treatment 5, indicating that increase in concentration of pressmud compost beyond 200 g/ kg of soil does not affect markedly the growth rate of the plant.

Table 34.5 represents interaction effect of *Rhizobium* with pressmud compost on the growth of gram (*Cicer arietinum*) in deep black soil. It is seen from the table that there is a increase in nodule number, nodule biomass, plant biomass, root growth and shoort growth of the gram plant by addition of *Rhizobium* and pressmud compost in the soil. It is also seen that nodule number, nodule biomass, plant biomass, root growth and shoort growth was maximum with treatment 5 followed by that with treatment 4 and treatment 3. But there is no marked difference between these values with treatment 4 and treatment 5, indicating that increase in concentration of pressmud compost beyond 200 g/kg of soil does not affect markedly the growth rate of the plant.

Prabhakaran *et al.* (1996) observed that application pressmud compost at the rate of 5 t/ha in the soil with *Rhizobium* has significantly increased growth, nodulation and grain yield of vamban–1 green gram in acid soils.

Table 34.6 represents the interaction effect of pressmud compost at the rate of 200 g/kg of soil and *Rhizobium* in deep black, medium black and light soils. It is seen from the table that the per cent values of increase in nodule number, nodule biomass and root growth are maximum in light soils, followed by that of medium black and deep black soils. It is already short (Table 34.2) that deep black soil are already rich in organic matter, total nitrogen and moisture whereas light soil is poor in organic matter, total nitrogen and moisture contents. It is clear that the addition of pressmud compost has enriched the light soils causing more effect on plant growth as compare to medium black and deep black soils. The pH of all the soil types is almost equal (Table 34.2), indicating that the pH of soil has not affected plant growth in different soil types. Thus it can be concluded that the pressmud compost promote the plant growth significantly in light soils.

Bharati Jadhav (1999) found that inoculation of pressmud compost in soil has significantly increased dry matter, nitrogen content of spinach plant. Vijayasarangan (2000) found that inoculation of composted sugarcane trash at the rate of 5 t/ha with *Azospirillum* has significantly increased grain yield in rice. Tapan Rai *et al.* (2000) found that crop residues compost has characteristics that can increase plant growth. They further observed that compost prepared posses large number of phospate solublizing bacteria. Balachandar and Kannaiyan (1996) have shown that compost prepared from cellulosic waste has improved significantly plant growth, grain yield in rice. Raman and Sundaram (1996) found that application of pressmud compost in the soil has significantly increased germination of tomato seedling.

Acknowledgement

The author is thankful to University Grants Commission for providing financial assistance under minor research project scheme.

References

Alexander, M. 1961. *Introduction to Soil Microbiology*. John Wiley, New York.

Bharati Jadhav. 1999. Microbial technology as a tool for utilization of waste and increasing the yield of green leafy vegetables In: *Abstract of State Level Symposines of Recent Trends in Biotechnology for Rural and Industrial Development*, Latur, pp. 27.

Balachandar, D. and Kannaiyan, S. 1996. Preparation of biomanure from viscose factory, waste and its effect on rice crap. In: *Abstract of 37th Annual Conference of the Association of Microbiologists of India*, Chennai, pp. 172.

Bergey. 1989. *Manual of Systematic Bacteriology*, Vol. 4. Williams and Wilkins, London.

Buchana, R.E. and N.E. Bibbons (Eds). 1974. *Bergey's Manual of Determinative Bacteriology*, 8th Ed. The Wiliams and Wilkins Co., Baltimore.

Deschetelets, L. and Yo, E.K.C. 1986. A simple pentose assay for biomass conversion studies. *Appl. Microbiol. Biotechnol.*, 24: 379–385.

Deshmukh, A.M. 1996. *Handbook of Media, Stains and Reagents in Microbiology*. Pama Publication, Karad.

Dubois, M., Gilles, K.A., Hamilton, J.K., Robers, P.A. and Smith, F. 1956. Colorimetric estimation of total sugars and related substances. *Anal. Chem.*, 28: 350–356.

Jackson, M.L. 1973. *Soil Chemical Analysis*. Prentice Hall of India Pvt. Ltd., New Delhi.

Juwarkar, A.S., Shede A., Thawale, P.R., Satyanarayanan, S., Deshbratar, P.B., Bal, A.S. and Juwarkar, A. 1994. Biological and industrial waste as source of plant nutrients. In: *Fertilizer, Organic Manure, Recyclable Waste and Biofertilizers*, (Ed.) Tandon, H.L.S., p. 72–90

Lowry, O.H., Rosebrongh, N.J., Farr, A.L. and Randall, R.J. 1951. Protein measurement with the folin phenol reagent. *J. Biol., Chem.*, 193: 265–275.

Methods of Wood Chemistry. 1967. Vol. 1. Interscience Publ., New York, U.S.A.

Miller, G.L. 1959. Estimation of reducing sugars by DNSA method. *Anal. Chem.*, 31: 426–428.

Piper, C.S. 1968. *Soil and Plant Analysis*. Reprint for Asia. Hans Publ. Hons, Bombay, p. 368.

Prabhakaran, J., Ravi, K.B. and Nadarajan, N. 1996. Influence of bio-digested slurry on the growth and yield of greengram in alfisol. In: *Abstract of 40th Annual Conference of Association of Microbiologists of India*, Bhubaneshwar, p. 119.

Raman, G. and Sundaram, M.D. 1996. Studies on the pretreatment of sugar factory wastes and the effect of sugar factroy wastes on the growth of tomato. In: *37th Annual Conference of the Association of Microbiologists of India*, p. 147.

Sharma, R.K., Ramamurthy, V. and Kothari, R.M. 1999. Soil conditioner a vital bioresource for sustainable agriculture and forestry. In: *Bioresource Biotechnology*, (Ed.) A. Kanniyan, TNAU, Coimbatore.

Subba Rao, N.S. 1988. *Biofertilizers in Agriculture*. Oxford and IBH Publ. Co. Pvt. Ltd., New Delhi, p. 189–203.

Tapan Rai, K.B. Munishamanna and D. Radhakrishna. 2000. Improvement of crop residue composting and its influence on crop growth. In: *Abstract of 41st Annual Conference of Association of Microbiologists of India*, Jaipur, p. 127.

Updegraff, D.M. 1969. Semi-micro determination of cellulose in biological material. *Anal. Biochem.*, 32: 420–424.

Chapter 35

Micropropagation of *Wedelia chinensis* through High Frequency Shoot Multiplication using Nodal Explants

Shally Sultana and P.J. Handique*

Department of Biotechnology, Gauhati University, Guwahati – 781014, Assam, India
**E-mail: pjhandique@sify.com*

ABSTRACT

An efficient micropropagation protocol was standardized for *Wedelia chinensis,* an important and commercially exploited medicinal plants species found in natural habitats of NE India. Formation of multiple shoots in high frequency was achieved in nodal explant culture. MS basal medium was used throughout the experiments with different combinations and concentrations of BAP, Kinetin and IAA. Combination of Kinetin (3 mg/l) and IAA (1 mg/l) in MS medium gave a maximum number of 15 shoots per explant. Individual shoots from about 7 weeks old multiple shoot cultures were separated and implanted in half-strength MS medium supplemented with IBA for rooting. Root initiation was noticed after 5–6 days and it became profuse after 10 weeks. Rooted plantlets were successfully established in soil after proper hardening. The survival percentage of rooted plants on soil was found to be 72 per cent. The results were interpreted and supported with necessary statistical analysis, figures and photographs.

Keywords: *Wedelia chinensis, Micropropagation, Medicinal plants, Nodal explants.*

---

*Author for Correspondence.

Introduction

Wedelia chinensis Merrill syn *W. calendulacea* Less (Family: Asteraceae) is a soft herb with considerable medicinal uses. It is commonly known as 'bhringaraj' (Assamese). The plant is semi-creeper in habit and attains a height of 1–3 feet with gorgeous growth. It was reported to be found in natural habitats of Bengal, Assam, Konkan and plain districts of Tamil Nadu (Hooker, 1879). *Wedelia chinensis* is extensively used in ayurvedic preparations, in modern drug industries and also in various traditional medicinal practices. Decoction of this herb is used in menorrhagia and uterine hemorrhages. Leaves are considered as tonic and used in cough, cephalagia, alopecia and skin diseases (Chopra *et al.*, 1956). Leaves are also used for dyeing hair and for promoting hair growth. Roots yield a black dye. Ethanolic extract of the herb inhibits the growth of Ehrlich ascites carcinoma (CSIR, 1992). In Assam, plant juice is taken orally for intestinal disorder. Paste of aerial portion is used as cooling agent in fever (Handique, 1994).

W. chinensis can be propagated by vegetative methods. However, due to habitat destruction and the use of the species in large quantities for medicinal purposes, it is becoming rare in natural habitat condition. Commercial scale cultivation of the species is also very much scanty in Northeast India though it has considerable demand in medicinal plant market. Further, no report is available on tissue culture of this plant species except a preliminary report published by the present authors (Bhuyan *et al.*, 2000). This paper describes a complete protocol for micropropagation of *W. chinensis.*

Materials and Methods

A few young plants of *Wedelia chinensis* were collected from their natural habitat, potted and maintained in the green house. Nodal explants were collected from the green house maintained plants. About 1.5 cm long stem segments which consist of a single node containing a small portion of the internodes on either side were excised and used as explants to initiate the culture. The explants were thoroughly washed in tap water and then dipped in 1 per cent Tween 20 for 30 minutes with occasional stirring. They were then washed thoroughly in running tap water for 10 minutes. These explants were then surface sterilized with 0.1 per cent $HgCl_2$ for 5 minutes followed by 70 per cent ethanol for 30 seconds. These explants were then thoroughly rinsed with sterile double distilled water and implanted vertically into the nutrient medium. MS basal medium (Murashige and Skoog, 1962) with 3 per cent sucrose and 0.8 per cent agar was used throughout the experiments. MS medium was supplemented with auxin and cytokinin either singly or in combination at various concentrations. The pH of the medium was adjusted to 5.8 before dispensing into 150 x 25 mm culture tubes and then autoclaved at 15 lb/inch^2 for 15 min. Cultures were incubated at 25±2° C under fluorescent light of about 2000 lux with 16-hour photoperiod. The cultures were maintained by repeated sub-culturing at 15 days interval and subsequently observations were recorded. Seven treatments were used with 10 replications in each and the experiment was repeated twice. *In vitro* grown shoots which were 2–3 cm long having at least four leaves and 2–3 nodes were excised from the clump and transferred to half strength MS medium containing IBA for induction of roots. After three weeks of root initiation, the rooted shoots were removed from culture tubes and washed free of agar. The plantlets were then potted in a sterilized mixture of garden soil, sand and compost (2 : 1 : 1) and covered with a plastic bag to maintain high humidity. A week later, the bags were removed and the plantlets were successfully established in field conditions.

Results and Discussion

The nodal explants started growing after about nine days of culture on MS medium producing green shoot buds. Initially establishment of culture was difficult due to a higher rate of fungal contamination, which was later eliminated by treatment of explants with 1 per cent Bavistin for 30 minutes followed by 0.1 per cent $HgCl_2$ for 5 minutes and 70 per cent ethanol for 30 seconds. Development of the first shoot was recorded after 9 days of inoculation in 85 per cent of the culture in MS medium supplemented with various concentrations of growth regulators. The control showed the development of first shoot after 31 days of culture with very little growth compared to other media formulations. This indicates that formation and development of new shoots solely depends on the optimum growth regulator level in the medium. Therefore, different combinations and concentrations of BAP, Kinetin and IAA were tried. MS medium supplemented with BAP (alone or in combination with IAA) showed insignificant results even after 45 days of culture. Hence, these were not taken into account. Whereas, MS medium supplemented with various concentrations of Kn alone or Kn and IAA in combination gave significant results. The effect of various hormones, either singly or in combinations, on number of shoots proliferated are presented below.

From the results presented in the Table 35.1(a) and Table 35.1(b), it is evident that a medium supplemented with Kn only at various concentrations produced a maximum of 5 shoots per explant which were clumped and reduced [Figure 35.2(a)]. These shoots although showed a normal induction, became reduced in size and lost the identity of individual shoots subsequently. But a medium containing Kn and IAA at different concentrations showed a better result. Kinetin at 2 mg/l and IAA at 0.5 mg/l in MS medium gave a maximum number of 9 shoots per explant [Tables 35.2(a) and 35.2(b); Figure 35.2(b)]. Further, Kn at 3 mg/l and IAA at 1 mg/l in MS medium gave a maximum number of 15 shoots per explant [Tables 35.3(a) and 35.3(b); Figure 35.2(c)]. This shows that addition of an auxin in a cytokinin-containing medium enhances shoot multiplication in *W. chinensis.* This is in contrast to earlier results in *Glehnia littoralis* (Hiraoka *et al.*, 1987). Chattopadhyay *et al.* (1990) also reported that auxin and cytokinin combination is essential for multiple shoot formation in mulberry. Sharma *et al.* (1991) reported a maximum average of 12.3 numbers of shoots at Kn (2.0 mg/l) and IAA (1.0 mg/l) in *Coleus forskohlii.*

Table 35.1(a): Effect of Kinetin on Number of Shoots*

Observation in Days	*Number of Shoots*					
	Kn Concentration in MS Medium (mg/l)					
	Control	*0.5*	*1.0*	*2.0*	*3.0*	*4.0*
15	0.0	0.5	1.0	1.0	2.0	0.0
30	0.0	2.0	2.8	3.0	3.0	1.5
45	1.0	2.0	3.0	3.5	4.0	2.0
60	1.0	3.0	3.5	4.8	5.0	3.0

*Values represented in the table are mean of ten replications.

Table 35.1(b): ANOVA Showing Effect of Kinetin on the Number of Shoots Proliferated *In vitro* of *Wedelia chinensis*

Source of Variation	*Degrees of Freedom*	*Sums of Squares*	*Mean Sums of Squares*	*Calculated Value of F*	*Table Value of F*
Time	3	22.031	7.344	38.25*	3.29
Treatment	5	23.688	4.738	24.67*	2.9
Error	15	2.879	0.192		

*Significant at 5 per cent level of significance.

Table 35.2(a): Effect of Kn + IAA on Number of Shoots*

Observation in Days	*Number of Shoots*				
	Kn + IAA Concentration in MS Medium (mg/l)				
	Control	*2.0+0.05*	*2.0+0.1*	*2.0+0.5*	*2.0+1.0*
15	0	0.9	1.0	1.5	0
30	0	3.0	4.0	3.8	2.5
45	1	5.8	6.0	6.5	3.0
60	1	7.0	7.5	9.0	4.5

*Values represented in the table are mean of ten replications.

Table 35.2(b): ANOVA Showing Effect of Kn + IAA on Number of Shoots Proliferated *in vitro*

Source of Variation	*Degrees of Freedom*	*Sums of Squares*	*Mean Sums of Squares*	*Calculated Value of F*	*Table Value of F*
Time	3	74.148	24.716	20.393*	3.49
Treatment	4	58.245	14.561	12.014*	3.26
Error	12	14.547	1.212		

*Significant at 5 per cent level of significance.

Table 35.3(a): Effect of Kn + IAA on Number of Shoots*

Observation in Days	*Number of Shoots*					
	Kn+IAA Concentration in MS Medium (mg/l)					
	Control	*3.0+0.05*	*3.0+0.1*	*3.0+0.5*	*3.0+1.0*	*3.0+1.5*
1	0	2.7	3.0	3.3	4.2	3.0
30	0	3.1	4.6	6.2	8.3	4.5
45	0	4.2	6.0	9.2	11.3	7.6
60	1	6.9	7.2	12.0	15.0	9.1

*Values represented in the table are mean of ten replications.

Table 35.3(b): ANOVA Showing Effect of Kn + IAA on Number of Shoots Proliferated *in vitro*

Source of Variation	*Degrees of Freedom*	*Sums of Squares*	*Mean Sums of Squares*	*Calculated Value of F*	*Table Value of F*
Time	3	115.395	38.465	16.573*	3.29
Treatment	5	197.473	39.495	17.016*	2.90
Error	15	34.81	2.321		

*Significant at 5 per cent level of significance.

The experiment further shows that a slight increase in the concentration of IAA (1.5 mg/l) with Kn (3 mg/l) the shoot number reduced significantly (Table 35.3). Thus, it can be stated that Kn

Figure 35.1: Stages of Micropropagation of *Wedelia chinensis* Merrill

35.1(a): Multiple Shooting in MS Medium Containing Kn (3.0 mg/l) + IAA (1.0 mg/l)

35.1(b): Rooting of Microshoots in MS Medium Containing IBA (0.2 mg/l) and IBA (0.5 mg/l)

35.1(c): Established Plantlets of *Wedelia chinensis* in Pots Containing Soil Mixture (Sand, Soil and Compost in 2 : 1 : 1)

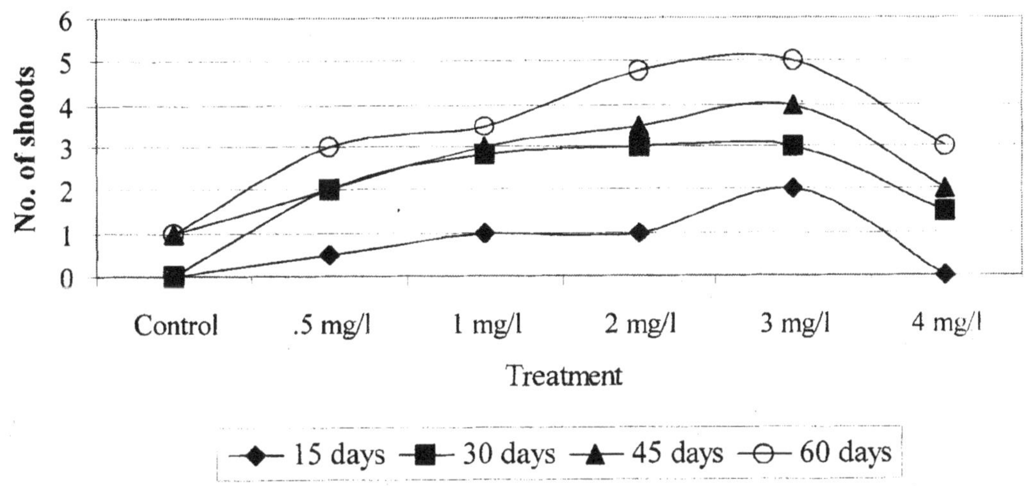

Figure 35.2(a): Effect of Kn on Number of Shoots

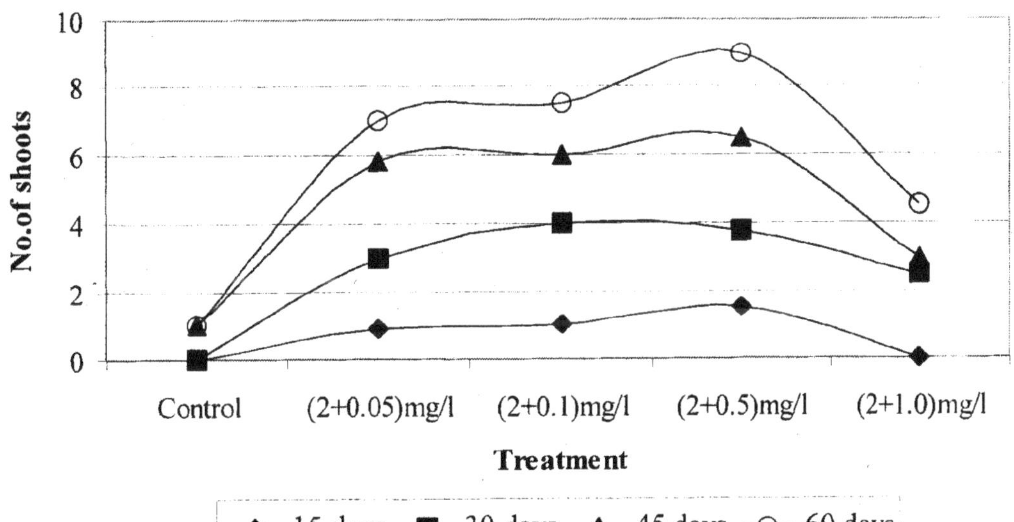

Figure 35.2(b): Effect of Kn + IAA on Number of Shoots

(3.0 mg/l) and IAA (1.0 mg/l) is optimum for high frequency shoot generation in *Wedelia chinensis* [Figure 35.1(a)]. After 7–8 weeks of culture, individual shoots of about 3 cm height were separated and implanted in half strength MS medium supplemented with IBA. Root initiation was noticed after 5–6 days and it became profuse after 10 weeks [Figure 35.1(b)]. MS medium supplemented with IBA proved to be better compared to IAA for rooting in *Wedelia chinensis*. Yadav *et al.* (1990) reported that

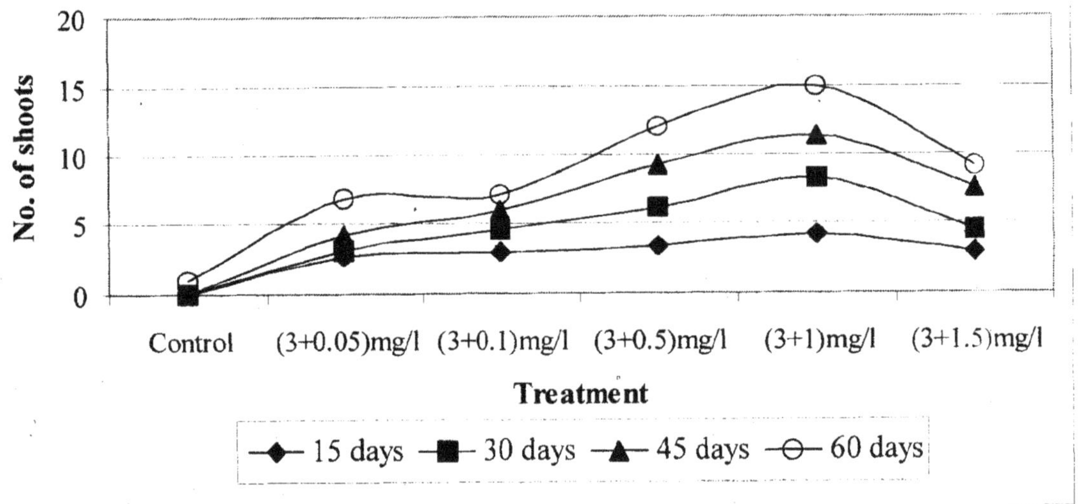

Figure 35.2(c): Effect of Kn + IAA on Number of Shoots

shoots rooted best in IBA containing media compared to IAA and NAA in *Morus nigra.* Similar observations were made in *Hemidesmus indicus* where IBA proved to be more effective than IAA (Patnaik and Debata, 1996). The rooted plantlets, after hardening, were successfully transferred initially into soil-pots containing garden soil, sand and compost (2 : 1 : 1) and then established in field [Figure 35.1(c)].

The protocol standardized through this study demonstrated the possibility of developing an efficient micropropagation system for successful mass propagation of *W. chinensis.*

Acknowledgement

The authors duly acknowledge the ICAR-NATP for financial assistance in terms of a ICAR-NATP (CGP) Project in the Department of Biotechnology, Gauhati University.

References

Bhuyan, P., Choudhury, S.S. and Handique, P.J. 2000. *In vitro* regeneration of an important medicinal plant–*Wedelia chinensis* Merrill. *J. Phyto. Res.,* 13(1): 57–59.

Chattopadhyay, S., Chattopadhyay, S. and Dutta, R.K. 1990. Quick *in vitro* production of mulberry (*Morus alba*) plantlets for commercial purposes. *Indian J. Exp. Biol.,* 28: 522–525.

Chopra, R.N., Nayar, S.L., and Chopra, I.C. 1956. In: *Glossary of Indian Medicinal Plants* (reprint 1986). CSIR, New Delhi.

CSIR. 1992. *The Useful Plants of India* (reprints). PID, New Delhi.

Handique, P.J. 1994. Ethno-medico botanical studies of the indigenous medicinal plants of Golaghat and Jorhat districts of Assam. *ASTEC Project Report,* Guwahati, p. 101.

Hiraoka, N. and Oyanagi, M. 1987. *Plant Cell Reports.*, 7: 39.

Murashige, T. and Skoog, F. 1962. A revised medium for rapid growth and bioassays with tobacco tissue cultures. *Physiol. Plant*, 15: 473–497

Patnaik, J. and Debata, B.K. 1996. Micropropagation of *Hemidesmus indicus* (L) R.Br. Through Axillary Bud Culture. *Plant Cell Reports*, 15: 427–430.

Sharma, N., Chandel, K.P.S. and Srivastava, V.K. 1991. *In vitro* propagation of *Coleus forskohlii* Briq., a threatened medicinal plant. *Plant Cell Reports*, 10: 67–70.

Yadav, U., Lal, M. and Jaiswal, V.S. 1990. Micropropagation of *Morus nigra* L. from shoot tip explants of mature trees. *Scientia Horticulturae*, 44: 61–67.

Chapter 36

Effect of Pesticides, Herbicides, Fumigants and Synthetic Fertilizers on the Nutrient Uptake of Rice

M.K. Mahesh and S.P. Hosmani

Department of Studies in Botany, University of Mysore, Manasagangothri, Mysore – 570 006

ABSTRACT

The effect of pesticides, herbicides, fumigants and synthetic fertilizers on the nutrient uptake of four cultivars of rice has been studied. The growth and yield of cultivar gowrisanna increases while in other three cultivars there is a significant decrease in growth. Nutrient uptake in shoots is reduced by the application of pesticides, herbicides and fumigants. Butachlor enhances uptake in shoots to a little extent. The application of these chemicals have a positive effect on the control of pests and harmful weeds, but the vigour of plants and quantitative yield of certain varieties of rice is reduced.

Introduction

Rice is the principal food of India and is cultivated in almost all states with above 85 million hectares of land under cultivation with an average rainfall varying between 14″ and 200″.

Since the plant requires much standing water, the environment is suitable for luxuriant growth of microbes, pests and weeds. These unwanted organisms often cause great damage to the crop and decreases its yield. In order to control such organisms the era of synthetic organic pesticides has progressed so rapidly that today at least 1000 pesticides are in common use in agriculture including 100 insecticides, and acaricides, 50 herbicides, 50 fungicides, 2 nematicides and 30 other chemicals of various properties. In agriculture it has been demonstrated that even after the effective use of these chemicals, pests still cause huge annual losses. Other deleterious effects on crop plants due to the residual remains of these on the surface of grains and other plant parts cause various diseases in consuming population, cause gene imbalance in crops. Combined bio-organic and chemical fertilizers show negative effects on physico-chemical properties, nitrogen transformation, macro and micro nutrient uptake, nutritional composition, production trends and dry matter yield if these chemicals are not treated in suitable proportion. Although a vast number of reports have appeared in the form of publications a few of utmost significance are quoted here. Singh and Bajwa (1987), Bhusan and Singh (1979), Biddappa *et al.* (1980), Mandal *et al.* (1980), Chaudhary *et al.* (1981), Chellaiah and Uthamasamy (1986), Dubey *et al.* (1987), Dyck and Orlido (1977), Gorantiwar *et al.* (1973), Gupta and Moolani (1971), Henisichs *et al.* (1979), Pandey and Thakur (1991), Prabhakara Rao *et al.* (1988), Prusty *et al.* (1989), Raman and Uthamasamy (1984), Ramu *et al.* (1988), Sai Ram *et al.* (1986), Sarvanan *et al.* (1987), Tripathy and Upadhyay (1989), Mahesh and Hosmani (2002), Mahesh and Hosmani (2001), Mahesh and Hosmani (2003).

The amount of work carried out on the effect of various pesticides, fumigants and synthetic fertilizers on crop plants seems to be enormous. However, when work relating to rice on the use of pesticides, fumigants and fertilizers from this part of the country is considered, it appears to be very meagre. There are a good number of high yielding rice varieties like Jaya, Gowrisanna, Prakash and I.R. 20 which are grown locally, but the improper use and the treatment methods of pesticides, fumigants and fertilizers have reduced their yield.

The aim of the present work was to determine the beneficial or harmful effects of some pesticides, fumigants and addition of excess fertilizers on the nutrient uptake in four varieties of rice.

Materials and Methods

The effect of Bavistin, Blitox, Emisan, Nuvacron, Butachlor, EDB, Aluminium phosphide and fertilizers like urea granules and the combination of N : P : K (15 : 15 : 15) on four varieties of rice namely Jaya, Prakash, IR-20 and Gowrisanna has been studied. The effect of these chemicals initially on the uptake of certain nutrients in roots as well as in shoots has been discussed.

Seed samples of rice varieties were obtained from local farmers (Jaya, Prakash, Gowrisanna and IR-20). Pesticides were obtained from local merchants and include Bavistin, Blitox, Emisan, Nuvacron, Butachlor and fumigants EDB and Aluminum phosphide from CFTRI, Mysore.

Bavistin

(Carbendazim) Methyl-2-benzimidazole Carbonate–4 gms was used to treat 1 kg wet seeds and shade dried; Blitox: (Agrimycin) copper oxychloride 1 kg of seeds are soaked in solution for 20–30 minutes and shade dried; Emisan: Methoxy Ethyl mercury chloride, 1 gm in 1 litre of water and used as wetting agent for 1 kg of rice.

Nuvacron

(0, 0-dimethyl, 1-0-(2 methyl-carbamyol-1-methyl-1-vinyl) phosphate, seeds soaked in 400 gms/l solution; Butachlor (Machete) 5 per cent is mixed in 100 litres of water and sufficient to spray 1 acre of land; Ethylene Dibromide and Aluminium phosphide are used for fumigation of seeds at time of processing. 3 ml of EDB is used to treat 100 kg of seeds for seven days, 2–3 tablets of aluminium phosphide are used to treat 1 ton of seeds for six days. The synthetic fertilizers added in excess for treatment of seeds were urea granules and N : P : K in the ratio of 15 : 15 : 15.

The details of the methods followed for treatment and cultivation of rice, methods of transplantation, drying and methods adopted for the analysis of various nutrients are as per the Laboratory Manual for Physiological Studies of Rice (Souchi Yoshida *et al.*, 1972). Sufficient sets of experiments were conducted and the data analyzed as per methods described by Gomez and Gomez (1984).

The experimental data indicates that at times one chemical is effective in reducing certain physiological parameters and at the same time it enhances another activity. Because of these variations it is difficult to recognise the direct economic benefit of the treatments on rice varieties. Generalized tables have been derived for discussion. Analysis of variance has been calculated for the treatments and the significance at 5 per cent levels discussed.

Results and Discussion

The study of the effect of different factors on growth and yield of cultivars of rice is indicative of the fact that Blitox has a retarding effect on cultivar Prakash and does not influence cultivar Jaya. The magnitude of significance is positive on IR-20 and Gowrisanna; Bavistin does not influence cultivar Prakash but retards the growth of other three cultivars. A positive effect on application of Emisan is noticed on cultivar Gowrisanna. EDB significantly retards the growth of all cultivars of rice. Cultivars IR-20 is greatly influenced by application of EDB. There is a significant improvement of cultivar Gowrisanna with a marked decrease in cultivar IR-20 on fumigation with Aluminium phosphide followed by cultivar Jaya. Its effect on cultivar Prakash is not marked. The effect of Nuvacron on these cultivars is also interesting with cultivar Prakash and Gowrisanna showing no response. Cultivars IR-20 and Jaya respond significantly to these treatments. The effect of Butachlor, which is negative is observed on Gowrisanna followed by IR-20. The effect reaches a positive magnitude in cultivar Jaya while there is no response on cultivar Prakash. Application of urea as an additional fertilizer has a positive influence on all parameters of all cultivars studied. Similarly the application of N:P:K also has a positive effect, except in cultivar Jaya where the response is slightly negative and in cultivar Prakash the response is almost nil. The conclusion derived from this data is that the growth and yield of cultivar Gowrisanna increases while in other three cultivars there is a significant decrease in growth.

The inference observed from the data of uptake of nutrients in roots is that all treatments show a retarding effect. A positive uptake of nutrients is observed where urea and NPK have been added in excess. Observation on individual cultivars also indicates that excess supply of urea and NPK only improve the uptake of nutrients like Calcium, Sodium and Potassium in roots. On the contrary cultivar Prakash responds negatively to treatments. Sodium and potassium uptake in roots is enhanced in cultivar IR-20 on addition of excess urea and NPK. Nitrogen uptake in all treatments is retarded. Magnesium, Phosphorus and Potassium uptake in roots is slightly influenced by application of Butachlor which is marked in treatments of urea and NPK.

A positive effect is observed in the uptake of nutrients when urea is applied in excess. Phosphorus uptake is enhanced in all four cultivars. Nitrogen uptake is not influenced very much while Iron uptake is significant in cultivars Prakash, IR-20 and Gowrisanna. In general uptake of nutrients is significantly enhanced by the application of additional urea.

The effect on the application of N : P : K to different cultivars of rice indicates that Phosphorus, Potassium and Sodium uptake is significantly enhanced in all the four cultivars whereas nitrogen uptake is not influenced. Jaya and IR-20 show positive uptake as compared to other two varieties (Table 36.1).

Table 36.1: Effect of Pesticides, Herbicides, Fumigants and Synthetic Fertilizers on the Nutrient Uptake of Roots

Rice Varieties	*Parameters*	*Blitox*	*Bavistin*	*Emisan*	*EDB*	*Al. Phosphate*	*Nuvacron*	*Butachlor*	*Urea*	*N : P : K*
	Nitrogen	−0.193	0.077	−0.563	1.407	1.259	−0.654	−0.803	1.771*	2.574**
	Iron	−0.975	0.378	−0.100	0.402	0.604	−2.158*	1.199	4.579**	4.277**
	Magnesium	0.079	2.101*	0.789	−3.137*	−3.308*	−2.101*	0.177	0.428	1.510
Jaya	Calcium	−0.459	1.014	0.000	1.155	1.525	0.000	3.043**	1.706*	0.192*
	Phosphorous	−0.320	1.090	4.797**	−1.151	−0.571	−0.796	−2.479*	2.815*	1.603
	Sodium	−0.885	−0.623	5.742**	3.115**	3.872**	−2.941**	0.598	3.269*	7.596**
	Potassium	−1.582	−0.460	5.906**	0.064	−3.308**	−2.101*	0.177	4.782**	10.864**
	Nitrogen	−1.120	−0.584	−0.565	0.263	0.328	0.000	0.613	1.004	1.775*
	Iron	−1.569	−0.572	−1.827*	−1.127	−1.189	1.824*	1.615	2.472*	1.608
	Magnesium	−0.077	−1.973*	−0.229	−5.497*	−1.501	−1.704*	1.893*	−0.623	−1.894*
Prakash	Calcium	−2.772**	−2.199*	−3.048**	−2.828**	−5.780*	−0.205	−2.745**	3.276	0.417
	Phosphorous	−1.033	−0.129*	6.855**	−3.956	−3.988**	0.447	04.783**	4.630**	2.399
	Sodium	−3.333**	−1.950*	3.737**	1.078	0.476	0.895	−0.458	0.895	1.473
	Potassium	−1.630	−0.483	2.396*	−2.845**	−1.501	−1.704*	1.893*	−1.507	0.169
	Nitrogen	0.071	−0.78	−0.763	−2.087*	−1.728*	0.823	0.439	1.412	0.967
	Iron	1.959*	−0.580	−1.387	−5.251**	−1.773*	1.322	0.733	1.286	0.481
	Magnesium	−3.817*	−1.562	−6.428*	−1.155	−1.525	−1.128	1.342	−0.970	2.009*
IR-20	Calcium	0.000	−0.603	1.069	−3.207**	−4.221	−0.61	2.683**	2.515**	3.411**
	Phosphorous	0.331	−0.047	−2.511**	−7.327**	−4.879**	0.750	0.190	4.769**	6.137**
	Sodium	2.399*	−0.808	1.208	−2.887**	3.121**	0.482	1.535	7.781**	4.024**
	Potassium	0.197	−1.301	−2.868**	−8.321**	−1.525	−1.128	1.342	5.669**	4.205**
	Nitrogen	−0.942	−1.399	−0.519	0.428	1.485	0.702	−0.062	1.785*	1.654
	Iron	0.946	−5.966**	−1.253	1.734*	2.482*	−2.079*	0.540	4.373**	0.519
	Magnesium	0.000	−1.055	0.363	0.654	0.960	−0.303	−0.312	0.446	2.449*
Gowrisanna	Calcium	−1.706*	−0.359	−0.359	0.000	−4.417	−0.320	−1.298	−1.960*	−0.555*
	Phosphorous	0.726	−0.603	−0.773	−1.680	0.026	−0.484	3.873*	2.714**	3.219**
	Sodium	−2.283*	−1.549	−2.698**	0.000	0.862	−0.550	2.072	2.942**	.3.112**
	Potassium	−0.497	0.804	0.920	−1.613	0.960	−0.303	−0.312	2.408*	2.246*

* Significant at 5%.

Table 36.2: Effect of Pesticides, Herbicides, Fumigants and Synthetic Fertilizers on the Nutrient Uptake of Shoots

Rice Varieties	*Parameters*	*Blitox*	*Bavistin*	*Emisan*	*EDB*	*Al. Phosphate*	*Nuvacron*	*Butachlor*	*Urea*	*N : P : K*
	Nitrogen	−0.038	−0.423	−0.294	−0.741	−1.224	−0.445	−0.865	−0.198	0.671
	Iron	−0.0887	−0.939	−2.545**	−2.049*	−0.923	−1.425	0.082	0.891	3.219**
	Magnesium	−1.083	−1.437	−0.626	−1.612	−1.912*	−1.079	−0.094	2.689**	0.810
Jaya	Calcium	−0.385	0.187	−1.623	0.000	0.000	−1.016	3.048**	1.688	1.601
	Phosphorous	−2.189*	−1.630	−5.239**	−.0297	−2.1201**	−2.116*	0.666	1.104	3.150**
	Sodium	−2.269*	−1.147	−6.099**	−2.396*	−0.853	−1.449	4.812**	1.265	2.727**
	Potassium	−0.617	−1.288	−5.826**	−3.358**	−0.100	−3.704**	4.232**	0.054	2.730**
	Nitrogen	−0.887	−0.182	−0.793	−0.966	−0.725	−0.867	0.667	0.920	0.897
	Iron	−1.560	−1.612	−1.507	−0.969	−0.652	−3.883**	1.001	2.094*	1.568
	Magnesium	−0.639	−0.932	−0.634	.0.760	0.215	−0.108	−0.446	4.667**	0.036
Prakash	Calcium	−1.809*	−1.414	−4.898**	−2.042*	−1.459	2.000*	0.894	1.052	1.155
	Phosphorous	−0.875	−0.333	−1.505	−2.209*	2.6039	−1.566	1.219	−0.096	4.180**
	Sodium	−1.466	−1.542	−0.669	−1.147	−1.518	−1.824*	2.498*	1.999*	1.839*
	Potassium	−1.512	1.332	−2.315*	−3.611**	−1.733*	−1.204	4.414**	2.396*	1.931*
	Nitrogen	0.311	−0.235	−0.988	−2.058*	−1.734*	−0.316	0.450	1.723*	1.692
	Iron	1.163	−9.958	−1.706*	−6.113**	−0.634	−1.307	1.054	1.478	0.921
	Magnesium	−1.670	0.159	−0.286	−1.155	−0.149	−0.381	1.765*	9.140**	3.579*
IR-20	Calcium	−0.717	−3.244**	−3.718*	−3.719*	02.000*	0.000	0.730	12.849**	3.411**
	Phosphorous	−1.206	−1.642	−1.204	−2.851**	0.9575	−1.008	0.856	−0.946	9.669**
	Sodium	−1.869*	−2.964**	−2.656**	−4.712**	0.738	−2.974**	−3.444**	2.582**	7.488**
	Potassium	0.004	−0.702	−0.887	−2.458	−0.585	−1.808*	2.677**	4.933**	3.031**
	Nitrogen	−0.951	−2.888**	−0.695	0.226	0.402	−1.417	1.282	2.474*	1.161
	Iron	0.313	−2.199**	0.197	0.563	1.167	−1.530	0.719	−0.661	0.142
	Magnesium	−0.933	−1.564	−1.070	−1.146	0.000	−0.326	0.922	4.261**	1.214
Gowrisanna	Calcium	−1.924*	−0.730	0.000	−1.376	−1.414	−1.960*	−0.381	3.483**	2.242*
	Phosphorous	−0.704	−1.322	−1.462	−0.982	−1.5977	−4.655**	3.982**	1.183	5.932**
	Sodium	−0.967	0.000	0.935	−1.541	−0.201	−0.928	1.471	0.000	1.713**
	Potassium	−1.522	−1.354	−0.802	−0.815	−1.114	−1.085	3.255**	0.752	1.347

* Significant at 5%.

The uptake of nutrients by shoots in cultivar Jaya indicates a positive uptake on application of Butachlor and NPK; while in Prakash and IR-20 positive uptake of nutrients is observed on application of urea. Butachlor and urea influence positively the uptake of nutrients in cultivar Gowrisanna. In general nutrient uptake in shoots is reduced by application of pesticides, herbicides and fumigants. Butachlor enhances uptake in shoots to a little extent (Table 36.2).

A generalized conclusion drawn from this study is that there is no doubt about the fact that application of pesticides, herbicides and fumigants control insect pests and harmful weeds but on the other hand the vigour of plants and many a times the quantitative yield of certain varieties of rice is reduced.

References

Bhusan, L.S. and Singh, G. 1979. Yield and nutrient uptake by rainfed direct sown rice as affected by nitrogen and phosphorus fertilization. *Fet. News*, 24: 25–27.

Biddappa, C.C., Sarkunam, V. and Prakasa Rao, P.S. 1980. Effect of urea insecticide mixtures on the yield of rice and nitrogen transformation in alluvial soil. *Oryza*, 17: 144–147.

Bikash Kumar Mandal, Ghorai, A.K. and Mandal, B.K. 1986. Relative efficiency of selected herbicides for controlling weeds in transplanted rice. *Oryza*, 23: 242–248.

Chaudhary, M.L. and Uppal, A.K. 1981. Accumulation of phosphorous by three varieties of rice and its relationship to phosphate fertilization. *Indian J. Agric. Res.*, 15: 145–148.

Chelliah, S. and Uthamasamy, S. 1986. Insecticide-induced resurgence of insect pests of rice. *Oryza*, 23: 71–82.

Dubey, R.S., Sharma, K.N. and Singh, B. 1987. Influence of phosphorus fertilization on accumulation of phytic acid, nucleic acid and total phosphorus in developing rice grains. *Oryza*, 24: 250–256.

Dyck, V.A. and Orlido, G.C. 1977. Control of brown plant-hopper Nilaparvata lugens by natural enemies and timely applications of narrow-spectrum insecticides. In: *The Rice Brown Plant Hopper*, FFTC, ASPAC, Taipei, p. 59–72.

Gorantiwar, S.M., Jaggi, I.K. and Khanna, S.S. 1973. Nutrient uptake by rice under different soil moisture regimes. *J. Indian Soc. Soil Sci.*, 21: 133–141.

Gupta, U.S. and Moolani, M.K. 1971. Effect of herbicides on the availability and uptake of nutrients. *Indian J. Weed Sci.*, 3: 51–61.

Mahesh, M.K. and Hosmani, S.P. 2001. Fumigation with aluminium phosphide and its effect on the nutrient uptake of four varieties of rice. *J. Env. and Poll.*, 8(1): 129–131.

Mahesh, M.K. and Hosmani, S.P. 2002. Effect of ethyl dibromide on the nutrient uptake by four varieties of rice. *Nature Env. and Poll. Tech.*, 1(3): 275–277.

Mahesh, M.K. and Hosmani, S.P. 2003. Changes in morphology and nutrient uptake in four varieties of rice treated with Blitox. *Asian J. Microbiol. Biotech. Env. Sci.*, 5(1): 99–101.

Mahesh, M.K. and Hosmani, S.P. 2004. Morphological changes and nutrient uptake in some cultivars of rice treated with Bavistin. *J. Ecotoxicol. and Env. Monit.*, 14(3): 195–198.

Prabhakara Rao, Y.P., Narasimha Rao, P.V., and Verma, S.C., 1988. Nutritional composition of rice as affected by phosphorous and potash application and seedling root dipping in nutrient solutions. *Oryza*, 25: 43–46.

Prusty, J.C., Belura, B. and A.K. Parida, 1989. Weed control in direct seeded shallow submerged lowland rice in wet season. *Oryza,* 26: 106–107.

Raman, K. and Uthamasamy, S. 1984. Effect of insecticide application on rice growth. *International Rice Res. Newsletter,* 9: 20.

Sairam, R.K., Tomar, P.S., and Horika, A.S., 1986. Effect of subherbicidal levels of 2, 4-D, iso proturan and atrazine on the nitrogen metabolism and growth of Oats (*Avena sativa* Linn.). *Indian J. Physiol.,* 29: 237–242.

Saravanan, A., Velu, V. and Ramanathan, K.M. 1987. Effect of combined application of bio-organic and chemical fertilizers on physico-chemical properties, nitrogen transformation and yield of rice in submerged soils of Cauvery delta. *Oryza,* 24: 1–6.

Singh, Balwinder and Bajwa, M.S. 1987. Efficiency of urea as affected by time and method of its application to rice grown in partially reclaimed sodic soils. *Oryza.* 24: 7–13.

Tripathy, B.N. and Upadhyay, R.M. 1989. Effect of biocide and fertilizer application on the nutritional status of rice. *Oryza,* 26: 43–47.

Yoshida, Shouichi; Douglas, A. Forno, James, H. Cock, Kwanchai A. Gomez. 1972. *Laboratory Manual for Physiological Studies of Rice,* 2nd Edition. The International Rice Research Institute, Phillipines, p. 70.

Chapter 37

Correlation and Path Analysis in Rice (*Oryza sativa* L.)

Purabi Das, Avijit Kundu*, Nirmal Mandal* and Indrani Dana***

Faculty of Agriculture, Bidhan Chandra Krishi Viswavidyalaya, Mohanpur – 741 252, West Bengal Chinsurah Rice Research Station, Hooghly – 712 102, West Bengal*

ABSTRACT

Genotypic correlation coefficient and path analysis carried out in fifty genotypes of rice showed that yield was significantly and positively correlated with number of panicles per plant, number of filled grains per panicle, grain density, 100-grain weight and 100-kernel weight. Plant height also had positive significant association with grain yield at genotypic level. Path analysis indicated that the direct effect of number of panicle per plant, number of filled grains per panicle and 100-grain weight were found to be negligible or negative while the correlation coefficients were positive. In such complex situation, the indirect causal factors also are to be considered simultaneously for efficient selection.

Keywords: *Genotypic correlation, Direct effect, Path analysis, Rice.*

Introduction

Grain yield is a dependent trait which is influenced by many independent traits. Studies on the correlation of traits and their relative direct and indirect effects on yield are important as they are helpful in selecting desirable traits. Hence, an attempt was made to study eighteen quantitative characters, their correlations and effects in fifty rice genotypes.

Materials and Methods

Fifty rice genotypes comprised of advances lines, released varieties and local cultivars were evaluated at Chinsurah Rice Research Station, Hooghly, West Bengal during *kharif* season of 2002–2003. The genotypes were grown in a Randomized Block Design with three replications. In each replication each of the genotypes were grown in a plot of 2 rows of 2 m length spaced 25 cm apart. 15 cm distance between plant to plant was maintained. Agronomical practices were adopted as per the recommended package of practices. Non-experimental rows were planted all round the experimental field to avoid any possible boarder effects. Data were collected on ten randomly selected plants for each genotype in each replication for days to 50 per cent flowering, plant height (cm), number of panicles per plant, panicle length (cm), number of filled grains per panicle, grain density, 100-grain weight, 100-kernel weight (g), kernel-grain weight ratio, grain length (mm), grain breadth (mm), grain L : B ratio, grain thickness (mm), kernel length (mm), kernel breadth (mm), kernel L : B ratio, kernel thickness (mm), grain yield per plant (g). The data were subjected to statistical analysis. Genotypic correlations were estimated by the usual procedure. The path analysis was carried out as per procedure prescribed by Deway and Lu (1959).

Results and Discussion

Yield of plant is generally associated with several other characters, expression of such characters by and large are influenced by the environment, which increases the complexity of inheritance pattern and in shaping such characters in breeding programmes. Hence, correlation studies provide better insight in understanding yield components, which help in choosing proper selection criteria (Robinson *et al.*, 1951; Johnson *et al.*, 1955).

Genotypic correlations were in most of the cases higher than their corresponding phenotypic correlations in this study (Data not shown), which indicated that there is a strong association between two characters genetically, but the phenotypic value reduced due to insignificant interaction of the genotype with environment. Johnson *et al.* (1955) also suggested that the low phenotypic correlations might be due to masking/modifying effect of the environment in genetic association between characters. Similar results were observed by Reddy *et al.* (1997). Grain yield showed significant positive association with plant height, number of panicles per plant, number of filled grains per panicle, grain density, 100-grain weight, grain length and 100-kernal weight (Table 37.1). These results indicate that increase in one variable will cause increase in the other and *vice versa.* Similar results were also reported by Li *et al.* (1991), Yadav (1992), He and Chen (1992), Rao and Saxena (1999) and Meenakshi *et al.* (1999) for number of panicles per plant; Bapu and Soundarapandian (1992) and Bai *et al.* (1992) for panicle length; Reddy and Raffia Chandraih (1990), Roy and Kar (1992) and Chauhan *et al.* (1993) for number of filled grains per panicle, Roy and Kar (199.2) and Chauhan *et al.* (1993) for grain density; Chauhan and Tandon (1984), Paramasivam (1991), Surek *et al.* (1998) and Bagali *et al.* (1999) for 100-grain weight. A positive correlation between important characters is desirable since it helps in simultaneous improvement of both the characters. Maximum value of correlaton of grain yield per plant was found with 100-grain weight followed by number of filled grains per panicle. Yuam *et al.* (1995), Deosarkar *et al.* (1989), Luzi-Kihipi also reported that (1998) the most important character affecting grain yield per plant was found to be the number of filled grains per panicle.

Path coefficient analysis measures the direct and indirect contributions of independent variables on dependent variables. Therefore, to understand the relationship between yield per plant and related characters, path analysis was computed using simple genotypic correlation coefficient values between characters and has been presented in Table 37.2. Maximum positive direct effect on grain yield per

Table 37.1: Genotypic Correlation Coefficients Among Yield Attributes and Gram Characters

Trait	*X2*	*X3*	*X4*	*X5*	*X6*	*X7*	*X8*	*X9*	*X10*	*X11*	*X12*	*X13*	*X14*	*X15*	*X16*	*X17*	*X18*
X1	0.634**	−0.259**	0.573**	0.308**	0.149	−0.043	−0.041	0.008	−0.159	0.335**	−0.338**	0.151	−0.285**	0.277**	−0.348**	0.094	0.120
X2		−0.091	0.348**	0.381**	0309**	−0.076	−0.165	−0.231**	−0.100	0.258**	−0.274**	0.005	−0.309**	0.121	−0.265**	−0.108	0.309**
X3			−0.253*	−0.015	0.100	0.077	0.086	−0.046	0.281**	−0.340**	0.405**	−0.179	0.311**	−0.266**	0.364**	−0.151	0.199*
X4				0.480**	0.215*	−0.180	−0.248**	−0.149	−0.028	0.109	−0.099	−0.033	−0.006	−0.002	−0.041	−0.146	0.166
X5					0.954**	−0.190	−0.221*	−0.120	−0.274**	0.098	−0.282**	−0.079	−0.365**	−0.043	−0.221*	−0.208*	0.316**
X6						−0.197*	−0.199*	−0.082	−0.295**	0.028	−0.249**	−0.163	−0.373**	−0.086	−0.198*	−0.237*	0.309**
X7							0.890**	−0.119	0.537**	0.583**	0.583**	−0.074	0.716**	0.428**	0.563**	−0.111	0.420**
X8								0.342**	0.511**	0.469**	0.005	0.674**	0.417**	0.506**	−0.073	0.709**	0.314**
X9									0.078	−0.119	0.164	0.004	0.069	−0.040	0.096	0.044	−0.379**
X10										−0.092	0.693**	0.311**	0.921**	−0.124	0.638**	0.137	0.207*
X11											−0.773**	0.783**	−0.242*	0.951**	−0.775**	0.792**	0.157
X12												−0.351**	0.759**	−0.740**	0.958**	−0.462**	0.013
X13													0.204*	0.763**	−0.375**	0.948**	0.203*
X14														−0.234**	0.761**	0.105	0.166
X15															−0.798**	0.865**	0.142
X16																−0.499**	−0.005
X17																	0.096

X1: Days to 50 per cent flowering; X2: Plant height; X3: Number of panicle per plant; X4: Panicle length; X5: Number of filled grain per panicle; X6: Grain density; X7: 100-grain weight; X8: 100-kernel weight; X9: Kernel-grain weight ratio; X10: Grain length; X11: Grain breadth; X12: Grain L : B ratio; X13: Grain thickness; X14: Kernal length; X15: Kernel breadth; X16: Kernel L : B ratio; X17: Kernel thickness; X18: Grain yield per plant.

*, **: Denote significance at 0.05 and 0.01 respectively.

Table 37.2: Direct and Indirect Effects of Component Characters on Grain Yield Per Plant Through Path Analysis

Trait	*X1*	*X2*	*X3*	*X4*	*X5*	*X6*	*X7*	*X8*	*X9*	*X10*	*X11*	*X12*	*X13*	*X14*	*X15*	*X16*	*X17*	*X18*
X1	– **0.207**	0.383	– 0.023	0.968	– 1.637	0.754	0.033	– 0.063	– 0.007	0.902	2.553	– 3.102	– 0.037	– 1.112	– 1.796	2.417	0.094	0.120
X2	– 0.131	**0.604**	– 0.008	0.588	– 2.030	1.560	0.058	– 0.251	0.203	0.567	1.965	– 2.514	– 0.001	– 1.205	– 0.784	1.845	– 0.108	0.359
X3	0.053	– 0.055	**0.092**	– 0.427	0.078	0.505	– 0.060	0.132	0.041	– 1.593	– 2.588	3.718	0.044	1.214	1.724	– 2.530	– 0.151	0.199
X4	– 0.119	0.210	– 0.023	**1.688**	– 2.552	1.087	0.139	– 0.378	0.131	0.158	0.834	– 0.910	0.008	– 0.256	0.011	0.282	– 0.146	0.166
X5	– 0.063	0.230	– 0.001	0.809	– **5.323**	4.822	0.147	– 0.338	0.105	1.555	0.745	– 2.586	0.024	– 1.425	0.282	1.540	– 0.209	0.316
X6	– 0.031	0.186	0.009	0.363	– 5.079	**5.053**	0.153	– 0.304	0.072	1.674	0.210	– 2.286	0.040	– 1.454	0.557	1.379	– 0.237	0.309
X7	0.009	– 0.045	0.007	– 0.303	1.009	– 0.997	– **0.777**	1.359	0.104	– 3.049	4.443	– 0.676	– 0.177	1.668	– 3.656	0.769	0.732	0.420
X8	0.008	– 0.099	0.007	– 0.418	1.178	– 1.007	– 0.692	**1.526**	– 0.302	– 2.900	3.576	0.045	– 0.167	1.625	– 3.286	0.509	0.710	0.314
X9	– 0.001	– 0.139	– 0.004	– 0.252	0.637	– 0.414	0.092	0.522	– **0.883**	– 0.445	– 0.903	1.507	– 0.001	0.270	0.262	– 0.669	0.043	– 0.379
X10	0.033	– 0.060	0.025	– 0.047	1.459	– 1.491	– 0.417	0.780	– 0.069	– **5.674**	– 0.703	6.356	– 0.077	3.591	0.802	– 4.436	0.137	0.207
X11	– 0.069	0.155	– 0.031	0.184	– 0.521	0.139	– 0.453	0.716	0.104	0.524	**7.620**	– 7.087	– 0.194	– 0.942	– 6.172	5.389	0.793	0.157
X12	0.070	– 0.165	0.037	– 0.167	1.501	– 1.259	0.057	0.007	– 0.145	– 3.931	– 5.887	**9.173**	0.087	2.960	4.803	– 6.664	– 0.463	0.013
X13	– 0.031	0.003	– 0.016	– 0.055	0.516	– 0.825	– 0.556	1.029	– 0.003	– 1.763	5.965	– 3.221	– **0.248**	0.794	– 4.948	2.611	0.950	0.203
X14	0.059	– 0.186	0.028	– 0.111	1.945	– 1.884	– 0.332	0.636	– 0.061	– 5.225	– 1.841	6.963	– 0.050	**3.900**	1.516	– 5.294	0.105	0.166
X15	– 0.057	0.073	– 0.024	– 0.003	0.231	– 0.434	– 0.438	0.773	0.035	0.701	7.247	– 6.789	– 0.189	– 0.911	– **6.489**	5.550	0.866	0.142
X16	0.072	– 0.160	0.033	– 0.068	1.179	– 1.002	0.086	– 0.111	– 0.085	– 3.620	– 5.906	8.791	0.093	2.969	5.179	– **6.954**	– 0.500	– 0.005
X17	– 0.019	– 0.065	– 0.013	– 0.247	1.110	– 1.197	– 0.568	1.081	– 0.034	– 0.777	6.032	– 4.235	– 0.235	0.408	5.610	3.471	**1.002**	0.096

X1: Days to 50 per cent flowering; X2: Plant height; X3: Number of panicle per plant; X4: Panicle length; X5: Number of filled grain per panicle; X6: Grain density; X7: 100-grain weight; X8: 100-kernel weight; X9: Kernel-grain weight ratio; X10: Grain length; X11: Grain breadth; X12: Grain L : B ratio; X13: Grain thickness; X14: Kernal length; X15: Kernel breadth; X16: Kernel L : B ratio; X17: Kernel thickness; X18: Grain yield per plant.

Bold figures indicate the direct effect. Residual effect = 0.4436.

plant was observed in grain L : B ratio (9.173) followed by grain breadth (7.620), grain density (5.053) and kernel length (3.90). The direct effect of these characters were very high and positive, but the correlation coefficients were found negligible. In such situation, a restricted simultaneous selection model is to be followed, *i.e.*, restrictions are to be imposed to nullify the indesirable effects in order to make use of the direct effect.

The direct effect of days to 50 per cent flowering (–0.207), number of filled grains per panicle (–5.323), 100-grain weight (–0.777), grain length (–5.674), grain thickness (–0.248) and kernal breadth (–6.489) were found to be negative, but the correlation coefficients were positive. In such complex situation, the indirect cause factors too are to be considered simultaneously for coefficient selection.

Residual effect indicates the involvement of the other characters in governing grain yield, which was not considered in this study. The residual effect was 44.36 per cent, so dependent variables could explain only 55.64 per cent of the effect for grain yield per plant. The reverse relationship between the magnitude of correlation coefficient and direct effect of path value revealed here may be due to the residual factor. Inclusion of some other additional characters might have covered the full range of variation for grain yield per plant.

References

Bagali, P.G., Hittalmani, S. and Shashidhar, H.E. 1999. Character association and path coefficient analysis in *indica* x *japanica* doubled haploid population of rice. *Oryza*, 36: 10–12.

Bai, N.R., Devika, R., Regina, A. and Joseph, C.A. 1992. Correlation of yield and components in medium duration rice cultivars. *Env. and Eco.*, 10: 469–470.

Bapu, J.R.K. and Soundarapandian. 1992. Genotypic association and path analysis in F_3 generation of rice crosses. *Madras Agri. J.*, 79: 619–623.

Chauhan, J.S., Chauhan, V.S. and Varira, M. 1993. Genetic variation and character association in rainfed upland rice. *Oryza*, 30: 116–119.

Chauhan, V.S. and Tandon, J.P. 1984. Genetic variability and character association in Hill Rice. *Oryza*, 21: 138–142.

Deosarkar, D.B., Mishal, M.B. and Nerker, Y.S. 1989. Variability and correlation studies for yield and yield contributing characters in breeding lines of upland rice. *J. Maharashtra Agril. Univ.*, 14: 28–29.

Dewey, D.R. and Lu, K.H. 1959. A correlation and path-coefficient analysis of components of crested wheat seed production. *Agron. J.*, 51: 515–518.

He, S.V. and Chen, S.J. 1992. Analysis on the effects of major characteristics on the yields of rice varieties using grey relational grade analysis. *Guongdong Agri. Sci.*, 1: 4–6.

Johnson, H.W., Robinson, H.F. and Comstock, R.E. 1955. Estimation of genetic variability and environments variability in soybean. *Agron. J.*, 47: 314–318.

Li, Q.L., Guo, G.T., Ji, G.C., Aug, G. and Zhu, U. 1991. Study on high yield breeding and genetic analysis of yield components of main rice cultivars in *jilin. J. South China Agric. Univ.*, 16: 3–6.

Luzi-Kihipi, A. 1998. Inter-relationship between yield and some selected agronomic characters in rice. *African Crop Sc. J.*, 6: 323–328.

Meenakshi, T., Amrita Deva Ratinam, A. and Backiyarari, S. 1999. Correlation and path analysis of yield and some physiological characters in rainfed rice. *Oryza*, 36: 154–156.

Paramasivam, K.S. 1991. Correlation studies in dwarf rice accession. *Oryza*, 28: 237–239.

Rao, S.S. and Saxena, R.R. 1999. Correlation and regression analysis in upland rice. *Oryza*, 36: 82–84.

Reddy, C.D.R. and Ramachandraih, D. 1990. Genetic correlation of components of yield in parents and their F_1s in rice. *Orissa J. Agri. Res.*, 3: 136–142.

Reddy, J.N., De, R.N. and Surya Rao, A.V. 1997. Correlation and path analysis in lowland rice under intermediate (0–50 cm) water depths. *Oryza*, 34: 187–190.

Robinson, H.F., Comstock, R.E. and Harvey, P.H. 1951. Genotypic and phenotypic correlations in corn and their implications in selection. *Agron. J.*, 43: 262–267.

Roy, A. and Kar, M.K. 1992. Heriability and correlation studies in upland rice. *Oryza*, 29: 195–199.

Surek, H., Karkut, Z. K. and Bilgin, O. 1998. Correlation and path analysis for yield and yield components in rice in a 8-parent harf diallel set of crosses. *Oryza*, 35: 15–18.

Yadav, R.K. 1992. Genetic variability, correlation studies and their implication in selection of high yielding genotypes of rice. *Advances in Plant Sci.*, 5: 306–312.

Yuan, Q., Quan, L.Y., Zhan, S.Z., Wu, Y., Wang, Y.B. and Zhang, G.Q. 1995. Analysis of heterosis, combining ability and path coefficient in *japonica* rice hybrids. *Acta Agriculturae, Shanghai*, 11: 23–27.

Chapter 38

Rapid *in vitro* Propagation of *Pogostemon cablin*: An Aromatic Plant Species with High Demand

Hemashree Deka*, H.K. Gogoi and P.J. Handique**

**Defence Research Laboratory, Post Bag No. 2, Tezpur – 784 001, Assam*

***Department of Biotechnology, Gauhati University, Guwahati – 781 014, Assam*

ABSTRACT

A protocol for rapid and large scale *in vitro* clonal propagation of an important aromatic plant *Pogostemon cablin,* has been developed. Nodal segments and leaf pieces were used as explants for callus induction. Earliest callusing was observed in MS medium containing NAA (3.0 mg/l) and BAP (0.5 mg/l). But earliest shoot initiation was recorded in MS medium with NAA (1.0 mg/l) and Kn (1.0 mg/l). Roots developed on full strength MS medium containing IBA (1.0 mg/l). Micropropagated plants were successfully acclimatized in garden soil were uniform without any morphological variation.

Keywords: *In vitro propagation, Pogostemon cablin, Aromatic plant.*

Introduction

Pogostemon cablin is a highly aromatic bushy under-shrub of Lamiaceae family. The highly fragrant leaves contain a very sweet smelling oil of lashing sticky odour. The oil is used in scenting of soaps, cosmetics, after-shave lotions, detergents and many fancy products. In very low concentrations

(2 ppm), the oil is used to flavour food, beverages, candy and baked products (Chadha, 2002). This important aromatic plant grows in mild summer and high rainfall, tropical and subtropical climate in foothills and has a good possibility of expansion in humid orchards. Moreover, North Eastern Development Finance Corporation Limited (NEDFi) has developed a large programme on export of Patchouli oil and its cultivation. Though the plant is propagated by terminal shoot cuttings but for large scale rapid propagation, *in vitro* culture is an option for multiplication of Patchouli. Again the oil can be directly extracted from the callus by suspension culture without plantlet formation and cultivation.

In view of its aromatic and commercial importance, there has been felt an urgent need to apply non-conventional propagation methods for conservation and commercial availability. Therefore, present study was designed to develop procedures which allow to obtain a system for callus production giving cell masses, for suspension culture as well as to establish an efficient protocol to regenerate plants from leaves and nodal segments of *Pogostemon cablin.*

Materials and Methods

Plant Material

Plant material of *Pogostemon cablin* were collected from nursery of DRL, Tezpur. Leaves and nodal segments were washed thoroughly for 30 mins under running tap water, then washed with commercial 'Extran' solution and again washed thoroughly under running tap water. In the laminar flow air cabinet they were surface sterilized 70 per cent aqueous ethanol (v/v) for 1 min followed by 0.1 per cent (w/v) Hg Cl_2 and rinsed thoroughly thrice with sterile double distilled water. Leaves and nodal segments were trimmed from all sides prior to inoculation onto culture media.

Initial explants consisted of leaf pieces (6–10 mm) and nodal segments. The callus initiation medium contained MS salts and vitamins of MS (Murashige and Skoog, 1962), 30 g^{-1} sucrose and 10 g Agar. The pH was adjusted to 5.7 before autoclaving at 121°C for 20 mins. The effect of growth regulators were tested using BAP (0.5 mg/l), NAA (1.0–3.0 mg/l) and Kinetin (1.0–2.0 mg/l). Cultures were maintained at 25±2°C and under 16 hrs photoperiod.

Results and Discussion

In *Pogostemon cablin,* preliminary experiments showed that both the type of explants-node and leaf, formed callus *in vitro*. These explants were inoculated on MS medium supplemented with various concentrations and combinations of NAA, BAP and Kin (Table 38.1). The earliest callus initiation occurred in NAA (3.0 mg/l) + BAP (0.5 mg/l) after 1 week of inoculation. Followed by treatment NAA (1.0) + Kin (1.0) after 12 days of inoculation. In other treatments callus initiation delayed with decreasing NAA concentration.

Higher concentration of NAA combined with BA was previously found to promote callus formation from leaf explant in *Centella asiatica* (Banerjee *et al.*, 1999; Tiwari *et al.*, 2000). Of the various treatments maximum amount of friable callus proliferation was observed in the medium of earliest callus initiation *i.e.*, NAA (3.0) + BAP (0.5) and declined with lower concentrations of hormones. Because the supply of exogenous growth regulators from the medium was essential for differentiation and did not occur in their absence. In the treatment NAA (1.0) and Kin (1.0 mg/l) yellowish friable callus was formed and in the same callus induction medium, shoot initiation and growth occurred. The mature dividing callus itself may establish the polarity in the surrounding cells and be the result of growth and the establishment of the shoot beds and subsequent shoots (Yao and Wang, 1987).

Table 38.1: Effect of Hormones/GR on Regeneration of Callus of *Pogostemon cablin* on MS Medium Growth Regulator Concentration (mg/l)

NAA	*BAP*	*Earliest (Days) Callus Initiation*	*Shoot Initiation (Days)*	*After 45 Days No. of Shoots*	*After 90 Days No. of Shoots*
0.5	0.5	22	60	–	5
1.0	0.5	18	66	–	7
2.0	0.5	15	70	–	7
3.0	0.5	6	80	–	8
NAA	KIN				
0.5	1.0	16	65	–	6
1.0	1.0	12	30	20	25

Table 38.2: Effect of Hormone on Rooting of Shoots of *P. cablin*

Hormone Concentration mg/l	*Root Length (cm)*	*Root No.*
IBA		
0.5	3.5	5
1.0	7	8
2.0	4.2	4
3.0	3.2	3

The callus in the NAA (1.0) + Kin (1.0) treatment showed shoot regenerative ability in the callus induction medium itself. Within 1 month of explant inoculation, multiple shoot formation was observed in this treatment.

Although after 80 days in the treatment BAP (0.5) + NAA (3.0 mg/l) multiple shoots were forming but early multiple shoot formation within 45 days were observed in the treatment NAA (1.0 mg/l) + Kin (1.0 mg/l). Average 20 nos of shoots having average 3.5 cm length were recorded where small shoots were still arising. Similar results were also seen in *Hyperium perforctum* upto 22 shoots per callus (Pretto and Santarem, 2000). Due to rapid multiple shoot formation in treatment NAA (1.0 mg/l) + Kin (1.0 mg/l), all the subsequent shoot regeneration from callus were done in the above treatment.

Rooting of Shoots

Although very thin roots were observed in the treatment of NAA (3.0 mg/l) + BAP (0.5 mg/l) in the prolonged culture of 5-6 months. But rooting of shoots was induced in different treatments by IBA. Different concentrations of IBA (0.5, 1.0, 2.0, 3.0 mg/l) were tested and earliest rooting (15 days) and maximum rooting (8 nos) was observed in the treatment of IBA (1.0 mg/l). Similar promontory effect of IBA in rooting was observed in *Centalia asiatica* (Beena *et al.*, 2002; Tiwari *et al.*, 2000).

In *Pogostemon cablin* elongated shoots obtained from callus were rooted in all media evaluated, although higher frequencies were obtained in the treatment IBA (1.0 mg/l) with higher mean root length (7 cm). With the increase in IBA concentration neither root length nor root no. increased. Hence, IBA (1.0 mg/l) was optimum for rooting of *P. cablin*. The footed plants were transferred from culture flasks into plastic cups containing sterilized vermiculite and covered with bell jars. After two weeks they were generally exposed to reduced relative humidity by progressively removing the glass cover over a period of two weeks. Then they were moved to green houses. The acclimatized plantlets were successfully established in the field. This finding may be helpful in rapid micropropagation and *ex situ* conservation of *P. cablin*. Further studies can be carried out focussing on the establishment of suspension cell cultures and on the analysis of their secondary constituents.

Acknowledgement

Authors are thankful to DRDO under Ministry of Defence for providing financial assistance as Junior Research Fellowship and to the Director of Defence Research Laboratory, Tezpur for providing laboratory facilities to carry out this investigation.

References

Banerjee, S., Zehra, M. and Kumar, S. 1999. *In vitro* multiplication of *Centella asiatica*, a medicinal herb from leaf explant. *Curr. Sci.*, 76: 147–148.

Beena, M.R., Martin, K.P., Kirti, P.B. and Hariharan, M. 2003. Rapid in *in vitro* propagation of medicinally important *Ceropegia candelabrum. Plant Cell, Tissue and Org. Cult.*, 72: 285–289.

Chadha, K.L. 2003. In: *Handbook of Horticulture*. Krishi Anusandhan Bhavan, ICAR, Pusa, New Delhi, p. 625–626.

Pretlo, F.R. and Santarem, E.R. 2000. Calluc formation and plant regeneration from *Hypersium perforatum* leaves. *Plant Cell, Tissue and Org. Cult.*, 62: 107–113.

Tiwari, K.N., Sharma, N.C., Tiwari, V. and Singh, B.D. 2000. Micropropagation of *Centella asiatica* (L), a valuable medicinal herb. *Plant Cell, Tissue and Org. Cult.*, 63: 179–185.

Yao, D.Y. and Wang, J.Z. 1987. *J Shandong Teacher's University. Natural Science Edition*, 2: 243–284.

Chapter 39

Combining Ability Studies in Sunflower (*Helianthus annuus* L.)

K. Venkata Siva Reddy and M.R. Manjare***

**Research Associate, **Senior Scientist (Breeding), Mahatma Phule Krishi Vidyapeeth, Department of Botany, Rahuri, Maharashtra*

ABSTRACT

An experiment was conducted to assess the magnitude and direction of combining ability of four cytoplasmic male sterile lines and seven fertility restorer lines in sunflower. The results revealed that the parental lines RHA-274, SS-56, Morden, CMS-343A and CMS-852A possessed genes for earliness, circumference of head, total number of seeds per head, number of filled seeds per head, hundred seed weight, per cent oil content and finally seed yield per plant. From the estimates of general combining ability it was observed that the cytoplasmic male sterile lines CMS-343A and CMS-852A found to be good combiners for most of the characters under study, among restorer lines RHA-274 and Morden were found to be good combiners for all the characters under study. The hybrid CMS-343A x Morden was the best cross combination identified with high significant sca effects for seed yield per plant, plant height at maturity, stem girth, circumference of head and hundred seed weight. The hybrid CMS-89A x RHA-348 was found to be the best combination for days to 50 per cent flowering. The hybrids CMS-852A x RHA-348 and CMS-89 x RHA-274 were found to be best cross combinations for per cent oil content and filled seeds per head respectively. For all the characters under study non-additive gene action was found to be predominant, indicating usefulness by straight exploitation of heterosis breeding.

Keywords: *Combining ability, Line x tester, Sunflower.*

Introduction

Sunflower is one of the most important oil seed crop in the world and in India. Sunflower is good substitute for other oilseed crops because of its favourable characters like suitability to late sown conditions, short duration, drought and saline tolerance, photo-insensitivity, high adaptability, good market price and oil content. The discovery of cytoplasmic male sterility (Leclercq, 1969) in France and fertility restoration (Kinman, 1970) in U.S.A had provided a break through for heterosis breeding in sunflower. As a result plant breeders have extensively used and exploited heterosis to improve seed and oil yields in sunflower. In the systematic breeding programme, it must be necessary to identify superior lines for hybridization and crosses to expand the variability reservoir for selecting superior genotypes. Combining ability analysis helps the breeders in the selection of desirable parents and provides information on the importance of GCA and SCA variances and effects for interpreting the genetic basis of important traits. In the present study an attempt was made to assess the combining ability and gene action governing the quantitative traits in sunflower for four cytoplasmic lines and seven restorer lines by employing line x tester analysis.

Materials and Methods

Twenty eight hybrids were obtained by crossing four cytoplasmic male sterile lines (CMS 86A, CMS 89A, CMS 343A and CMS 852A) and seven fertility restorer lines (RHA 274, RHA 278, RHA 296, RHA 297, RHA 348, SS 56 and Morden) in line x tester fashion. Twenty eight crosses along with their eleven parents were sown in a randomized block design with three replications. The experiment was conducted at Post Graduate Institute Farm, Department of Botany, Mahatma Phule Krishi Vidyapeeth, Rahuri, Maharashtra. The crossing was carried out during summer season of 2000 and the evaluation of the crosses along with their parents was done during *kharif*, 2000. Each entry was sown in a single row of 5 m length, with a spacing of 60 cm between rows and 30 cm between plants with in a row. Standard package of practices were followed for raising the crop. Observations were recorded on five random competitive selected plants in each treatment, replications wise for ten characters under study *i.e.*, days to 50 per cent flowering, days to maturity, plant height at maturity (cm), stem girth (cm), circumference of head (cm), total number of seeds per head, number of filled seeds per head, hundred seed weight (g), per cent oil content and seed yield per plant (g). Mean values of all the characters for different entries were subjected to line x tester analysis (Kempthorne, 1957) to estimate general combining ability (GCA) and specific combining ability (SCA) effects and variances.

Results and Discussion

Analysis of variance (Table 39.1) revealed that the cytoplasmic male sterile lines differed significantly for days to 50 per cent flowering, plant height at maturity (cm), stem girth (cm), circumference of head (cm), total number of seeds per head, number of filled seeds per head, hundred seed weight (gm) and seed yield per plant (gm). Fertility restorer lines differed significantly for days to 50 per cent flowering, days to maturity, plant height at maturity (cm), stem girth (cm), circumference of head (cm), total number of seeds per head, number of filled seeds per head, hundred seed weight (gm) and seed yield per plant (gm). The line x testers showed significant differences for all the characters under study. The mean sum of squares due to crosses was highly significant for all the characters under study. It could be because of diverse nature of lines and testers and the significant interaction between lines and testers.

The estimates of GCA and SCA effects of 11 parents and 28 hybrids respectively, for ten characters are furnished in Table 39.2. Days to 50 per cent flowering, number of filled seed per head, oil content

Table 39.1: Analysis of Variance for Combining Ability

Sl.No.	Source of Variation	DF	Mean Sum of Squares									
			Days to 50% Flowering	Days to Maturity	Plant Height at Maturity (cm)	Stem Girth (cm)	Circumference of Head (cm)	Total Number of Seeds per Head	Number of Filled Seeds per Head	Hundred Seed Weight (g)	Per cent Oil Content	Seed Yield per Plant (g)
1.	Replication	2	0.062	12.937	124.312	0.134**	81.796**	9704.00	5806.00	0.616**	9.898	36.459
2.	Female	3	63.218**	10.708	1187.75**	0.228**	85.281**	126869.3**	71940.00**	4.963**	84.468**	795.294*
3.	Male	6	17.052**	50.375**	2611.18**	0.196**	162.117**	167604.0**	131735.3**	3.0173**	50.776**	451.046
4.	Female x male	18	8.967**	18.572**	611.389**	0.101**	92.984**	90991.11**	60223.33**	3.0226**	23.551**	264.78**
5.	Error	54	2.689	5.483	65.592	0.023	12.508	16711.410	9883.408	0.119	5.361	25.153

**: Significant at 1 per cent level of significance; *: Significant at 5 per cent level of significance.

Table 39.2: Estimate of General Combining Ability Effects of 11 Parents for 10 Characters

Sl.No.	Parents	Days to 50% Flowering	Days to Maturity	Plant Height at Maturity (cm)	Stem Girth (cm)	Circumference of Head (cm)	Total Number of Seeds per Head	Number of Filled Seeds per Head	Hundred Seed Weight (g)	Per cent Oil Content	Seed Yield per Plant (g)
Females											
1.	86A	−0.06	−0.44	−10.78**	−0.14**	−0.93	9.50	7.45	0.09	1.35**	−2.42*
2.	89A	2.22**	0.61	5.36**	0.04	−0.36	32.40	12.78	−0.68**	−2.08**	−7.48**
3.	343A	−0.15	−0.77	0.60	0.10**	2.92**	68.78*	60.07**	0.46**	2.04**	6.08**
4.	852A	−2.01**	0.61	4.31*	−0.004	−1.61*	−110.69**	−80.31**	0.13	−1.30*	3.82**
	S.E. $(g_i)\pm$	0.35	0.51	1 76	0.03	0.77	28.20	21.69	0.07	0.50	1.09
	CD at 5%	0.71	1.02	3 54	0.06	1.54	56.58	43.51	0.15	1.01	2.19
	CD at 1%	0.95	1.36	4 72	0.08	2.06	75.37	57.96	0.20	1.35	2.92
Males											
1.	RHA–274	−1.64**	−1.77*	16.24**	0.16**	3.71**	171.89**	144.63**	0.65**	2.98**	−0.19
2.	RHA–278	−0.39	3.06**	−20.29**	−0.20**	−3.77**	−146.60**	−120.78**	−0.20*	−1.77*	−1.91
3.	RHA–296	2.10**	1.56*	3.11	−0.12**	−4.97**	−145.35**	−141.86**	−0.72**	−3.08**	−5.56**
4.	RHA–297	0.52	−2.52**	4.16	−0.004	−0.95	104.31**	97.96**	−0.30**	1.75*	−8.99**
5.	RHA–348	−0.72	−1.69*	3.54	0.009	0.13	−15.85	15.71	−0.25.	0.62	2.13
6.	SS–56	0.52	0.89	13.46**	0.04	0.67	15.06	−11.61	0.35**	−0.60	7.30**
7.	Morden	−0.39	0.47	−20.23**	0.12**	5.18**	16.56	15.96	0.47**	0.09	7.21**
	S.E. $(g_i)\pm$	0.47	0.67	2.33	0.04	1.02	37.31	28.69	0.09	0.66	1.44
	CD at 5%	0.94	1.35	4.68	0.08	2.04	74.85	57.56	0.19	1.34	2.90
	CD at 1%	1.26	1.80	6.24	0.11	2.72	99.71	76.68	0.26	1.78	3.86

**: Significant at 1 per cent level of significance; *: Significant at 5 per cent level of significance.

Table 39.2(b): Estimate of Specific Combining Ability Effects of 28 Hybrids for 10 Characters

Sl.No.	Hybrids	Days to 50% Flowering	Days to Maturity	Plant Height at Maturity (cm)	Stem Girth (cm)	Circumference of Head (cm)	Total Number of Seeds per Head	Number of Filled Seeds per Head	Hundred Seed Weight (g)	Per cent Oil Content	Seed Yield per Plant (g)
1.	86A x RHA-274	−0.02	2.44	10.43*	−0.02	−2.27	138.91	113.13	0.36	−2.64	−1.19
2.	86A x RHA–278	0.72	−1.39	746	−0.05	4.12*	194.41*	128.21*	1.19**	−0.76	11.25**
3.	86A x RHA-296	−1.10	0.77	−15.50**	−0.26**	−7.95**	−135.83	−138.36*	−0.88**	1.79	−9.16**
4.	86A x RHA-297	−0.85	−0.14	7.35	0.07	−0.004	−35.16	25.13	0.76**	−1.10	1.55
5.	86A x RHA-348	3.06**	0.02	20.28**	0.27**	5.14*	41.66	−20.28	0.69**	0.09	5.94*
6.	86A x SS-56	−0.85	−1.56	7.08	0.07	2.93	17.41	16.71	−1.11**	0.59	−2.53
7.	86A x Morden	−0.94	−0.14	−37.82**	−0.07	−1.97	−221.41**	−124.53*	−1.01**	2.01	−5.84*
8.	89A x RHA-274	0.35	0.72	−3.01	0.10	1.35	262.34**	229.46**	−0.05	1.78	−8.58*
9.	89A x RHA-278	0.10	−2.10	2.58	0.18*	5.95**	−24.15	18.88	0.41*	2.42	6.28*
10.	89A x RHA-296	0.27	−2.94*	8.27	0.13	3.60	51.26	−25.03	0.58**	2.15	8.24**
11.	89A x RHA-297	0.85	0.47	−7.13	−0.09	−2.70	−70.73	−43.53	−0.67**	1.94	−2.87
12.	89A x RHA-348	−2.89**	1.31	−3.31	0.004	0.70	−83.23	35.71	0.52*	−2.42	3.86
13.	89A x SS-56	−1.14	−0.94	−3.23	−0.07	−1.93	−224.15**	−104.95	0.60**	−0.35	4.16
14.	89A x Morden	2.44*	3.47*	5.82	−0.25**	−6.98**	88.67	−110.53	−1.39**	−5.53**	−11.11**
15.	343A x RHA-274	0.07	−0.56	−5.31	−0.09	−1.39	−244.70**	−128.15*	−0.54**	2.65	3.67
16.	343A x RHA-278	0.15	2.60	−11.51*	−0.27**	−11.07**	−297.20**	−248.40**	−1.45**	0.53	−20.56**
17.	343A x RHA-296	1.65	0.10	9.17	0.06	6.418**	166.54*	180.01**	−0.35	−2.00	7.11*
18.	343A x RHA-297	−0.76	0.85	2.29	0.002	−0.66	107.54	19.84	0.60**	−0.98	−3.97
19.	343A x RHA-348	1.48	2.02	−5.91	−0.02	−0.84	71.04	74.09	−0.54**	−2.48	1.73
20.	343A x SS-56	−1.09	−2.22	−9.00	−0.01	−2.42	149.79*	9.09	0.31	1.81	−1.67
21.	343A x Morden	−1.51	−2.81*	21.00**	0.34**	9.99**	46.96	93.51	1.98**	0.47	13.70**
22.	852A x RHA-274	−0.40	−2.60	−2.10	0.02	2.31	−156.56*	−214.44**	0.23	−1.79	6.10*
23.	852A x RHA-278	−0.98	0.89	1.46	0.14	1.00	126.94	101.31	−0.15	−2.20	3.02
24.	852A x RHA-296	−0.82	2.06	−1.94	0.05	−2.07	−81.97	−16.60	0.65**	−1.94	−6.19*
25.	852A x RHA-297	0.76	−1.19	−2.51	0.01	3.37	−1.64	−1.44	−0.69**	0.13	5.29
26.	852A x RHA-348	−1.65	−3.35*	−11.76*	−0.25**	−5.00*	−29.47	−89.52	−0.67**	4.81**	−11.53**
27.	852A x SS-56	3.09**	4.72**	5.15	0.01	1.41	56.94	79.14	0.20	−2.05	0.04
28.	852A x Morden	0.01	−0.52	11.70*	−0.01	−1.03	85.77	141.56*	0.42*	3.04*	3.26
	S.E. (Sij)±	0.94	1.35	4.67	0.08	2.04	74.63	57.39	0.19	1.33	2.89
	CD at 5%	1.89	2.71	9.37	0.17	4.09	149.71	115.13	0.39	2.68	5.80
	CD at 1%	2.52	3.61	12.49	0.23	5.45	199.42	153.36	0.53	3.57	7.73

**: Significant at 1 per cent level of significance; *: Significant at 5 per cent level of significance.

and seed yield per plant are of much practical significance in sunflower and the discussion is mainly around these characters. Among parents CMS line 852A and restorer RHA-274 exhibited significant GCA effects for days to 50 per cent flowering in negative direction which is a desirable one. CMS line 343A and restorer line RHA-274 exhibited significant GCA effects for number of filled seeds per head, stem girth, circumference of head, hundred seed weight and per cent oil content. This indicates that this particular line RHA-274 seemed to posses decreasing alleles for days to 50 per cent flowering and increasing alleles with non-additive effects for number of filled seeds per head, circumference of head, hundred seed weight and per cent oil content. Similar results were also reported (Alone *et al.*, 1996; Naware, 1999). For the character plant height at maturity, two CMS lines *viz.*, CMS-89A and CMS-852A and two restorers RHA-274 and SS-56 expressed significant positive GCA effects. Further, the lines CMS-343A, SS-56 and Morden appeared to posses favourable alleles with non-additive effects for the character seed yield per plant.

Table 39.3: Estimate of General and Specific Combining Ability Variances and Gene Action

Sl.No.	*Characters*	σ^2_g	σ^2_s	σ^2_g/σ^2_s
1.	Days to 50% flowering	1.889	2.093	0.903
2.	Days to maturity	0.725	4.363	0.166
3.	Plant height at maturity	78.065	181.932	0.429
4.	Stem girth	0.0068	0.025	0.272
5.	Circumference of head	1.861	26.825	0.069
6.	Total number of seeds per head	3408.82	24759.9	0.137
7.	Number of filled seeds per head	2522.08	16779.97	0.150
8.	Hundred seed weight	0.058	0.967	0.059
9.	Per cent oil content	2.67	6.063	0.441
10.	Seed yield per plant	21.72	79.875	0.271

σ^2_g: Variance due to gca; σ^2_s: Variance due to sca.

In general no single line or tester exhibited significant GCA effects for all the characters under study. But, the line CMS-343A and tester RHA-274 recorded significant GCA effects for maximum number of characters under study. The restorer lines SS-56 and Morden which recorded highest GCA effect for seed yield per plant seemed to possess unfavourable alleles for oil content. Oil yield is the function of both seed yield and oil content. In this regard, although the lines CMS-852A, SS-56 and Morden which did not posses favourable genes for oil content, by virtue of their good combining ability for seed yield, expressed a fairly high magnitude of GCA effects for oil yield. Significant positive GCA effects for oil yield has been reported by earlier workers (Kumar *et al.*, 1998; Shekar *et al.*, 1998).

Out of 28 cross combinations, three crosses *viz.*, CMS-89A x RHA-348, CMS-852A x RHA-348 and CMS-343A x Morden, exhibited significant negative SCA effects for days to 50 per cent flowering. The parental lines included in the former cross combinations did not seem to possess favourable genes for earliness. The negative SCA effects of these combinations may be attributed to the favourable epistatic interaction of genes from these two lines. The cross CMS-89A x RHA-274 had high SCA effects for total number of seeds per head and number of filled seeds per head. Of the parental lines involved in the former combination, CMS-89A had significant negative GCA effects, where as RHA-274 had significant positive GCA effects, inspite of the GCA effects of these two parental lines in opposite direction, the significant SCA effects of this combination in the desirable direction may be due to high frequency of dominant alleles for total number of seeds per head and number of filled seeds per head. Only four hybrids, CMS-343A x Morden, CMS-86A x RHA-348, CMS-343A x RHA-348 and CMS-343A x RHA-274 recorded significant positive SCA effects for seed yield per plant. In addition to this the hybrid CMS-343A x Morden exhibited significant positive SCA effects for the traits plant height at maturity, stem girth, circumference of head and hundred seed weight. From this it appears that the significant positive SCA effects for yield could be attributed to the significant positive SCA effects either for circumference of head, hundred seed weight plant height at maturity or stem girth. In addition to this both the parents in the cross combination, CMS-343A and Morden had significant positive GCA effect for all the characters in the same direction thus contributing for high SCA effect for these characters. The cross CMS-852A x RHA-348 had high SCA effects for per cent oil content. Both the parental lines in the former combination did not seem to possess favourable genes for earliness. The negative SCA effects of this combination may be attributed to the favourable epistatic interaction of genes from these two lines.

The parental lines RHA-274 and CMS-852A were found to possess genes for earliness as evident from their significant negative GCA effects for days to 50 per cent flowering. The hybrid CMS-89A x RHA-348 was adjudged to be the best cross combination for earliness. The lines CMS-343A and CMS-852A and the testers SS-56 and Morden appeared to transmit non-additive genes for higher seed yield in their progenies along with other characters *viz.*, circumference of head, total number of seeds per head, number of filled seeds per head, hundred seed weight and hence could be used as parents advantageously in hybridization programme to obtain better yields. The cross combination CMS-343A x Morden registered the highest positive SCA effect for seed yield and other parameters contributing seed yield. The parental lines CMS-343A and RHA-274 were found to possess genes for total number of seeds per head and number of filled seeds per head. The hybrid CMS-89A x RHA-274 was found to be the best specific combination for total number of seeds per head and number of filled seeds per head. The lines CMS-86A, CMS-343A, RHA-274 and RHA-297 were found to transmit genes for higher oil content. The hybrid CMS-852A x RHA-348 was found to be the best as it expressed the highest positive SCA effects for per cent oil content.

Predominance of non-additive gene action was evident in the inheritance of most of the quantitative characters from the ratio of GCA : SCA variance (< 1) for all the characters. Non-additive gene action had been reported by earlier workers (Govindaraju *et al.*, 1992; Alone *et al.*, 1996; Lande *et al.*, 1997; Limbore *et al.*, 1997; Shekar *et al.*, 1998; Singh *et al.*, 1998). However additive gene action for these traits has also been reported (Shankar, 1981; Singh *et al.*, 1989; Khurana and Bhatia, 1996; Bajaj *et al.*, 1997; Kumar *et al.*, 1998).

Most interesting point to note that the cytoplasmic male sterile line CMS-343A and restorer line Morden involved as a common female and male parent in most hybrids which exhibited significant SCA effects in favourable direction for plant height at maturity, stem girth, circumference of head, hundred seed weight and seed yield per plant. From this, it could be ascertained that these two lines are the most versatile in their ability to combine among the parental lines. From the present study it could be concluded that the lines CMS-343A, CMS-852A, RHA-274, SS-56 and Morden, may be further evaluated for the confirmation of their superiority with respect to their general combining ability for seed yield and oil content. The hybrid CMS-343A x Morden which exhibited the highest SCA effect for seed yield and other important quantitative parameters need to be critically evaluated over different seasons and locations to confirm its superiority and stability.

References

Alone, R.K., Patil, R.D., Mate, S.N. and Manjare, M.R. 1996. Combining ability in sunflower. *Indian J. Agric. Res.*, 30(3–4): 215–220.

Ashok Kumar, A., Ganesh, M. and Janila, P. 1998. Combining ability analysis for yield and yield contributing characters in sunflower. *Ann. Agric. Res.*, 19(4): 437–440.

Bajaj, R.K., Aujla, K.K. and Chahal, G.S. 1997. Combining ability studies in sunflower. *Crop Improvement*, 24(1): 50–54.

Castiglioni, V.B.R., Olieveira, M.F., Arias, C.A.A. and Olieveira, H.F. 1999. Combining ability analysis among inbred lines of sunflower. *Pesquisa Agropecuaria Brasilieira*, 34(6): 981–988.

Gangappa, E., Channa Krishnaiah, K.M., Harini, H.S. and Ramesh, S. 1997b. Studies on combining ability in sunflower. *Helia*, 20(27): 73–84.

Govindaraju, T.A. and Sindagi, S.S. 1992. Combining ability for achene yield and its attributes in sunflower. *J. Oilseeds Res.*, 9(2): 314–319.

Javed, N. and Aslam, M. 1995. Combining ability effects in sunflower F_1 hybrids. *Helia*, 18(23): 41–45.

Kandalkar, B.S. 1997. Combining ability analysis in sunflower. *Bhartiya Krishi Anusandhan Patrika*, 12(2): 77–80.

Kempthorne, O. 1957. *An Introduction to Genetical Statistics*. John Wiley and Sons, Inc., New York, p. 217–568.

Khurana, S.R. and Bhatia, N. 1996. Combining ability studies in sunflower. *Indian J. Genet.*, 56: 533–539.

Leclercq, P. 1969. Une sterilite male cytoplasmique, chezle tounerol. *Annls. Amel. P.*, 12: 99–106.

Limbore, A.R., Weginwar, D.G., Gite, B.D. and Ghorade, R.B. 1997. Combining ability in sunflower. *J. Soils and Crops*, 7(1): 39–42.

Shekar, G.C., Jayaramaiah, H., Virupakshappa, K. and Jagadesh, B.N. 1998. Combining ability of high oleic acid in sunflower. *Helia*, 21(28): 7–14.

Singh, D.P., Singh, S.B. and Raheja, R.K. 1999. Combining ability analysis for seed yield, oil and oil quality in sunflower. *J. Oilseed Res.*, 16: 38–42.

Vanishree, G., Anantha Sayana, K., Nagbhushanam, G.V.S. and Jagdish, C.A. 1988. Studies on combining ability in sunflower. *J. Oilseeds Res.*, 5: 52–61.

Chapter 40

Effect of Planting Varying Number of Seedlings per Hill on Growth and Yield of Some Rice Varieties During Dry Season in West Bengal

B. Mitra, S. Sinha, S. Basu and R.L. Nayak

Department, of Agronomy, Faculty of Agriculture, Bidhan Chandra Krishi Viswavidyalaya, Mohanpur, Nadia – 741 252, West Bengal

ABSTRACT

A field experiment, conducted at Teaching Farm, Mondouri, BCKV during dry season of 2000–2001 in Factorial Randomised Block Design with twelve treatment combinations (4 varieties and 3 different number of seedlings per hill) on clay loam soil having neutral pH, revealed that the hybrid varieties especially Pro-Agro 6201 performed better than high yielding variety Khitish (IET 4094) in respect of different growth attributes (plant height, number of tillers/m^2 and dry matter accumulation) as well as grain yield. Hybrid varieties exhibited better performance in terms of growth attributes and grain yield at 2 seedlings/ hill than at 1 seedling/hill. However, there was no gain in yield by planting 3 seedlings/hill over 2 seedlings/hill in case of hybrid varieties, but for high yielding varieties, there was increase in grain yield.

Keywords: *Hybrid varieties, HYVs, Rice, Growth, Yield.*

Introduction

Rice (*Oryza sativa* L.) is the most important cereal crop in the developing world and is the staple food over half of the world's population. It is generally considered as a semi-aquatic annual grass plant.

The most important challenge in India is to produce sufficient foodgrain for growing population of more than 100 crores of people. Rice will have to share the major portion of additional foodgrain production in the coming years. As there is no scope for increasing area under cultivation; therefore, we have to increase our foodgrain production per unit area per unit time through introduction of modern, promising high yielding varieties as well as hybrid rice which will be considered as an important step towards augmentation of yield of rice.

Materials and Methods

The field experiment was conducted during dry season of 2000–2001 at Teaching Farm, Mondouri, BCKV, West Bengal. The soil of the experimental field was clay loam in texture and neutral in reaction. In this experiment, 3 hybrids and 1 high yielding variety of rice were taken to find out the varietal performance with varying number of seedlings/hill during dry season in West Bengal.

The experiment was laid out in factorial RBD with 12 treatment combinations of 2 factor.

Factor A: Varieties (4)

V_1–Pro-Agro 6201

V_2–CNHR-3

V_3–PAC-801

V_4–IET 4094 (Khitish)

Factor B: Number of Seedlings/Hill (3)

T_1–1 seedling/hill

T_2–2 seedlings/hill

T_3–3 seedlings/hill

Proper care was taken for management of all the experimental plots right from transplanting to maturity. Periodical observations on some growth parameters and finally the yield were recorded accordingly.

Results and Discussion

Effect of Variety

From Table 40.1, it was clear that the plant height of different varieties differed significantly at all stages of growth. Hybrid varieties showed lower plant height than HYV upto panicle initiation stage, then all three rice hybrid surpassed the HYV because of more culm elongation in hybrid than in HYV. At heading stage, hybrid Pro-Agro 6201 gave rise to the significantly tallest plant (92.29 cm).

The varietal differences on tiller production were found to be significant at all the stages of growth. At early growth stage (22 DAT), Khitish produced maximum tiller number/m^2 (343.17) followed by Pro-Agro 6201 (308.44), CNHR-3 (304.00) and PAC-801 (301.51). However, at and after maximum tillering stage (45 DAT), hybrids produced higher number of tillers/m^2 than HYV due to profuse

Table 40.1: Effect of Different Planting Varying Number of Seedlings/Hill on Plant Height (cm), Number of Tillers/m² and Dry Matter Accumulation (g/m²) of Different Rice Varieties

Treatments	*Plant Height (cm)*				*Number of Tillers/m²*				*Dry Matter Accumulation (g/m²)*			
	AT	*MT*	*PI*	*H*	*AT*	*MT*	*PI*	*H*	*AT*	*MT*	*PI*	*H*
Variety												
Pro-Agro 6210 (V_1)	41.88	65.14	68.86	92.29	308.44	586.50	573.50	535.67	127.78	791.07	841.35	1264.89
CNHR-3 (V_2)	39.44	64.34	68.08	87.77	304.00	571.33	554.83	531.36	126.74	775.22	828.17	1256.19
PAC-801 (V_3)	40.00	64.59	68.12	86.76	301.51	568.27	550.67	530.66	126.27	769.17	820.20	1248.58
Khitish (V_4)	45.69	68.49	72.76	83.98	343.17	491.67	474.67	438.00	170.66	738.41	772.85	1053.92
S.Em (±)	1.101	1.013	1.246	1.154	1.433	1.758	1.909	1.626	3.075	2.980	4.038	4.296
CD at 5%	3.229	2.971	3.654	3.384	4.202	5.156	5.598	4.768	9.018	8.739	11.842	12.598
Seedlings Numbers												
1 Seedling/hill (T_1)	41.36	64.82	68.58	87.24	302.84	544.77	531.34	498.00	134.48	780.44	820.84	1195.55
2 Seedlings/hill (T_2)	41.86	65.41	69.67	87.63	317.63	557.00	540.67	513.13	139.35	785.18	831.92	1212.36
3 Seedlings/hill (T_3)	42.06	65.95	70.12	88.24	322.38	561.56	543.34	515.65	141.79	787.54	834.57	1216.03
S.Em (±)	0.953	0.877	1.079	0.998	1.241	1.634	1.653	1.408	1.603	1.581	2.497	3.721
CD at 5%	NS	NS	NS	NS	3.639	4.792	4.848	4.129	4.701	4.636	7.323	10.912
Interaction	NS				NS				NS			

AT: Active tillering; MT: Maximum tillering; PI: Panicle initiation; H: Heading; NS: Not significant.

tillering habit. After maximum tillering, tiller number of all the varieties declined because of tiller mortility.

At early growth stage (22 DAT), Khitish produced the highest amount of dry matter due to higher number of tillers/m^2. However, at and after maximum tillering stage, hybrids produced significantly higher dry matter as compared to Khitish.

All the hybrid varieties namely Pro-Agro 6201 (7.91 t/ha), CNHR-3 (7.58 t/ ha) and PAC-801 (7.54 t/ha) produced significantly higher grain yield as compared to Khitish (5.52 t/ha). The yield advantage of Pro-Agro 6201 over Khitish was 2.39 t/ha, *i.e.*, 43.29 per cent (Table 40.2).

Table 40.2: Effect of Number of Seedlings/Hill on Grain Yield (t/ha) of Different Rice Varieties

Number of Seedlings per Hill	Variety				
	V_1	V_2	V_3	V_4	Mean
T_1	7.42	7.17	7.14	5.01	6.69
T_2	8.20	7.80	7.75	5.72	7.37
T_3	8.11	7.76	7.74	5.84	7.36
Mean	7.91	7.58	7.54	5.52	

	Variety	Number of seedlings/hill	Interaction
S.Em (±)	0.030	0.026	0.052
CD at 5 per cent	0.088	0.076	0.153

Effect of Number of Seedlings/Hill

The number of seedlings/hill on plant height was not found to be significant at any stage of growth, but with increasing number of seedlings from 1 to 3/hill, the height tended to increase (Table 40.1).

It was observed that at maximum tillering stage, P.I. and heading stage, tiller number/m^2 at 2 seedlings/hill was significantly higher than 1 seedling/hill, but there was no significant increment in tiller number/m^2 by planting 3 seedlings/hill. At all stages of growth, dry matter produced at 2 and 3 seedlings/hill were statistically at par.

When the number of seedlings/hill was increased from 1 to 2, the grain yield increased significantly. But there was no significant difference in grain yield when number of seedlings were increased from 2 to 3 per hill (Table 40.2). The yield increase at 2 seedlings/hill over 1 seedling/hill was estimated to be 10.16 per cent.

Effect of Interaction

Interaction effect of variety and number of seedlings/hill on plant height, number of tillers/m^2 and dry matter accumulation was found to be not statistically significant. However, the interaction effect was significant for grain yield. All the hybrid rice cultivars gave significantly higher yield by transplanting 2 seedlings/hill. For HYV Khitish, it was found that with increasing number of seedlings from 1 to 3/hill, the grain yield increased significantly. All these findings were in agreement with Annonymous (1997a, 1997b), Paroda and Siddique (1996).

References

Annonymous.1997a. Annual Report, Rice Research Station, Chinsurah, Hooghly, West Bengal.

Annonymous. 1997b. Progress Report, All India Co-ordinated Rice Improvement Project, DRR, Hyderabad.

Paroda, R.S. and Siddique, E.A. 1996. Current Status of hybrid rice research and development in India. In: *Hybrid Rice Technology*, (Eds.) M. Rangaswamy, P. Rangaswamy, K. Thiyagarajan and H. S. Ponnuswamy, p. 18–30.

Chapter 41

Effect of Sowing Direction and Planning Pattern of Raya Intercropping on Wheat Yield under Rainfed Conditions

Sukhvinder Singh, Parvender Sheoran, D.S. Rana and B.S. Sidhu

PAU Zonal Research Station for Kandi Area, Ballowal Saunkhri, District Nawanshahr – 144 521, Punjab

ABSTRACT

A field experiment was conducted at farmer's field during *rabi* season as a part of Operational Research Project (ORP) for Dryland Agriculture for two years and it was observed that sowing of raya as an intercrop with is a useful tool for increasing and assuring the per unit crop productivity under rainfed conditions. The per cent increase in mean WEY and net returns due to raya intercropping with wheat was 5.41 per cent (1.77 q/ha) and 18.23 per cent (1335 Rs/ha), respectively over sole wheat yield. It was also found that wheat sowing in east-west direction and raya rows across the direction of wheat sowing *i.e.*, in north-south direction resulted in highest mean WEY (36.09 q/ha) as compared to the rest of the treatments. The results also demonstrate the impact of extent and distribution of rainfall particularly under rainfall deficient situations by testifying the superiority of raya intercropping with wheat.

Keywords: *Wheat, Raya intercropping, Sowing direction, Planting pattern.*

Introduction

Selection of crops for a particular region is determined not only by physical, biological and economic factors but social factors like personal taste, tradition and ethical code also plays their role.

Sowing of rainfed *rabi* crop (wheat) is generally done under deficient seeding zone moisture as upper soil profile dries up particularly due to early withdrawl of monsoon in this region. Therefore, there is a general perception among farmers that growing of raya as an intercrop gives an additional yield advantage without affecting the main crop's yield, clearly elucidating the beneficial effect of intercropping under rainfed conditions (Parkash *et al.*, 1986). This is mainly due to the fact that raya is very remunerative crop having very low water requirement and will give certain yield in case of complete failure of main crop due to drought. In addition to this, it was also realized that proper sowing direction and planting pattern might have an additional advantage on account of its common prevalence among the farmers of this region. Keeping these facts in view, a research experiment was carried out at farmer's fields in kandi belt of Punjab.

Materials and Methods

A field experiment was conducted at farmer's field during *rabi* season at village Dada (Nara-Dada watershed), District Hoshiarpur as a part of All India Coordinated Operational Research Project for Dryland Agriculture for two years. The soil of experimental field was slightly alkaline in reaction (pH 8.2), silty loam in texture (sand –68.4 per cent, silt –26.9 per cent and clay –4.7 per cent), low in available nitrogen and medium in available phosphorus and potassium having 0.63 per cent organic carbon and 0.21 ds/m electrical conductivity. The experiment was laid out in split plot design having two sowing direction (north-south and east-west) as main plot treatments and three planting patterns (sole wheat, raya rows along and across the direction of wheat sowing) as sub-plot treatments and each treatment being replicated thrice every year.

Rainfed wheat variety (PBW 175) was sown in lines 30 cm apart by using 100 kg seed rate in the first week of November and was fertilized with 80 kg N and 40 kg P_2O_5 per hectare. Full dose of P and half dose of N were applied before sowing of wheat by placing it below seeding zone and the remaining dose of N was top-dressed after the receipt of winter rains in the months of December and early January. After the completion of wheat sowing, raya (RLM 619) was drilled in between wheat rows 3 m apart in such a way that the first and last raya row remains at one-half distance from the field bunds. Wheat seed was treated with chlorpyriphos @ 4 ml/kg seed before sowing for termite control and raya rows were sprayed with metasystox 25 EC @ 625 ml/ha for the control of aphids. The raya rows and wheat crop was harvested in the first and third week of April, respectively. The total rainfall received during the crop growth season was 285.9 and 210.2 mm in the first and second year, respectively.

Results and Discussion

Effect of Sowing Direction

There was a non-significant effect of sowing direction on wheat and raya yields during both the years (Table 41.1). The data when converted into wheat equivalent yield (WEY) showed that nearly similar levels of WEY were obtained whether the crop is sown in east-west or north-south direction. Although, it was observed that wheat sowing in east-west direction recorded 1.32 q/ha (3.97 per cent) and 663 Rs/ha (7.84 per cent) higher mean WEY and net returns, respectively over its sowing in north-south direction.

Effect of Planting Pattern

The data in Table 41.1 revealed that intercropping of raya with wheat (irrespective of planting pattern) resulted in significantly higher mean WEY as compared to sole wheat crop. The per cent increase in mean WEY and net returns due to raya intercropping with wheat was 5.41 per cent (1.77 q/ha) and 18.23 per cent (1335 Rs/ha), respectively over sole wheat crop, hence, showing the beneficial effect of intercropping under rainfed conditions (Singh *et al.*, 2000). Further, it was observed that raya rows across the direction of wheat sowing recorded significantly higher mean WEY over its sowing along the direction of wheat. The per cent increase in mean WEY and net returns due to sowing of raya rows across the direction of wheat was 5.06 per cent (1.70 q/ha) and 7.04 per cent (589 Rs/ha), respectively over raya rows along the direction of wheat sowing. This might be ascribed to shading effect of raya rows on wheat resulting in its lower grain yield (Lal *et al.*, 1998).

Table 41.1: Effect of Sowing Direction and Planting Pattern of Raya Intercropping on the Grain Yield of Wheat Under Rainfed Conditions

Treatments	*Yield (q/ha)*						*Mean WEY*	*Net Returns (Rs/ha)*	*B : C Ratio*
	First Year			*Second Year*					
	Wheat	*Raya*	*WEY*	*Wheat*	*Raya*	*WEY*			
Sowing Direction									
North-south	31.73	3.61	35.49	27.80	2.70	30.93	33.21	8907	1.66
East-west	33.23	3.23	37.43	28.70	2.51	31.63	34.53	9570	1.70
CD (P = 0.05)	NS	NS	NS	NS	NS	NS	NS	–	–
Planting Pattern									
Sole wheat	35.61	–	35.61	29.77	–	29.77	32.69	7322	1.56
Raya rows along the direction of wheat sowing	30.13	3.46	36.18	26.43	2.64	31.04	33.61	8362	1.62
Raya rows across the direction of wheat sowing	31.70	3.38	37.60	28.54	2.57	33.03	35.31	8951	1.66
CD (P = 0.05)	2.54	NS	1.07	2.76	NS	1.46	0.66	–	–

WEY: Wheat equivalent yield.

Normal price (Rs/qtl.): Wheat–620/-, Raya–1100/-.

Interaction Effect

The data in Table 41.2 elucidate the interaction effect of sowing direction and planting pattern and it was observed that wheat sowing in east-west direction and raya rows across the wheat sowing *i.e.*, north-south direction resulted in significantly higher mean WEY (36.09 q/ha) as compared to other treatments under study.

Table 41.2: Interactive Effect of Sowing Direction and Planting Pattern of Raya Intercropping on Wheat Equivalent Yield (Mean of 2 Years)

Planting Pattern	*Sowing Direction*		
	North-south	*East-west*	*Mean*
Sole wheat	32.17	33.22	32.69
Raya rows along the direction of wheat sowing	32.92	34.29	33.61
Raya rows across the direction of wheat sowing	34.54	36.09	35.31
Mean	33.21	34.53	–

CD (P = 0.05): 1.16.

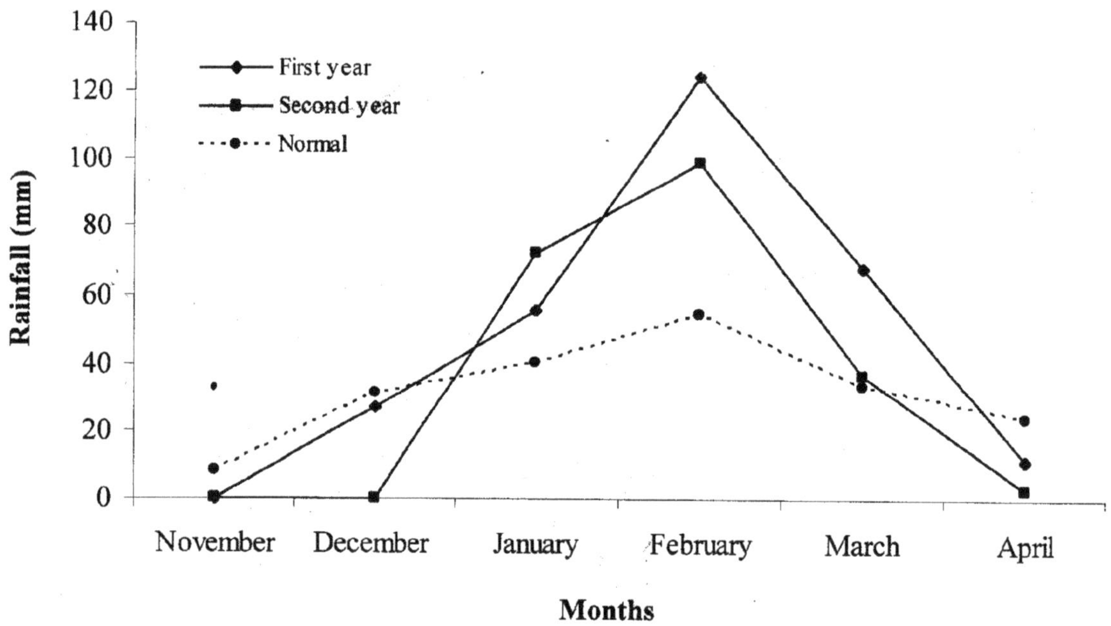

Figure 41.1: Monthly Rainfall Recorded During Crop Growth Period at ORP Village

Effect of Rainfall

The critical analysis of the rainfall data indicated that in the first year, the overall crop yield was higher (irrespective of treatments) as compared to the second year's yield (Table 41.1 and Figure 41.1). This might be due to better residual profile moisture and commencement of winter rains in December, coinciding with the critical growth stage *i.e.*, tillering followed by heavy rains during reproductive phase in the first year as compared to second year (Verma *et al.*, 2000).

Acknowledgement

The financial support provided by All India Coordinated Research Project for Dryland Agriculture (ICAR), Hyderabad to carry out this study is sincerely acknowledged.

References

Lal, R.B., Verma, A.K. and Ahuja, K.N. 1998. Intercropping of oilseed and pulses crops in wheat (*Triticum aestivum*) under fertilizer and water-constraints situations. *Indian J. Agronomy*, 43: 253–255.

Parkash, V., Tandon, J.P. and Prasad, K. 1986. Studies on intercropping in rainfed wheat. *Ann. Agric. Res.*, 7: 258–262.

Singh, Sukhvinder, Verma, H.N., Gill, M.S. and Narang, R.S. 2000. Performance of wheat-raya intercropping under rainfed conditions of *Kandi* areas of Punjab. *J. Res. PAU*, 37(1–2): 15–19.

Verma, H.N., Sukhvinder Singh and Narang, R.S. 1999. Yield of rainfed wheat (*Triticum aestivum*) in relation to seeding zone moisture and seed. *Indian J. Agronomy*, 45(3): 548–553.

Chapter 42

Influence of Some Cereals Diets on Breeding of *Corcyra cephalonica* Statinton

J.R. Kadam, A.P. Chavan, S.R. Parate, D.B. Kadam and B.M. Mhaske
Mahatma Phule Krishi Vidyapeeth, Rahuri – 413 722, India

ABSTRACT

The results of the study to find out most suitable diet for high eggs yield of *C. cephalonica* revealed that out of 17 dietary treatments sorghum 2 kg + yeast 2 gm + Vitamin E 0.2 gm emerged as the best diet for getting high yield of *Corcyra* eggs. It recorded highest (342 + 48 eggs/female) fecundity. The next highly promising and at par treatments were Bajara 2 kg + yeast 2 gm + Vitamin E w2 gm (322±33 eggs) and Bajara 1 kg + maize 1 kg + yeast 2 g + Vitamin 32 g (319±18 eggs). The cereals fortified with yeast and Vitamin E reduced duration of life cycle, increased moth emergence and sex ration of the factious host *C. cephalonica.*

Keywords: *Bio agents, Corcyra cephalonica, Factitious host, Fecundity.*

Introduction

Pest suppression is a vital aspect of successful crop production. Though chemical method of pest control is easy, economic and most popular its hidden dangers are becoming visible over a period of time especially due to undermined and excess use of pesticides in agro ecosystem of major crops.

The bio control agents are required to be mass-multiplicated for field releases on some alternate and easily culturable laboratory host under introduction and inundation programme culturability is an important character of ideal parasitoid. Mass multiplication of the parasitoids could be achieved easily and fastly if culturable factitious host is available. The eggs of rice moth, *C. cephalonica* are used as a factitious host for the multiplication of some predators and parasitoids. Among Trhchogrammatids, *Trichogramma aexiguum* (Maninder and Verma, 1982) were reared on eggs of rice moth. The key objective in all such programmes is that the cost of mass production of hosts and natural enemies should not exceed the cost of accruing benefits and progeny of alternate host should be healthy and fertile. The main aim of work carried out on suitability of hosts is getting healthy and fertile eggs, larvae and pupae and pupae for mass rearing of different parasitoids and predators and to minimise cost of their breeding. Therefore an attempt was made to study the suitability of different combinations of dietary ingredients with yeast and Vitamin E for *C. cephalonica* along with the aspect of fecundity and egg storability of factitious host for a Trichogrammtid production.

Materials and Methods

A laboratory experiment was conducted during the 1997–1998 in the Department of Agricultural Entomology, Mahatma Phule Krishi Vidyapeeth, Rahuri, District Ahmednagar (Maharashtra, India). Initial pure culture of the test insect (*C. cephalonica* stainton) was obtained from the bio control laboratory of the department. Major part of the ingredients were cereal grains. Jowar (M-35-1) 2 kg/tray, Bajra (wee-75) 1 kg/tray and Maize (Vijay) 0.667 kg/tray supplemented with yeast 2 gm, 4 gm/tray and Vitamin 'E' (Evion capsule 250 mg) 0.4 gm, 0.8 gm/tray to enhance insect growth were used to study the suitability of combinations of dietary ingredients to the host insect. These cereals yeast and Vitamin 'E' were made available from the local market. Experiment was conducted in randomized block design with three replications and seventeen treatment releasing 3600 eggs per box of size 38 × 22 × 12 cm having capacity 3.5 kg.

The cereal grains were fumigated (separately) in air tight containers by ethylene dibromide for 24 hours for destruction of eggs and hibernating stages of other stored grain pests. The insect free grains were coarsely ground into small pieces in grinder mill to ensure better development of the host larvae (Seshagiri Rao, 1954). The wooden boxes filled with the ingredients as per treatments. Male and female moths emerged from each box during one month. Percentage of moths emerged per treatment was calculated.

The male and females were identified on the basis of morphological character. The males were smaller in size as compared to females. The female abdomen is larger and broader.

Results and Discussion

Results of the study considering influence of dietary ingredients on various aspects of potential of rice moth, *C. cephalonica* as factitious host for bio agents are presented below.

Fecundity

The fecundity of the rice moth in various dietary treatments varied from 233±14.15 to 342±47.81 eggs/female. $JMY_1 E_1 (T_5)$ registered significantly highest fecundity of 342±48 eggs. The next promising and at par treatments were $BY_1 E_1$ resulting in 322±33 and 319±18 eggs/female, respectively. Among cereals bajara alone of fortified it with yeast and Vitamin E were better than correspondents jowar and maize treatments. Mixtures of jowar + bajara were significantly superior to those of containing maize. Rose *et al.* (1982) observed lesser duration 40 to 50 days on sorghum fortified with 5 per cent yeast than the duration of 44 to 52 days on unfortified sorghum. This finding confirmed the trend of results on the duration of life cycle in present study.

Eggs Production/2 kg Diet

The estimated number of eggs in the treatment was 81.2 to 137.6 thousands, JMY_1E_1 was the best diet for *Corcyra* which gave significantly highest yield of 137.6 thousand eggs at par treatment of followed by BY_1E_1 (136.6 thousand). BMY_1E_1 (132.6 thousand) was the next highly promising treatment. Satisfactory yield (110.6 to 115.8 thousand) of the factitious host was registered in at par treatments in descending order of superiority were BY_2D_2, JBY_1E_1, $B_1BMY_2E_2$ and JYE_1.

Duration of *Corcyra* Life Cycle

The duration in the treatments was narrowly varied from 453.70 in BY_2E_2 to 47.50 in M treatment. Bajara was the best diet for *Corcyra* among the 3 cereals. It recorded minimum duration (44.50 days) of the host. However, it was on par to jowar which was better than bajara. Most of treatments fortified with yeast and Vitamin 'E' reduced duration of life cycle as compared that in alone respective cereal treatments.

Mean Number of Emergence of *Corcyra* Moth/Tray

It ranged from 927 to 1056 BY_2E_2 resulted in highest (1056) number of moth emergence. It was on par to the treatments with BY_1E_1 and B only. The next highly promising treatments were JBY_2E_2 (1034), BMY_2E_1 (1024) and JBY_1E_1 (1027). Later was at par to that with BYM_1E_1 (1019 moth/tray).

Sex Ratio

Male to female was most appropriate (1 : 0 : 76) in BY_1E_{12} followed by that in BMY_1E_1 (1 : 0 : 69), JMY_2E_2 (1 : 0 : 66) and BMY_2E_2 (1 : 0 : 63). Jowar and maize diet showed almost equal sex ratio of host while it was lowest (1 : 0 : 50) in only bajara diet. However, all the three test cereals increased male to female ratio when fortified with yeast (Y) and Vitamin 'E' (E). The male *Corcyra* moth outnumbered female moth in all the treatments. The sex ratio recorded by Aware (1993) was 1 : 0 : 96 for *Corcyra* reared on Sorghum CSH-1 in summer. Low sex ratio might attributed to the winter season of present study. The result on fecundity, egg, yield and moth emerged could not be compared for lack of literature on such type study.

Table 42.1: Influence of the Dietary Ingredients on Potentiality of *C. cephalonica* as Factitious Host for Bioagents Production

Sl.No.	*Treatments**	*Fecundity/ Female (No. of Eggs)*	*Estimated Egg Production per 2 kg Diet (000)***	*Duration of Life Cycle (Days) Egg to Adult Death*	*Mean Number of Moth Emerged*	*Sex Ratio (M : F)*
1.	JY_1E_1	277±39	110.6	45.40	990	1 : 0.62
2.	BY_1E_1	322±55	136.6	43.85	1055	1 : 0.76
3.	MY_1E_1	268±47	90.8	47.25	940	1 : 0.59
4.	JBY_1E_1	297±33	113.6	44.50	1027	1 : 0.65
5.	JMY_1E_1	342±48	137.6	45.50	950	1 : 0.68
6.	BMY_1E_1	319±18	132.6	46.30	1019	1 : 0.69

Contd...

Table 42.1–Contd...

Sl.No.	Treatments*	Fecundity/ Female (No. of Eggs)	Estimated Egg Production per 2 kg Diet (000)**	Duration of Life Cycle (Days) Egg to Adult Death	Mean Number of Moth Emerged	Sex Ratio (M : F)
7.	$JBMY_1E_1$	272±29	102.2	47.00	969	1 : 0.64
8.	JY_2E_2	283±25	103.4	44.60	1000	1 : 0.59
9.	BY_2E_2	294±24	115.8	43.70	1056	I : 0.60
10.	MY_2E_2	268±34	92.4	46.50	949	1 : 0.58
11.	JBY_2E_2	289±26	99.0	44.10	1034	1 : 0.51
12.	JMY_2E_2	223±14	87.8	45.00	964	1 : 0.66
13.	BMY_2E_2	277±27	110.8	45.75	1024	1 : 0.63
14.	$JBMY_2E_2$	234±16	81.2	45.30	980	1 : 0.56
15.	J	267±39	96.6	46.10	990	1 : 0.56
16.	B	302±33	112.8	44.50	1050	1: 0.50
17.	M	256±23	87.4	47.50	927	1 : 0.57
	Mean	280.30	106.54	45.28	995.00	–
	SE±	1.9	1.4	0.50	4	–
	CD (P = 0.05)	6.0	4.4	1.70	12	–

J: Jowar; B: Bajara; M: Maize; Y_1: Yeast 2 gm; Y_2: Yeast 4 gm; E_1: Vitamin E 0.2 g and E_2: 0.4 g/kg cereal diet.

**Based on No. of females emerged per tray X mean fecundity per treatment.

References

Aware, B.Y. 1993. Suitability of some sorghum varieties and hybrids to rice moth, *Corcyra Cephalonica* stainton. *M.Sc. (Agri.) Thesis*, Marathwada Agril. Univ., Parbhani, M.S.

Maninder, S. and G.C. Verma. 1982. The growth of population of *Trichogramma exigum* Perkins. *Indian J. Ent.*, 44(2): 179–181.

Ray, R. 1994. Life table and intrinsic rate of increase of *Corcyra cephalonica* stainton in maize. *Env. and Ecol.*, 12(1): 202–204.

Ray, R., N. Shi and H.P. Patanik 1990. The life table and intrinsic rate of increase of *Corcyra cephalonica* stainton on jowar. *Orissa J. Agric. Res.*, 3(1): 22–26.

Rose, H.S., Saini, D.K. and Dhillon, S.S. 1982. Food preference behaviour of the larvae of *Corcyra cephalonica* stainton. *Bull. Grain Tech.*, 20(2): 137–139.

Seshagiri Rao, D.S. 1954. Notes on rice moth, *Corcyra cephalonica* stainton. *Indian J. Ent.*, 16(2): 95–114.

Chapter 43

Preliminary Field Evaluation of Ready Mix Sherlone 24 EC for Control of Sucking Pest Complex of Chilli

Panduran B. Mohite and Namdeo Patil***

**Biotechnology Centre, Mahatma Phule Krishi Vidyapeeth, Rahuri – 413 722, District Ahmednagar*
***Department of Agricultural Entomology, College of Agriculture, Kolhapur*

ABSTRACT

A field experiment was carried out during 1999–2000 and 2000–2001 to evaluate the bio-efficacy of Sherlone 24 EC ready mix alongwith commonly used insecticides against the sucking pest complex of chilli. The Sherlone 24 EC at 1.5 lit/ha was significantly effective in reducing the survival population of aphids, thrips and whiteflies followed by Spark 36 EC. Sherlone 24 EC at 1.5 lit/ha exhibited phytotoxic effect which resulted in to higher (4.98 MT/ha) yield.

Keywords: *Sherlone, Efficacy, Pest complex, Phytotoxic effect.*

Introduction

Chilli, *Capsicum annum* L. is important vegetable spice crop growth in India earning valuable foreign exchange for the country. The main constraints in the production and productivity is the damage caused by sucking pest *viz.*, thrips, *Scirtothrips dorsalis*, aphids, *Aphis gossypii* and whiteflies *Bemasia tabaci* and also reported to transmit leaf curl disease. Several insecticides including organophosphates have been evaluated against these pests (Rajasri *et al.*, 1991 and Rai and Singh, 1992) and has been found to give effective control. There is however, little information about efficacy recently introduced ready mix Sherlone 24 EC (Cypermiethrin 5 per cent + Phosalone 24 per cent) developed by Phone-poulenc Agrochemicals and hence, the present study reports on performance of Sherlone 24 EC in controlling the sucking pest complex of chilli.

Materials and Methods

Field experiment was carried out at farmers' field Rukadi, Tal. Hat Kanangale, Distt. Kolhapur (M.S.) during 1999–2000 and 2000–2001 to test the bioefficacy of ready mix Sherlone 24 EC (Cypermethrin 5 per cent + Phosalone 24 per cent) for controlling the sucking pest complex of chilli. The experiment was conducted in Randomized Block Design with three replications. Sherlone was tested at four different doses *i.e.*, 1.00, 1.25, 1.50 and 1.75 lit/ha and compared with endosulfan 35 EC at 2.00, Cypermethrin 10 EC at 0.60 and Spark 36 EC at 1.00 lit/ha along with an untreated check. Insecticides were sprayed five times at 15 days interval starting from 15 days after transplanting using high volume knapsack sprayer. Survival population of thrips, aphids and whiteflies on three selected leaves (top, middle and bottom) from each five randomly selected plants from each plot were assessed prior to each spray. The data on survival population and yield of green chilli were analysed statistically by analysis of variance.

Results and Discussion

The data revealed that Sherlone 24 EC at a dose of 1.5 lit/ha was significantly effective in reducing the mean survival population of thrips (2.87), aphids (3.70) and whiteflies (1.97) per leaf as compared to their population 11.90 to 17.30 in untreated control, however, it was on par with the higher dose of 1.75 lit/ha (Table 43.1). Of remaining three products spark gave good control of thrips, aphids and whiteflies. The Sherlone at 1.50 lit/ha exhibited phytotoxic effect which resulted in to higher yield (12.300 kg/plot *i.e.*, 4.98 MT/ha).

Table 43.1: Effect of Sherlone on Average Number of Sucking Insects on Three Leaves per Plant

Sl.No.	*Treatments*	*Dose lit./h*	*Mean Survival Population**			*Yield*
			Thrips	*Aphids*	*Whiteflies*	*kg/Plot MT/ha*
1.	Sherlone 24 EC	1.00	4.77 (2.29)	3.77 (2.52)	3.77 (2.06)	8.400[a] 3.42[b]
2.	Sherlone 24 EC	1.25	4.53 (2.24)	5.83 (2.52)	4.10 (2.10)	9.370 9.370
3.	Sherlone 24 EC	1.50	2.87 (1.84)	3.70 (2.04)	1.97 (1.47)	12.300 (4.98)
4.	Sherlone 24 EC	1.75	3.47 (1.95)	7.37 (2.68)	4.93 (2.25)	9.610 (3.89)
5.	Endosulfan 35 EC	2.00	5.53 (2.43)	5.50 (2.43)	4.70 (2.27)	9.470 (3.84)
6.	Cypermethrin 10 EC	0.60	6.10 (2.57)	6.33 (2.61)	5.83 (2.90)	7.380 (8.930)
7.	Spark 36 EC	1.00	5.67 (2.48)	5.10 (2.32)	4.67 (2.23)	8.930 (3.62)
8.	Untreated control		11.90 (3.52)	15.40 (3.98)	17.30 (4.16)	6.290 (2.55)
	S.E.±		0.079	0.22	0.23	0.44
	C.D. at (P = 0.05)		0.17	0.46	0.50	0.94

* Mean of five sprayings (Pooled analysis of two years); Figures in parentheses are $\sqrt{x + 0.5}$ transformation.

[a]: yield kg/plot; [b]: MT/ha.

References

Rai, P.C. and N.N. Singh. 1992. Determination of most profitable rate of spraying of some insecticides against thrips in relation to the yield of green chilli. *J. Insect Sci.*, 925(1): 109–110.

Rajasri, M., Reddy, G.P.V., Krishnamurthy, M.N. and Prasad, U.P. 1991. Bioefficacy of certain new insecticides including neem products against chilli pest complex. *Indian Coconut, Arecanut and Spices Journal*, 15(2): 40–42.

Chapter 44

Effect of Thiourea on the Germination of Three Varieties of *Vigna radiata* (L.) Wilczek

Arvind Kumar

Department of Botany, T.N.B. College, T.M. Bhagalpur University – 812 007, Bihar, India

ABSTRACT

The present paper deals with the effect of Thiourea on the germination of three varieties of *Vigna radiata* (L.) wilczek *viz.*, SUNAINA ML-65 and PANT MOONG-2 ML-65 shows tolerant as compared to SUNAINA and PANT MOONG-2. PANT MOONG-2 shows maximum sensitivity.

Keywords: *Thiourea, Germination, Tolerancy, Sensitivity.*

Introduction

Germinability of a seed depends on many factors substances which inhibit us germination of seeds are called inhibitor. Natural inhibitor have been found in the pulp of fruits, seed coats, embryos and endosperm. Certain inhibitor check the inhibition of water by having high osmotic pressure while other bring about dormancy due to their interference in the normal metabolism of seed for germination.

In the present investigation, the effect of different concentrations of Thiourea on the germination and early seeding growth of SUNAINA, ML-65 and PANT MOONG-2 varieties of *Vigna radiata* (L.) wilczek have been studied.

Materials and Methods

Germination experiments were conducted is sterlised petridishes containing thin cotton pads covered over by the filter circles. Effects of Thiourea was tested in 0.125 per cent, 0.25 per cent, 0.5 per cent, 1 per cent and 2 per cent. Thiourea solutions were prepared in distilled water. Respective concentrations were used to soaking the seeds and germination medium except control where only distilled water was used. Three replicates of 10 seeds were maintained at room temperature. Germination percentage was calculated on the basis of mean replicates. Length and fresh weight of radicles and hypocotyls were measured separately after 96 hours.

Results and Discussion

The effect of Thiourea on germination pattern was towards retardation is all the varieties at every concentrations up to 1 per cent. In 2 per cent Thiourea, seedling growth did not take place. The decreasing trend with increasing concentration of Thiourea was observed in all the three varieties. ML-65 shows Thiourea tolerance in comparison to SUNAINA and PANT MOONG-2 in respect to seedling length and its fresh weight after 96 hours. PANT MOONG-2 shows maximum sensitivity.

Table 44.1: Effect of Thiourea on Germination Percentage of Different Varieties of *Vigna radiata* (L.) Wilczek

Attributes Treatment	*Germination Percentage*		
	SUNAINA	*ML-65*	*PANT MOONG-2*
Control	95	90	80
0.125%	75	80	80
0.25%	75	80	70
0.5%	65	85	75
1%	55	75	70
2%	–	–	–

Table 43.2: Effect of Thiourea on Seedling Length of Different Varieties of *Vigna radiata* (L.) Wilczek

Attributes Treatment	*Seedling Length After 96 Hours (in cm)*								
	SUNAINA			*ML-65*			*PANT MOONG-2*		
	R	*H*	*Total*	*R*	*H*	*Total*	*R*	*H*	*Total*
Control	3.60	2.58	6.18	2.55	3.84	6.39	4.02	3.77	7.79
0.125%	1.74	2.14	3.88	2.05	2.07	4.12	1.82	1.56	3.38
0.25%	1.04	2.97	4.01	1.53	1.8	3.33	1.07	1.4	2.47
0.5%	0.41	0.83	1.24	0.52	0.87	1.39	0.53	0.9	1.43
1%	0.4	0.36	0.76	0.38	0.5	0.88	0.21	0.37	0.58
2%	–	–	–	–	–	–	–	–	–

R: Radicle; H: Hypocotyl.

Table 43.3: Effect of Thiourea on Seedling Fresh Weight of Different Varieties of *Vigna radiata* (L.) Wilczek

Attributes Treatment	*Seedling Fresh Weight After 96 Hours (in mg)*								
	SUNAINA			*ML-65*			*PANT MOONG-2*		
	R	*H*	*Total*	*R*	*H*	*Total*	*R*	*H*	*Total*
Control	34.4	117.77	152.17	15	117.22	132.22	20	150.71	170.71
0.125%	16	142	158	13.12	132.5	145.62	11.11	105.55	116.66
0.250%	7.14	115.71	122.85	11.11	110.0	121.11	10.00	130	140
0.5%	3	96	99	5	98	103	2.75	95	97.75
1%	2	94	96	2	100	102	1.11	93.33	94.44
2%	–	–	–	–	–	–	–	–	–

R: Radicle; H: Hypocotyl.

Acknowledgement

The author thanks to Dr. M.M. Prasad, Head, Department of Botany, T.N.B. College, Bhagalpur, Dr. M.M. Sharma, Head of Department of Botany, Kisan College and Dr. Damodar Singh, Head of Department of Botany, Nalanda College, Bihar Sharif for their constant encouragement and help.

References

Agarwal, S.K. 1971. Effect of Thiourea and ascorbic acid on seed germination. *J. Indian. Bot. Soc.*, 50: 374–376.

Mayer, A.M. and Poljakoff-Mayber, A. 1982. *The Germination of Seeds*, 3rd ed. Pergamon Press, Oxford, London.

Sen, D.N. and Chawan, D.D. 1971. Action of growth retardants on the seed germination and early seedling growth in *Leptadaenia pyrotechnica* (For SK.) *Decne. Biochem. Physiol. Pflanzen* (BPP), 162: 484–494.

Vyas, L.N. and Shrimal, R.L. 1973. Studies in the effect of Thiourea, IAA and Gibberallic acid on the germination of *Celosia argentea* L. *Acta. Bot. Sci. Hung.*, 18: 423–430.

Chapter 45

Reaction of Blackgram Genotypes Against Major Insect Pests

Devendra Prasad*, Dharmjeet Kumar*, Rabindra Prasad* and Santosh Sahay**

**Department of Entomology, **Department of Soil Science and Agricultural Chemistry, Birsa Agricultural University, Kanke, Ranchi – 834 006, Jharkhand*

ABSTRACT

A field experiment was conducted in the randomised block design with three replications in the Research Farm of Birsa Agricultural University, Kanke, Ranchi to find out reaction of different entries of blackgram [*Vigna mungo* (L.) Hepper] against major insect pests during *kharif* 2000. Out of 12 cultivars/entries of blackgram, Pant U-19 received minimum incidence of major insect pests *viz.*, leaf miner. *Acrocercops phaeospora* Meyr., Bihar hairy caterpillar, *Spilosoma obliqua* Walker., leaf hopper, *Empoasca kerri* Pruthi., foliage feeder, *Spodoptera litura* Fab., thrips, *Megalurothrips distalis* Karney, pod boring insects, *Catochrysops cnejus* Fab, *Lampides boeticus* L. and *Helicoverpa armigera* (Hub.) resulted the highest grain yield (799.40 kg/ha). TU-99-852 and BAU-1 remained almost at par with Pant U-19 with regard to the pests incidence but resulted to the lower grain yields of 569.70 and 452.66 kg/ha, respectively.

Keywords: *Blackgram, Genotypes, Reaction, Insect pests.*

Introduction

Blackgram [*Vigna mungo* (L.) Hepper] is one of the important pulse crop grown in India. The crop is heavily damaged by several insect pests. The loss due to insect pests were estimated upto 23 per cent in Bihar (Hameed and Dubey, 1983) and 60 per cent in Madhya Pradesh (Rawat and Kaushik, 1983).

Among the various methods of pest control, use of resistant varieties is considered to be highly effective and ideal for raising the pulse crops with minimum loss (Sachan, 1992). With this objective, the present investigation was undertaken with a view to search out the resistant/tolerant/less susceptible genotypes of blackgram against the pest species.

Materials and Methods

Altogether 12 entries were taken to test their susceptibility/resistance against the insect pest species in the representative agroclimatic condition of Ranchi (Jharkhand). Seeds of different genotypes of blackgram received from the Indian Institute of Pulses Research (IIPR), Kanpur, were sown on 11th July, 2000 in a randomised block design with three replications in the plot size of 4 m x 1.2 m with a buffer strip of 0.6 meter. The recommended packages of practices were applied without any protection measures. Ten randomly selected plants were tagged in each plot on which the incidence of insect pests were recorded at weekly intervals during 7.00 to 11.00 am. Absolute populations per plants were counted in the case of thrips, pupea of leaf miner, nymph of leaf hopper, larvae of semilooper, *Spodoptera litura*, Bihar hairy caterpillar and other countable insect pests. Number of total pods as well as damaged pods were counted at the time of maturity for calculating percentage of pod damage. Grains yield were recorded plot wise and converted into kg/ha. Data on the incidence of major insect pests and grain yield were subjected to the statistical analysis for drawing the inferences.

Results and Discussion

Blackgram was found to be infested as many as nineteen insect pests during 2000. Among them, eight insect pests were categorised as major pests (Table 45.1). Out of eight major insect pests occurring on the crop plants, pod boring insects *viz., Catochrysops cnejus* Fab., *Lampides boeticus* L. and *Helicoverpa armigera* (Hub.) were found to be highly injurious pests and hence were categorised as key pest of blackgram.

Table 45.1: Major Insect Pests Infesting Blackgram During 2000

Sl.No.	*Common Name*	*Scientific Name*	*Pest Status*
1.	Leaf hopper	*Empoasca kerri* Pruthi	Major
2.	Leaf miner	*Acrocercops phaeospora* Meyr	Major
3.	Foliage feeder	*Spodoptera litura* Fab.	Major
4.	Bihar hairy caterpillar	*Spilosoma obliqua* Wlk.	Major
5.	Thrips	*Megalurothrips distalis* Karney	Major
6.	Pod borer	*Catochrysops cnejus* Fab.	Key
7.	Pod borer	*Lampides boeticus* (L.)	Key
8.	Pod borer	*Helicoverpa armigera* (Hub.)	Key

Incidence of Insect Pests

Leafhopper

The lowest population of *Empoasca kerri* (2.80 jassids/plant) was recorded in Pant U-19 which remained at par with TU-99-852 and BAU-1 (3.00 jassids/plant). The highest incidence (7.70 jassids/plant) was recorded in the case of OBG-16 (Table 45.2) which did not differ significantly with UG-774 (7.50 jassids/plant) and KU-96-3 (7.20 jassids/plant).

Table 45.2: Reaction of Different Genotypes of Blackgram Against Insect Pests

Genotypes	*Number of Insects per Plant at 40 DAS*					*Pod Damage (%)*	*Yield (kg/ha)*
	Leaf Hopper	*Leaf Miner*	*Foliage Feeder*	*Bihar Hairy Caterpilla*	*Thrips*		
Pant U-19	2.80	1.10	0.20	1.00	1.40	6.85	796.40
	(1.82)	(1.26)	(0.84)	(1.22)	(1.38)	(15.27)	
KU-96-5	6.14	2.10	1.50	2.00	4.80	17.62	535.65
	(2.58)	(1.61)	(1.41)	(1.58)	(2.30)	(24.84)	
OBG-17	5.00	3.20	2.70	3.00	3.20	11.96	570.52
	(2.35)	(1.92)	(1.79)	(1.87)	(1.92)	(20.08)	
UG-774	7.50	3.80	2.60	3.60	5.40	19.88	468.88
	(2.83)	(2.07)	(1.76)	(2.02)	(2.43)	(26.53)	
TU-99-852	3.00	1.50	0.20	1.10	1.80	14.87	569.70
	(1.87)	(1.41)	(0.84)	(1.26)	(1.52)	(22.75)	
BAU-1	3.00	1.20	3.10	1.10	2.00	13.27	452.66
	(1.87)	(1.30)	(1.90)	(1.26)	(1.58)	(21.31)	
KU-96-3	7.20	3.10	4.50	2.60	5.00	29.06	504.76
	(2.77)	(1.90)	(2.24)	(1.76)	(2.35)	(32.46)	
KU-99-1	5.20	3.10	4.60	2.80	3.80	13.65	506.15
	(2.39)	(1.90)	(2.26)	(1.81)	(2.07)	(21.72)	
UG-950	6.32	3.40	5.60	3.00	3.80	16.72	435.65
	(2.61)	(1.97)	(2.47)	(1.87)	(2.07)	(24.00)	
OBG-16	7.70	3.20	5.70	3.80	5.60	26.14	468.34
	(2.86)	(1.92)	(2.49)	(2.07)	(2.47)	(30.67)	
USJD-102	6.26	1.80	2.30	1.50	4.20	22.74	603.67
	(2.60)	(1.52)	(1.67)	(1.41)	(2.17)	(28.47)	
OBG-19	5.20	3.10	3.20	2.80	3.10	27.85	568.33
	(2.39)	(1.90)	(1.92)	(1.82)	(1.90)	(32.01)	
S.E.m (±)	0.07	0.06	0.04	0.15	0.05	0.82	22.57
C.D. (P = 0.05)	0.20	0.19	0.12	0.44	0.15	2.42	66.20
C.V. (%)	10.09	6.55	6.73	9.97	7.52	5.73	7.02

Data transformed into $\sqrt{x + 0.5}$ for statistical analysis.

Figures in parentheses correspond to transformed value.

Percentage date converted into arcsine √percentage.

Leaf Miner

The lowest incidence of 1.10 pupae of leaf miner/plant was observed in Pant U -19 which were not different significantly from the population recorded in TU-99-852 (1.50 pupae/plant) and BAU-1 (1.20 pupae/plant). The highest incidence (3.80 pupae/plant) was noticed in UG-774 (Table 45.2), which remained at par with OBG-16 and OBG-17 (3.20 pupae/plant), KU-96-3, KU-99-1 and OBG-19 (3.10 pupae/plant) and UG-950 (3.40 pupae/plant).

Foliage Feeder

The highest larval population of *Spodoptera litura* (5.70 larvae/plant) was obtained in OBG-16 (Table 45.2) which was significantly not different from UG-950 (5.60 larvae/plant). The lowest population of 0.20 larva per plant was observed in Pant U-19 and TU-99-852.

Bihar Hairy Caterpillar

The highest larval population of *Spilosoma obliqua* (3.80 larvae/plant) was noted in OBG-16 (Table 45.2) which remained at par with UG-774 (3.60 larvae/plant), OBG-17 and UG-950 (3.00 larvae/plant), KU-99-1 and OBG-19 (2.80 larvae/plant) and KU-99-3 (2.60 larvae/plant). The lowest larval population of the pest (1.00 larvae/plant) was recorded in Pant U-19 which did not differ significantly with the population recorded in TU-99-852 and BAU-1 (1.10 larvae/plant), USJD-102 (1.50 larvae/plant) and KU-96-5 (2.00 larvae/plant).

Thrips

OBG-16 received the highest population of *Megalurothrips distalis* (Karney) (5.60/plant) which was closely followed by UG-774 (5.40/plant), KU-96-3 (5.00/plant) and KU-96-5 (4.80/plant). The lowest incidence of 1.40 thrips/plant was recorded in Pant U-19 (Table 45.2) which remained at par with TU-99-852 (1.80/ plant).

Incidence of Pod Borers

The lowest pod damage (6.85 per cent) caused by *Catochrysops cnejus* (Fab.), *Lampides boeticus* (L.) and *Helicoverpa armigera* (Hub.) was registered in Pant U-19 (Table 45.2). The highest incidence of pod borers was recorded in KU-96-3 (29.06 per cent) which was not different from the pod damage obtained in OBG-19 (27.85 per cent) and OBG-16 (26.14 per cent).

Grain Yield

Pant U-19 was also responsible for producing the highest grain yield of 796.40 kg/ha. The lowest grain yield of 435.65 kg/ha was obtained in UG-950 which remained at par with the yield recorded in OBG-16, BAU-1 and UG-774 (Table 45.2).

A perusal of Table 45.2 indicated that Pant U-19 was found to be responsible for receiving the lowest populations of leaf hopper, *E. kerri;* leaf miner, *A. phaeospora;* defoliator, *S. litura;* Bihar hairy caterpillar, *S. obliqua* and *M. distalis* at flowering stage of the crop. However the population of these insect pests were not significantly different from the populations recorded in TU-99-852. The populations of leaf hopper, leaf miner, Bihar hairy caterpillar and thrips obtained on BAU-1 did not differ from TU-99-852. The higher populations of these insect pests were observed in OBG-16 followed by KU-96-3 and KU 99-1. The highest yield as well as lowest pod damage caused by *C. cnejus, L. boeticus* and *H. armigera* were recorded in Pant U-19.

Not much information on varietal screening of blackgram against the insect pests is available in the literature so far, however Teotia (1978) suggested that the use of plant varieties less susceptible to insect attack could help in reducing the seriousness of the pest problems. Several entomologists obtained less susceptible/ tolerant/ resistant varieties of blackgram (Thakur *et al.*, 1988; Sahoo *et al.*, 1989; Yadav *et al.*, 1992; Chhabra *et al.*, 1993; Chauhan and Dahiya, 1997; Ujagir and Sehgal, 1997; Katare *et al.*, 1998; Sahasrabuddhe and Patel, 2000. Pant U-19 was found to be least susceptible to thrips (Katare *et al.*, 1998) and Bihar hairy caterpillar (Yadav *et al.*, 1992). Resistant reaction against whitefly by Chauhan and Dahiya, 1997 and jassids and thrips by Ujagir and Sehgal, 1997 were obtained in Pant U-19. Similar results were also obtained in the present investigation where lowest incidence of leafhopper, leaf miner, *S. litura,* Bihar hairy caterpillar, thrips and pod borers as well as higher grain yield of blackgram (cv Pant U-19) were registered.

It may be concluded that pant U-19 carried minimum incidence of insect pests *viz.*, leaf miner, *A. phaeospora;* Bihar hairy caterpillar, *S. obliqua;* leaf hopper, thrips and pod boring insects which in turn resulted in highest grain yield (796.40 q/ha).

Acknowledgements

The authors are grateful to the Director, Indian Institute of Pulses Research, Kanpur who made available the seeds of different entries for evaluation against the pest species. The authors are also thankful to the Director (Research) BAU, Ranchi for providing the facilities for conducting the trial.

References

Chauhan, R. and Dahiya, B. 1997. Insect Pest Management in Pulse crops. In: *Advanced Training Course on Insect Pest Management,* (Eds.), Naresh, J.S., Singh, R. and Rohilla, H.S. Department of Entomology, CSSHAU, Hissar, p. 123–124.

Chhabra, K.S. and Kooner, B.S. 1981. Field resistance in blackgram, *Vigna mungo* L. against insect pests and yellow mosaic virus. *Indian J. Ent.,* 43: 288–293.

Chhabra, K.S., Kooner, B.S., Sharma, A.K. and Saxena, A.K. 1993. Screening of blackgram genotypes against whitefly, jassids and yellow mosaic virus–role of phytochemicals in resistance. *Indian. J. Pulses Res.,* 6(1): 76–81.

Hameed, S.F. and Dubey, S.L. 1983. Losses due to insect pests in North Bihar. Special Issue: *Crop losses due to Insect Pests. Indian J. Ent.,* 1: 136–146.

Katare, Subhash, Deshpande, R.R., Bhadauria, N.S., Bhadauria, N.K.S. and Katare, S. 1998. Response of blackgram cultivars against their major insect pests. *Bhartiya Krishi Anusandhan Patrika,* 13(1–2): 89–91.

Rawat, R.R. and Kaushik, U.K. 1983. Crop losses due to insect pests in North-East Madhya Pradesh. Special Issue: *Crop Losses Due to Insect Pests. Indian. J. Ent.,* 1: 73–83.

Sachan, J.N. 1992. Present status of *Helicoverpa armigera* in pulses and strategies for its management. In: *Helicoverpa Management: Current Status and Future Strategies,* (Ed.) Sachan, J.N. Proceedings of 1st National Workshop held at Directorate of Pulses Research, Kanpur, India, 30–31st August, 1990, p. 7–23.

Sahasrabuddhe, A.G. and Patil, P.D. 2000. Evaluation of some promising blackgram cultivars against some major sucking pests. *Pestology,* 24(7): 43–44.

Sahoo, B.K., Sontakke, B.K. and Rath, L.K. 1989. Varietal susceptibility of different greengram and blackgram cultivars to the leaf beetles and pod borer complex. *Env. and Ecol.,* 7(2): 345–347.

Teotia, T.P.S. 1978. Presidential address: Role of Entomologists in Agriculture, Rural Development and Rural Health Care. *Indian J. Ent.,* 40(4): 428–432.

Ujagir, R. and Sehgal, V.K. 1997. Insect Pest Management in pigeon pea and *Vigna* crops. In: *Recent Advances in Insect Pest Management in Major Crops,* (Eds.), Sehgal, V.K., Pathak, P.K., Sachan, G.C. and Kanaujia, K.R. Department of Entomology, G.B. Pant University of Agriculture and Technology, Pantnagar, pp. 90.

Yadav, R.P., Kumar, N., Yazdani, S.S. and Goel, S.C. 1992. Effect of insecticides on field populations of Bihar hairy caterpillar, *Spilosoma obliqua* Walker and its naturally occurring parasitoids in Bihar. In: *Bioecology and Control of Insect Pests.* Proceedings of National Symposium on Growth, Development and Control Technology of Insect Pests, p. 192–199.

Chapter 46

Survey of Fungal Diseases of Economically Important Crops from Ahmednagar District

S.K. Aher, R.K. Aher, S.L. Khapke and R.N. Deshmukh

Department of Botany, New Arts, Commerce and Science College, Parner, District Ahmednagar, M.S., India

ABSTRACT

This article deals with screening of some fungal diseases of economically important crops from Ahmednagar District with their causal organism, host etc. During the present investigation about 83 fungal diseases caused to 35 economically important plants were recorded. The important symptoms caused by the pathogens are rust, anthracnose, leaf spots, mildew, damping off, wilting and leaf blight etc. The aim behind this investigation is an awakening in the minds of farmers towards crop diseases will certainly help to apply proper pesticides or fungicides to control the diseases.

Keywords: Fungal pathogens, Economically important plants.

Introduction

Today at the national and state levels, there is great emphasis on developing a more productive agriculture by transforming an ancient agriculture into a modern agricultural practices. Various diseases caused to economically important crops is one of the serious problem. It can cause severe deformation of foliage, stunted growth of plants, and premature fruit fall. This resulted in poorer yield. Morever premature leaf and fruit dropping caused by severe fungal diseases resulted heavy losses to the farmers (Rangaswami and Mahadevan, 1999; Sangwan and Mehta, 2001). To control these diseases farmers uses heavy doses of fungicides which causes soil, air and water pollution. Considering the seriousness of the fungal diseases in Ahmednagar District, the present study was conducted.

Ahmednagar District has an area of 17,345 sq. km. The District has hilly as well as plain area where various cereals (wheat, maize, jawor, bajara, rice), pulses (pigeon pea, bengal gram, greengram, blackgram) oil seed crops (groundnut) vegetables, fibre yielding plants, fruit plants (banana, grapes) as well as cash crops like sugarcane, pea, tomato, potato, etc. are grown. The economically important crops are subjected to various fungal diseases resulting in poor yield.

Material and Method

The comprehensive study of diseases of crop plants is examined in the field. The fields with diseased crops visited frequently to observe developmental stages of plant growth as well as diseases. Crops grown in different seasons and under different agronomic and cultural conditions observed periodically by arranging field tours. To obtain a clear knowledge of disease symptoms, the plants examined carefully in the field, and a description of the disease, recorded in a note book. Some diseases easily identified in the field while unidentified diseases along with plant material brought to the laboratory, sorted out and spread out on sheets of paper or blotter and pressed systematically. The blotter changed at frequently. The dried specimens used for identification. Diseases identified by using the methods of culturing of pathogenic organisms, preparation of culture media, isolation of plant pathogens, purification of the culture and identification of plant pathogens was done by refering standared references like Thomson and Raper (1945); Ellis (1976); Barnet (1972).

Results and Discussion

A total of 83 fungal diseases caused to economically important crops was recorded during the period of present investigation. There are certain fungal diseases of crop plants occuring year after year in Ahmednagar areas to cause severe damages. Important among them are listed in Table 46.1.

The aim behind this work is an awakening in the minds of farmers towards crop diseases and their causal organism in particular. A perfect knowledge of the disease will certainly help to apply proper pesticides or fungicides to control the diseases.

Of the 83 fungal pathogens, a vast majority are causing leaf and stem infections. The important genera causing these diseases are *Alternaria, Cercospora, Fusarium, Helminthosporium, Oidium, Pyricularia* etc. belonging to Deuteromycotina, *Plasmopara, Albugo, Perenospora, Sclerospora* etc. from Phycomycotina, rust and smut fungi *Puccinia* and *Ustilago* respectively from Basidiomycotina and *Erysiphae, Claviceps* etc. from Ascomycotina. Important symptoms caused by the pathogens are rust, anthracnose, leaf spots, mildew, damping off, smut, wilting, blight etc.

Because of the nature of their spread, mode of infection and establishment of disease symptoms on the host, these diseases cause severe damages to the crops. There are certain diseases of crop plants occurring year after year. Important among them are anthracnose of citrus, powdery mildew, downy mildew and rust of grapes, leaf and stem rust of wheat, smut of sorghum, red rot and smut of sugarcane, tikka disease of groundnut, rust of pea, sunflower etc.

There is an urgent need to study the problems and develop suitable techniques to control the plant diseases.

Table 46.1: List of Diseases and Causal Organism of Economically Important Crops of Ahmednagar District

Sl.No.	*Name of the Host*	*Disease Caused*	*Causal Organism*
1.	*Vitis vinifera*, L. (Grapes)	Powdery mildew	*Uncinula necator*
		Downy mildew	*Plasmopara viticola*
		Anthracnose	*Elsinoe viticola*
		Rust	*Phakosporw vitis*
		Black rot	*Guignardia bidwelli*
2.	*Psidium guajava* L. (Guava)	Guava wilt	*Fusarium oxysporum*
3.	*Mangifera indica* L. (Mango)	Powdery mildew	*Erysiphae cichoracearum*
		Downy mildew	*Oidium mangiferae*
4.	*Citrus aurantifolia* (Kagadi line)	Anthracnose	*Colletotrichum*
5.	*Cucumis sativus* L. (Khira)	Powdery mildew	*Erysiphae sphaerotheca*
		Downy mildew	*Perenospora cubensis*
6.	*Oryza sativa* L. (Rice)	Blast	*Pyriculara oryzae*
		Brown spot	*Helminthosporium oryzae*
7.	*Triticum aestivum*, L. (Wheat)	Leaf, brown rust	*Puccinia recondita*
		Black or stem rust	*P. graminis tritici*
		Loose smut	*Ustilago nuda* var. *tritici*
		Foot rot	*Puccinia graminicolum*
		Leaf blight	*Alternaria triticina*
8.	*Sorghum vulgare*, Pers (Sorghum)	Grain smut	*Sphacelotheca sorghi*
		Loose smut	*Sphacelotheca cruenta*
		Head smut	*Sphacelotheca reiliana*
		Rust	*Puccinia purpurea*
		Downy mildew	*Sclerospora sorghi*
		Leaf spot	*Cercospora sorghi*
		Anthracnose	*Colletotrichum graminicolum*
		Charcoal rot	*Macrophomia phaseolina*
9.	*Pennisetum typhoides*, (Burm. F) Stapf and Hubb. (Bajra)	Downy mildew	*Sclerospora graminicola*
		Rust	*Puccinia penniseti*
		Smut	*Tolyposporium penicillariae*
		Ergot	*Claviceps microcephala*
10.	*Zea mays*, L. (Maize)	Smut	*Ustilago maydis*
		Rust	*Puccinia sorghi*
11.	*Saccharum officinarum*, L. (Sugarcane)	Red rot	*Colletotrichum falcatum*
		Smut	*Ustilago scitaminea*
12.	*Gossypium herbaceum*, L.	Wilt	*Fusarium oxysporum*
		Grey mildew	*Ramularia areola*
13.	*Arachys hypogaea*, L. (Groundnut)	Tikka disease	*Cercosporidium arachidicola*
		Rust	*Puccinia arachidis*
14.	*Brassica oleracea* var. capitata, L. (Cabbage)	Damping off	*Pythium aphanidermatus*
		Club root	*Plasmodiophora brassicae*
15.	*Raphanus sativus*, L. (Radish)	White rust	*Albugo candida*
16.	*Beta vulgaris* L. (Beet root)	Leaf spot	*Cercospora beticola*
17.	*Ipomea batatas*, L. (Sweet potato)	White rust	*Albugo ipomoeae-panduranae*
18.	*Lycopersicon esculentum*, Mill. (Tomato)	Damping off	*Pythium aphanidermatum*
		Late Blight	*Phytophthora infestans*
		Early Blight	*Alternaria solani*

Contd...

Table 46.1–Contd...

Sl.No.	*Name of the Host*	*Disease Caused*	*Causal Organism*
19.	*Solanum melogena*, L. (Brinjal)	Leaf spot	*Alternaria solani* *A. solani–melongenae*
20.	*Capsicum annum*, L. (Chilli)	Anthracnose Leaf spot Damping off	*Cercospora capsici* *Pythium aphanidermatum*
21.	*Hibiscus esculentus*, L. (Bhendi)	Powdery mildew Rust	*Erysiphe cichoracearum* *Uromyces heterogeneus*
22.	*Solanum tuberosum* L. (Potato)	Late Blight Early Blight	*Phytophthora infestans* *Alternaria solani*
23.	*Helianthus annus*, L. (Sunflower)	Rust Rust	*Puccinia helianthi*
24.	*Pisum sativum*, L. (Pea)	Rust Downy mildew Leaf spot	*Eromyces fabae* *Peronospora pisi* *Cercospora cruenta*
25.	*Phaseolus aureus*, Roxb. (Green Gram)	Powdery mildew Leaf spot	*Erysiphae polygoni* *Cercospora canenscens*
26.	*Phaseolus mungo* var. *radiatus*, L. (Black Gram)	Powdery mildew Rust	*Erysiphae polygony* *Uromyces phaseoli*
27.	*Allium cepa*, L. (Onion)	Smut Rust	*Urocystis cepulae* *Puccinia porri*
28.	*Ricinus communis*, L. (Castor)	Rust	*Melampsora ricini* *Cercospora ricinella*
29.	*Brassica campestris*, L. (Mustard)	White rust Downy mildew Powdery mildew	*Albugo candida* *Peronospora brassicae* *Erysiphe polygoni*
30.	*Annona squamosa*, L. (Custard apple)	Leaf spot	*Cercospora annonae*
31.	*Musa paradisica*, L. (Banana)	Wilt Leaf spot	*Fusarium oxysporum* *Diectoniella torulosum*
32.	*Carthamus tinctorius*, L. (Safflower)	Rust	*Puccinia calcitrapae* var. *Centaureae*
33.	*Cajanus cajan*, L. (Pigeon pea)	Wilt Leaf spot Powdery mildew	*Fusarium oxysporum* *Cercospora indica* *Leveillula taurica*
34.	*Cicer arietium*, L. (Bengal Gram)	Wilt Rust Powdery mildew	*Fusarium oxysporum* var. *orthoceros* *Uromyces ciceris-arietini* *Erysiphae polygoni*
35.	*Daucas carota*, L. (Carrot)	Powdery mildew Leaf blight Leaf spot	*Erysiphe polygoni* *Alternaria radicina* *Cercospora apii*

References

Barenett, H.L. and B.B. Hunter. 1972. *Illustrated Genera of Imperfect Fungi*, 3rd edn. Burgess Publishing Company, Minnespolis, Minnesota.

Ellis, M.B. 1976. *More Dematiaceous Hyphomycetes*. Commonwealth Mycological Institute, Kew, England, pp. 607.

Rangaswami, G. and A. Mahadevan. 1999. *Diseases of Crop Plants in India*, 4th edition. Prentice Hall of India Private Limited, New Delhi.

Sangwan, M.S. and N.K. Mehta. 2001. Diseases and their control. *Indian Farming*, 51: 17–21.

Thompson, C. and Raper, C.B. 1945. *A Manual of Aspergilli.* The Willians and Wilkins Co.

Chapter 47

Genetic Architecture of Yield and its Component Traits in Rice

Purabi Das*, Avijit Kundu*, Nirmal Mandal* and Indrani Dana**

**Faculty of Agriculture, Bidhan Chandra Krishi Viswavidyalaya, Mohanpur – 741 252, West Bengal*
***Chinsurah Rice Research Station, Hooghly – 712 102, West Bengal*

ABSTRACT

Different genetic parameters and combining ability were studied for grain yield and its component traits in diallel crosses involving ten diverse parents of rice (*Oryza sativa* L.) following Griffing's Numerical Approach, Method II, Model I. The general and specific combining ability were significant for almost all the characters studied except sca for grain breadth, grain thickness and grain L : B ratio. The predictability ratio indicated importance additive genetic variance for days to 50 per cent flowering, plant height, panicle length, grain length, grain L : B ratio, 100-grain weight and grain yield per plant. The importance of non-additive genetic variance was found in number of panicles per plant, number of filled grains per panicle and spikelet sterility. While for grain density and grain breadth equal importance of both additive and non-additive gene effects were observed. Days to 50 per cent flowering, plant height, number of filled grains per panicle, spikelet sterility, grain length, L : B ratio and grain yield per plant showed high heritability and genetic advance that suggest the feasibility of genetic improvement of these characters under low selection pressure. Among the parents Pokkali and Annada were found to be good general combiners for grain yield and its components whereas, Pusa Basmati, Tarori Basmati and Karnal local-95 showed high *per se* peformances and high gca effects too for grain characters. While amongst the crosses Pokkali x Annada, IR 72 x Annada, IR43 x Annada and Tulsari x Pusa Basmati were the best combinations showing significant sca effects for grain yield, quality and its components. These crosses could be exploited to improve grain yield in rice.

Keywords: *Combining ability, Diallel analysis, Rice.*

Introduction

Success of a breeding programme in self pollinated crop like rice depends on the choice of parents for hybridization and identification of superior recombinants. Among the different methods to assess combining ability of parents, the diallel cross technique is a systematic method available to the breeders for discriminating parents as well as identifying the cross combinations having probability of throwing transgressive segregants. Furthermore, contribution of additive and non-additive gene actions in expression of quantitative characters are also gleaned from such analysis. The present investigation was conducted with 10 × 10 half diallel cross to generate informations on the relative magnitude of additive and dominance components in the genetic control of characters in planning suitable, economic and efficient breeding strategies to develop high yielding varieties with reasonably good productivity potential and quality.

Materials and Methods

The parental material comprising of ten diverse rice genotypes *viz.*, Mansarovar, Pokkali, IR 72, IR 43, Annada, IR 50, Tulsari, Pusa Basmati, Tarori Basmati and Karnal local-95 were crossed in a diallel fashion (excluding reciprocals) during *kharif* 2002. Ten parents and 45 F_1S were grown in a Randomized Block Design with three replications during wet season of 2003 at Chinsurah Rice Research Station, Hooghly. In each replication, each of the homogeneous generations, *i.e.*, parents and F_1s were grown in a plot of 2 rows of 2 m length spaced 25 cm apart. 15 cm distance between plant to plant was maintained in each plot. Non-experimental rows were planted all around the experimental material to avoid any possible boarder effects. Data were collected on 10 plants selected randomly per plot in each parent or F_1. The observations were recorded on days to 50 per cent flowering, plant height (cm), number of panicles per plant, panicle length (cm), number of filled grains per panicle, spikelet sterility (per cent), grain density grain length (mm), grain breadth (mm), grain L : B ratio, grain thickness (mm), grain type, 100-grain weight (g) and grain yield per plant (g). Plot means were used for statistical analysis. The analysis of variance was done following standard statistical method (Panse and Sukhatme, 1978). In the present investigation combining ability of parents and crosses were compute following Griffing's Numerical Approch, Method II, Model I (Griffing, 1956).

Results and Discussion

Analysis of variance revealed significant differences among the parents and F_1 hybrids in respect of all the characters under study. This indicates the existence of ample genetic differences in the cross combinations and parents (Table 47.1). The variances due to gca were found to be highly significant for all the characters which too demonstrates availability of significant variability among the selected parents. Whereas, insignificant sca variances were observed in grain breadth, grain thickness and grain L : B ratio only.

The estimates of relative proportion of gca and sca as proposed by Baker (1978) used to assess additive and non-additive gene actions in governing yield contributing characters. The low ratio of gca : sca variance (< 1) was recorded for number of panicles per plant, panicle length, number of filled grains per panicle and spikelet sterility indicating that those characters were principally governed by non-additive gene action. Similar result was obtained by Sharma and Koranne (1995) for 100-grain weight and spikelet sterility; Sarma *et al.* (1996), Ganesan and Rangaswamy (1998) for number of tillers per plant and panicle length. The operation of non-additive gene action in governing the characters offers ample scope for exploitation of hybrid vigour in rice. Operation of additive gene action was observed for days to 50 per cent flowering, plant height, grain density, 100-grain weight,

Table 47.1: Mean Sum of Squares in Respect of Yield and Yield Component Characters in Parents and F_1 Hybrids of Rice

Source	Mean Sum of Squares (MSS)													
	d.f.	*Days to 50% Flowering*	*Plant Height*	*No. of Panicle/ Plant*	*Panicle Length*	*No. of Filled Grain Panicle*	*Spikelet Sterility (%)*	*Grain Density*	*Grain Length*	*Grain Breadth*	*Grain Thick-ness*	*L : B Ratio*	*100-grain Weight*	*Grain Yield/ Plant*
Replication	2	2.59	0.15	1.73	1.03	90.74	8.07	0.27	0.0100	0.0008	0.0019	0.0009	0.004	33.62
Treatment	54	326.89**	1053.54*	25.76**	18.06**	2735.16**	304.55**	3.65**	2.2600**	0.2142**	0.0951**	0.9260**	0.26**	432.42**
Error	164	3.66	11.32	3.79	1.25	288.90	16.55	0.48	0.0100	0.0009	0.0104	0.0224	0.01	42.03
gca	9	175.65**	1342.66*	7.67**	6.40**	914.53**	90.28**	1.34**	2.6968**	0.0928**	0.0836**	1.0716**	1.63**	159.25**
sca	45	98.44**	153.08**	8.79**	5.49**	1518.34**	105.38.*	1.18**	0.3662**	0.0667	0.0218	0.1716	0.06**	141.17**
Error	108	1.22	3.77	1.06	0.41	96.30	5.48	0.16	0.0025	0.0356	0.0161	0.1611	0.003	14.00
gca : sca	–	1.78	8.77	0.87	0.64	0.60	0.85	1.95	7.3642	1.3913	3.8348	6.2447	21.15	1.12
Range	–	73.00 – 128.00	68.87– 152.07	4.67– 16.07	20.87– 33.50	65.93– 201.6	10.51– 58.47	2.62– 7.65	8.03– 11.22	2.27– 3.43	1.53– 2.73	2.38– 4.78	1.62– 3.23	12.51– 63.31
Grand mean	–	89.66	115.18	9.71	25.93	118.60	25.57	4.58	9.3710	2.7650	2.0310	3.4410	2.52	28.95
S.E	–	1.56	2.74	1.58	0.91	13.87	3.31	0.56	0.0817	0.0245	0.0832	0.0149	0.83	5.29
C.D.	–	3.06	5.38	3.11	1.78	27.20	6.49	1.10	0.1600	0.0480	0.1632	0.0292	1.63	10.37

* and ** denote significance at 0.01 and 0.05, respectively.

grain length, grain thickness and grain L : B ratio. Preponderance of additive gene action was also reported by Srivastava and Seshu (1983), Dhaliwal and Sharma (1990), Geetha *et al.* (1994) and Ganesan *et al.* (1997), for plant height; Sarawgi *et al.* (1991) for grain density and grain length; Ghosh (1993) and Munhot *et al.* (2000) for grain L : B ratio. The gca : sca ratio for grain yield per plant was found to be close to unity, which demonstrate almost equal importance of both additive and non-additive gene actions in manifesting those characters. Li (1975) and Mahapatra and Mohanty (1985) also indicated the existance of similar gene action for grain yield per plant. Since both additive and non-additive gene effects played pivotal role in the inheritance of grain yield and its components, their simultaneous improvement through adoption of biparental mating would be logical to implement with check (Chakraborty and Hazarika, 1994). Comstock *et al.* (1949) suggested that the use of reciprocal recurrent selection which would be more effective when both additive and dominant gene actions are involved in the expression of a character. Obviously use of component characters in improving grain yield largely depends upon their heritability values or relative magnitude of the additive genetic variance and its correlation with yield largely (Falconer, 1960).

High genotypic variance (σ^2_g) were observed as compared to phenotypic variance (σ^2_g) for days to 50 per cent flowering, plant height, number of productive tillers per plant, number of filled grains per panicle, spikelet sterility, grain length, grain L : B ratio, 100-grain weight and grain yield per plant (Table 47.2) which elucidates more involvement of genetic control in governing these characters. As genetic variability remains unaffected by environmental conditions this type of variability is more useful for genetic improvisation though selection or hybridization. The presence of the non-additive genetic variance offers immense opportunity for exploitation of heterosis.

High additive genetic variance was observed for days to 50 per cent flowering, plant height, number of filled grains per panicle, spikelet sterility and grain yield per plant as compared to dominance varince suggesting strong genetic control in governance of these characters. The additive genetic variance is associated with homozygosity, and it is fixable. Therefore, selection of characters governed by such variance would be very effective, may be used extensively for selection because this is the only variance which readily responds to selection efficiently. It is mentionable that breeding value of an individual is measured directly by the additive gene effects general combining ability (gca) effect of a parent.

Low dominance variance was observed for number of panicles per plant, panicle length and spikelet sterility. The dominance variance is associated with heterozygosity and generally non-fixable. Therefore, selection of those characters controlled by such variance would not be effective. Involvement of heterosis breeding for exploitation of heterosis may be advocated in such situation.

Low magnitude of differences were found between genotypic coefficient of variance (GCV) and phenotypic coefficient of variance (PCV) for days to 50 per cent flowering, plant height, panicle length, grain length, grain breadth, grain thickness and grain L : B ratio. It indicates less control of environmental influence i~ governing these characters. However, PCV estimates were found to be higher than GCV estimate for number of filled grains per panicle, grain density, spikelet sterility, 100-grain weight and grain yield per plant. It suggests the involvement of environmental influence in expression of those characters.

Heritability in broad sense (h^2a) ranged between 65.89 (number of panicles per plant) and 98.75 (grain breadth) (Table 47.2). High genetic advance (GA) was observed for number of filled grains per panicle (50.54) followed by plant height (37.78), days to 50 per cent flowering (21.03) and grain yield per plant (20.43). Days to 50 per cent flowering, plant height, number of filled grains per panicle, grain length, grain L : B ratio and grain yield per plant showed high h^2a and appreciable magnitude of GA

Table 47.2: Genetic Parameter Profile for Grain Yield and Yield Attributing Characters in Rice

Character Genetic Parameter	*Mean Sum of Squares (MSS)*												
	Days to 50% Flowering	*Plant Height*	*No. of Productive Tillers/ Plant*	*Panicle Length*	*No. of Filled Grain Panicle*	*Spikelet Sterility (%)*	*Grain Density*	*Grain Length*	*Grain Breadth*	*Grain Thick-ness*	*L : B Ratio*	*100-grain Weight*	*Grain Yield/ Plant*
Genotypic variance (σ^2g)	107.74	347.4.1	7.32	5.60	815.42	96.03	1.05	0.1942	0.0220	0.0051	0.0750	4.33	130.13
Phenotypic variance (σ^2p)	111.40	358.73	11.11	6.85	1104.32	112.48	1.53	0.1967	0.0576	0.0212	0.2361	5.38	172.16
Environmental variance (σ^2e)	3.66	11.32	3.79	1.25	288.90	16.45	0.48	0.0025	0.0356	0.0161	0.1611	0.01	42.03
Dominance variance (σ^2d)	97.21	149.31	7.73	9.90	1422.03	99.89	1.18	0.3637	0.0311	0.0057	0.0105	0.06	69.50
Additive variance (σ^2a)	215.49	694.81	14.64	11.20	1630.84	192.06	2.11	0.3884	0.0044	0.0.102	0.1500	8.67	260.26
Phenotypic coefficient of variance (PCV)	11.77	16.44	29.99	10.09	28.01	41.47	33.49	9.3028	9.7045	9.6743	16.5317	91.86	45.3.1
Genotypic coefficient of variance (GCV)	11.58	16.18	36.95	9.13	24.07	38.33	23.03	8.0033	9.6437	8.2703	15.9493	82.42	39.39
Heritability in broad sense (h^2a)	96.71	96.84	65.89	81.18	73.84	85.37	68.76	98.6482	98.7505	73.0802	93.0778	80.50	75.58
Heritability in narrow sense (h^2b)	68.11	81.32	56.15	50.12	48.80	62.27	63.98	51.4709	6.1884	31.8750	46.6418	19.14	70.07
Genetic advance (GA)	21.03	37.78	4.53	4.37	50.54	18.65	1.77	1.7721	0.5459	0.2959	1.0907	3.84	20.43
Genetic advance as % of mean (GAM)	23.45	32.80	46.56	16.87	42.61	72.54	38.63	18.9109	19.7461	14.364	31.6981	152.34	70.55

that suggest the feasibility of genetic improvement of these characters under low selection pressure. Manuel and Palanisamy (1989) and Yadav *et al.* (1999) also reported performance of non-additive gene effects in governing grain yield and advocated indispensability of high selection pressure to improve upon those characters.

Pusa Basmati, Tarori Basmati and Karnal local-95 (scented rice) showed high *per se* performances and high gca effects too for grain characters, which offers ample scope to develop desirable segregates for improvement of grain quality characters. Pokkali and Annada were found to be better general combiners for all the characters (Table not shown). IR 72, IR 43, Annada, IR 50 and Pusa Basmati showed negative gca values for days to 50 per cent flowering suggesting their breeding values may be utilized to develop early varieties by reducing crop duration. Similarly IR 72, IR 43 and IR 50 may reduce plant height of recombined through hybridization. Positive gca values were observed in Pokkali and Annada at signficant levels for grain density and grain yield per plant.

Table 47.3: sca Performance of Superior Cross Combination in Comparison with gca Values for Grain Yield Per Plant in Rice

Cross Combination	*sca Effect*	*gca Effect*	
		Female	*Male*
IR 72/Annada	32.55**	1.66	5.13**
Pokkali/Annada	28.09**	4.59**	5.13**
Tulsari/Pusa Basmati	25.53**	– 1.61	– 0.66
IR 72/Tulsari	21.97**	1.66	– 1.61
IR 43/Annada	20.45**	– 1.12	5.13**
Mansarovar/Pusa Basmati	8.16**	1.13	– 0.66
Pokkali/Tarorai Basmati	7.69*	4.59**	– 6.31**
Mansarovar/IR 72	7.58*	1.13	1.66
Tulsari/Karnal local-95	6.70*	– 1.61	– 4.55**
Pokkali/IR 43	6.48*	4.59**	– 1.12

* and ** denote significance at 0.01 and 0.05, respectively.

The ranking of F_1 hybrids based on *per se* performance and sca effects are presented in Table 47.3. High sca effects results mostly from the dominance component and interaction effects of the hybridizing parents. In the present study, significant sca effects were found to be exhibited by five crosses for grain yield per plant. They are IR 72 x Annada, Pokkali x Annada, Tulsari x Pusa Basmati, IR 72 x Tulsari and IR 43 x Annada. The ranking of F_1 hybrids based on *per se* performance and sca effects showed that most of the crosses with significant sca effects also had superior *per se* performance. This result is in agreement with the findings of Mandal and Banerjee (1997) in safflower and Munhot *et al.* (2000) in rice, where strong positive correlation between gca and *per se* performance was observed. Whereas, it is in contradiction with the finding of Peng and Viramani (1990), Sreemachandran Babu *et al.* (2000), Lavanya (2000) who suggested that selection of hybrids should be based on *per se* performances coupled with sca effects. The cross IR 50 x Karnal local 95 showed very high mean performance for grain L : B ratio with insignificant sca effect. IR 43 x Tarori Basmati recorded very high sca effect but did not rank top in respect of grain L : B ratio. On the other hand, crosses *viz.*, Tarori Basmati x Karnal local 95 and Pusa Basmati x Torori Basmati showed very high grain length with negatively significant gca and significantly positive sca values. Similar trend was also observed for grain breadth and thickness, it is interesting to note that crosses with insignificant sca effects performed better than some of the crosses, which showed significant sca effects. These results clearly indicate that superior performance of the hybrids need not be the ones with high sca effects *vice versa*. Similar result was also obtained by Rahman *et al.* (1981). It is, therefore, concluded that selection should be made on the basis of *per se* performance and sca effects concurrently.

References

Baker, R.J. 1978. Issues in diallel analysis. *Crop Sci.*, 18: 533–536.

Chakraborty, S. and Hazarika, M.H. 1994. Association analysis between grain yield and its components in rice. *Oryza*, 31: 263–265.

Comstock, R.E., Robinson, H.F. and Harrey, P.H. 1949. A breeding procedure design to make maximum use of both general and specific combining ability. *Oryza*, 30: 275–279.

Dhaliwal, T.S. and Sharma, H.L. 1990. Combining ability and maternal effects for agronomic and grain characters in rice. *Oryza*, 27: 122–128.

Falconer, D.S. 1960. *Introduction to Quantitative Genetics*. The Ronald Press Company, New York.

Gaesan, K.N., Wilfred, W., Manuel Vivekanandan, P. and Pillai Arumugan, M. 1997. Combining ability heterosis and inbreeding depression for quantitative traits in rice. *Oryza*, 34: 13–18.

Ganesan, K.N. and Rangaswamy, M. 1998. Combining ability studies in rice hybrids involving wild abortive (WA) and *Oryza* perennis sources of CMS lines. *Oryza*, 35: 113–116.

Geetha, S., Soundaray, A.P.M.K., Palanisamy, S. and Kareem, A.A. 1994. Combining ability analysis and gene action relating to grain character among medium duration rice genotypes. *Crop Res.*, 7: 239–242.

Ghosh, A. 1993. Combining ability for yield and its related traits in upland rice. *Oryza*, 30: 275–279.

Griffing, B. 1956. Concept of general and specific combining ability in relation to diallel crossing system. *Aust. J. Bio. Sci.*, 9: 463–493.

Lavanya, C. 2000. Combining ability for yield and its components in hybrid rice. *Oryza*, 37: 11–14.

Li, C.L. 1975. Diallel analysis for yield and its component in rice (*Oryza sativa* L.). *J. Agric. Assoc. China*, 92: 41–56.

Mahapatra, K.C. and Mohanty, H.K. 1985. Inheritance of some quantitative characters including heterosis in rice by combining ability analysis. *In: Proc. Int. Rice. Genetic Symposium* (Rice Genetic–I), 27–31 May, 1985, p. 579–591.

Manuel, W. and Palanisamy, S. 1989. Line x tester analysis of combing ability in rice. *Oryza*, 26: 27–32.

Mondal, A.B. and Banerjee, S.P. 1997. Diallel analysis of yield and yield components in safflower. *J. Genet. and Breed.*, 51: 211–215.

Munhot, K.K., Sarawgi, A.K. and. Rastogi, N.K. 2000. Gene action and combining ability for yield, grain quality and other related characters in rice. *Oryza*, 37: 16.

Panse, V.G. and Sukhatme, P.V. 1978. *Statistical Methods for Agricultural Workers*, 3rd Revised Edition. ICAR, New Delhi.

Peng, Y.J. and Virmani, S.S. 1990. Combining ability for yield and four related traits in relation to breeding in rice. *Oryza*, 27: 1–10.

Rahman, M., Pataerary, A.K. and Miah, A.J. 1981. Combining ability in rice. *Indian J. Agric. Sci.*, 51: 543–546.

Sarawgi, A., Srivastava, M.N. and Chowdhary, B.P. 1991. Partial diallel cross analysis of yield and its related characters in rice under irrigated and rainfed situations. *Indian J. Genet.*, 51: 30–36.

Sarma, M.K., Richharia, A.K. and Agarwal, R.K. 1996. Variability, heritability, genetic advance and genetic divergence in upland rice. *Int. Rice. Res. Notes*, 21: 25–26.

Sharma, R.K. and Koranne, K.D. 1995. Line x tester analysis for yield and yield components in *indica x japonica* crosses of rice. *Oryza*, 32: 234–238.

Sreemachandran Babu, M., Satyanarayana, P.V., Madhuri, J. and Kumar, R.V. 2000. Combining ability analysis for identifying elite parents for heterotic rice hybrids. *Oryza*, 37: 19–22.

Srivastava, M.N. and Seshu, D.V. 1983. Combining ability for yield and associated characters in rice. *Crop Sci.*, 23: 741–744.

Yadav, L.S., Maurya, D.M., Giri, S.P. and Singh, S.B. 1999. Combining ability analysis for yield and its components in rice. *Oryza*, 36: 208–210.

Chapter 48

Effect of Soil Solarization and Herbicides on Nutrients Uptake by Soybean and Associated Weeds

T.K. Tiwari, V.S. Pawar, P.V. Mahatale and A.V. Patil

Department of Agronomy, Post Graduate Institute, M.P.K.V., Rahuri – 413 722, Maharashtra, India

ABSTRACT

Field experiment was conducted during *Kharif* season of 2000–2001 to assess the nutrient uptake by soybean and loss caused by weeds. The results revealed that, the total NPK uptake by crop was found to be significantly highest (168.84, 14.17 and 22.49 kg NPK ha^{-1}) under soil solarization for 45 days, closely followed by weed free treatment, odyssey 2.0 L ha^{-1}, and soil solarization for 30 days. The increase in nutrient uptake due to soil solarization for 45 days was 53.70 per cent N, 58.73 per cent P, and 31.84 per cent K.

The minimum NPK uptake by weeds was observed under weed free treatment, soil solarization for 45 days (0.18, 0.27, 0.21 kg NPK ha^{-1}), odyssey 2.0 L ha^{-1} and pursuit 100 g a.i. ha^{-1}.

Keywords: *Soybean, Weed, Nutrient uptake, Soil solarization.*

Introduction

Soybean [*Glycine max* (L) Merill.] serves as an important fat and protein source for large population residing in Asia and American continents. In *kharif* season, favourable conditions of humid climate and intensive agricultural pattern provide congenial atmosphere for abundant population and growth of weeds. This ubiquitous organism causes enormous losses by diverting the energy from the direction desired by man. Weeds cause the losses in soybean production to the tune 38 per cent (Singh and Kolar, 1994). Therefore, efforts were made to study the relative efficacy of various herbicides and soil solarization treatments in controlling the weeds and to see their effect on mortality and assess the loss of nutrients caused by weeds.

Materials and Methods

Field study was conducted on soybean in medium deep black soil at Post Graduate Institute Research Farm, M.P.K.V., Rahuri District Ahmednagar, M.S., during *kharif* 2000–2001. The soil of the experimental site was alkaline (pH 8.3), with medium N (263.43 kg ha^{-1}) low P (7.84 kg ha^{-1}) and very high K (496.0 kg ha) contents. Fifteen treatments (Table 48.1) were tested in a randomized block design with three replications.

Table 48.1: Effect of Soil Solarization and Herbicides on the Uptake of Nitrogen, Phosphorus and Potassium (kg ha^{-1}) by Weeds in Soybean

Treatments	*Dry Matter of Weeds (q ha^{-1})*	*Nutrient Uptake by Weeds (kg ha^{-1})*		
		Nitrogen	*Phosphorus*	*Potassium*
Raptor 20 g a.i. ha^{-1}	4.10	3.10	1.75	2.13
Raptor 30 g a.i. ha^{-1}	4.08	2.46	1.39	1.63
Raptor 40 g a.i. ha^{-1}	4.50	2.31	1.50	1.83
Raptor 80 g a.i. ha^{-1}	3.34	2.04	1.14	1.48
Pursuit 75 g a.i. ha^{-1}	3.82	2.66	1.22	1.49
Pursuit 100 g a.i. ha^{-1}	2.16	1.13	0.90	0.67
Odyssey 0.8 Lha^{-1} (FP)	3.91	1.84	1.07	1.18
Odyssey 1.0 Lha^{-1} (FP)	2.94	1.46	0.86	1.14
Odyssey 2.0 Lha^{-1} (FP)	1.72	0.80	0.67	0.54
Oxyfluorfen 0.15 kg a.i. ha^{-1}	2.99	1.41	0.82	0.72
Soil solarization for 30 days	2.33	1.59	1.07	1.04
Soil Solarization for 45 days	0.68	0.18	0.27	0.21
Stale seedbed technique	7.93	4.42	2.64	2.94
Weedy Check	14.79	7.26	5.01	6.49
Weed free	0.00	0.00	0.00	0.00
C.D. at 5%	2.72	1.15	0.56	0.84

The JS-335 variety was sown at a spacing of 30 cm × 10 cm and uniformly fertilized with 50 kg N and 75 kg P_2O_5 ha as basal application.

For soil solarization, in summer months, the plots were covered with polyethylene sheet (0.014 mm) for 30 and 45 days following the procedure of Chauhan *et al.* (1988). All such plots were uniformly irrigated to field capacity one day prior to imposing the treatments.

30 days prior to sowing of soybean, the finally prepared plot was uniformly irrigated to field capacity. A flush of young weed seedlings appeared was removed by manual weeding in stale seed

bed technique. Oxyfluorfen was applied as pre-emergence spray on third day after sowing. Raptor, a 12 per cent water soluble concentrate of imazamox, whereas, persuit, a 10 per cent water soluble concentrate of imazethapyr and odyssey, a formulated product of imazamox and imazethapyr (2.5 per cent each) were applied as early post applications at two to three leaf stage of weeds (15–20 DAS). The quantity of water and the herbicides to be applied per plot were calculated and calibrated on area basis before spraying.

Species-wise weed population (per m^2, their weights and nutrient uptake by weeds and soybean were recorded at harvest. The nitrogen content in soybean plant and weed samples were estimated by Micro-Kjeldahl's method after destroying the organic matter by H_2O_2 digestion. Phosphorus was determined by vanadomolybdate phosphoric yellow colour colourimentric method in nitric acid system and potassium by flame photometer method (Chapman and Pratt, 1961). The uptake of NPK by crop and weeds were computed by multiplying per cent content and dry weight.

Results and Discussion

Various weed control practice significantly reduced the weed dry weight at harvest over weedy check (Table 48.1). The dry matter of weeds was significantly lowest under weed free treatment indicating superiority of this treatment for efficient weed control which was at par with those of pursuit 100 g a.i. ha (2.16 q ha^{-1}), odyssey 2 L ha^{-1} (1.72 q ha^{-1}), soil solarization for 30 days (2.33 q ha^{-1}) and soil solarization for 45 days (0.68 q ha^{-1}). The dry matter of weeds was significantly the highest (1.4.79 q ha^{-1}) under weedy check followed by stale seed bed technique (7.93 q ha^{-1}), indicating that this technique was not so effective for controlling weeds. These results were in conformity with those reported by Rana and Angiras (1996) and Singh *et al.* (2000).

Table 48.2: Effect of Soil Solarization and Herbicides on Nutrient Uptake by Soybean

Treatments	*Nitrogen*	*Phosphorus*	*Potassium*
Raptor 20 g. a.i. ha^{-1}	140.37	11.24	18.04
Raptor 30 g a.i. ha^{-1}	141.39	11.75	18.50
Raptor 40 g a.i. ha^{-1}	148.10	12.22	19.38
Raptor 80 g a.i. ha^{-1}	157.06	12.97	20.65
Pursuit 75 g a.i. ha^{-1}	142.21	11.66	18.69
Pursuit 100 g a.i. ha^{-1}	154.31	12.69	20.17
Odyssey 0.8 Lha^{-1} (FP)	158.82	13.11	20.96
Odyssey 1.0 Lha^{-1} (FP)	165.88	13.51	21.79
Odyssey 2.0 Lha^{-1} (FP)	167.44	13.79	21.71
Oxyfluorfen 0.15 kg a.i. ha^{-1}	144.64	11.75	18.60
Soil solarization for 30 days	165.47	13.66	21.75
Soil Solarization for 45 days	168.84	14.17	22.49
Stale seedbed technique	140.01	11.14	18.00
Weedy Check	106.31	8.40	13.73
Weed free	167.98	14.10	22.02
C.D. at 5%	7.32	0.60	0.95

Nutrient Uptake by Weeds

The uptake of N, P and K was significantly lowest in weed free treatment, soil solarization for 45 days, odyssey 2.0 L ha^{-1} and pursuit 100 g a.i. ha^{-1}. The uptake was found to be significantly higher under weedy check followed by stale seed bed technique because of more dry matter production (14.79 and 7.93 q ha^{-1} respectively) of weeds (Table 48.1). Maurya *et al.* (1990) also reported a reduction in nutrient removal by weeds with their control.

Nutrient Uptake by Soybean

The NPK uptake by soybean in soil solarized plots for 45 days was significantly highest (168.84, 14.17, 22.49 kg NPK ha^{-1} respectively) and it was at par with those observed in weed free treatment odyssey 2 L ha^{-1}, odyssey 1 L ha^{-1} and soil solarization for 30 days (Table 48.2). The superiority in nutrient uptake by soybean under soil solarization for 45 days may be attributed to the effective control of weeds and availability of more nutrients to soybean and enhanced nutrient concentration and the yield, thereby resulting in higher uptake of nutrients.

References

Chapman, H.D. and Pratt, P.F. 1961. *Methods of Analysis of Soil, Plant and Water*. Division of Agri. Sci., University of California.

Chauhan, Y.S., Nene, Y.L., Johansen, C., Haware, M.P., Saxena, N.P., Singh, S., Sharma, S.B. Sahrawat, R.L., Burford, J.R., Rupela, O.P., Kumar Rao, J.V.D.K. and Sithanantham, S. 1988. Effects of soil solarization on pigeonpea and chickpea. *Res. Bull. II, ICRISAT,* Patancheru, pp. 16.

Maurya, B.M., Gogulwar, N.M. and Tiwari, J.P. 1990. Herbicidal weed control efficiency and nutrient removal by weeds in soybean [*Glycine max* (L.) Merill.]. *Indian J. Weed Sci.,* 22(3–4): 51–56.

Rana, M.C. and Angiras, N.N. 1994. Dynamics of weed flora in soybean (*Glycine max*) and wheat (*Triticum aestivum*) + pea (*Pisum sativum*) due to imazethapyr application. *Indian J. Weed Sci.,* 26 (3&4): 64–70.

Singh, H. and Kolar, J.S. 1994. Crop weed competition studies in soybean [*Glycine max* (L). Merill.]. *Indian J. Weed Sci.,* 26(1&2): 83–86.

Singh, V.P., Mishra, J.S. and Bhan, V.M. 2000. Effects of varieties, row spacing and weed control measures on the weed growth and grain yield of soybean (*Glycine max* (L) Merill.]. *Pestology,* 24(11): 22–24.

Chapter 49

Long-term Influence of Organic and Inorganic Fertilization on the C/N Ratio of Alfisol Under Maize-Wheat Cropping Sequence

Santosh Sahay, B.P. Singh*, Birendra Kumar** and Dharmjeet Kumar****

Department of Soil Science and Agricultural Chemistry, **Department of Biotechnology, * Department of Entomology, Birsa Agricultural University, Kanke, Ranchi – 834 006*

ABSTRACT

Effect of long-term application of organic and inorganic fertilizers was studied on C/N ratio of soil after maize and wheat harvest. In general, C/N ratio lowered down with the continuous application of fertilizers and manure. Mean C/N ratio were varied from 5.40 to 10.56 and 5.19 to 12.14 after maize and wheat harvest respectively. Plot receiving continuous application of manures had wider C/N ratio than in plots receiving continuous application of fertilizers but the value was found less in all the treatments over control.

Keywords: *Long-term, C/N ratio, Fertilization.*

Introduction

Immobilization and mineralization depends on the C/N ratio of organic matter. Generally, application of organic matter having C/N ratio more than 30 causes immobilization of soil N at the early phase of decomposition. Lower total N content in soil or higher C/N ratio slows down the decomposition process. However, not much information is available on the influence of continuous use of manure and fertilizers on C/N ratio of alfisol and the present study is an attempt in this direction.

Materials and Methods

Soil samples (0–15) were collected after maize and wheat harvest (2002–2003) from different treatments of the permanent manurial trial experiments progressing since 1956 at Birsa Agricultural University Farm, Ranchi with maize-wheat cropping. The soil was initially clay loam with organic carbon (0.529 per cent) and total N (0.067 per cent). C/N ratio was calculated from the value of total organic carbon and total nitrogen determined in each soil sample. Organic carbon was determined by Walkley and Black rapid titration method as described by Piper (1950) and total N by Kjeldahl's wet digestion method as described by Jackson (1967). Treatment details are as follows:

T_1; Control: No manure and fertilizer. T_2; N: 110 kg ha^{-1} as urea. T_3; NP_xK_y: 110 kg N ha^{-1} as urea + P_2O_5 present in full dose of FYM, *i.e.*, X and Y kg respectively. T_4; FYM: 110 kg N ha^{-1} through farmyard manure (full dose of FYM @ 220 q ha^{-1}). T_5; ½ (FYM + N) + $P_{X/2}$ + $K_{Y/2}$ – 55 kg N ha^{-1} as FYM + 55 kg N ha^{-1} as urea + X/2 kg P_2O_5 ha^{-1} as SSP + Y/2 kg K_2O ha^{-1} as MOP. T_6; NP: 110 kg N ha^{-1} as urea + 90 kg P_2O_5 ha^{-1} as SSP. T_7; FYM + P (90-X) – T_4 + (90-X) kg P_2O_5 ha^{-1} as SSP. T_8; ½ (FYM + N) + P (90-X/2) – 55 kg N ha^{-1} as FYM + 55 kg N ha^{-1} as urea + (90-X/2) kg P_2O_5 ha^{-1} as SSP. T_9; NPK: 110 kg N ha^{-1} as urea + 90 kg P_2O_5 ha^{-1} as SSP + 70 kg K_2O ha^{-1} as MOP. T_{10}; FYM + P (90-X) + K (70-Y) – T_7 + K (70-Y) kg K_2O ha^{-1} as MOP. T_{11}; ½ (FYM + N) + P (90-X/2) + K (70-Y/2) – T_8 + (70-Y/2) kg K_2O ha^{-1} as MOP. T_{12}; Lime + NPK – Lime + T_9. T_{13}: Lime + FYM + P (90-X) + K (70-Y) – Lime + T_{10}. T_{14}; Lime + N – Lime + T_2. Where X, Y represent the amount of P_2O_5 and K_2O present in full dose of farmyard manure (FYM) required to supply 110 kg nitrogen.

Results and Discussion

Influence of long-term organic and inorganic fertilization on organic carbon, total nitrogen and C/N ratio are given in Table 49.1.

Application of nitrogenous fertilizers alone (T_2) recorded lowest organic carbon content which may be attributed to very poor crop growth and poor plant residues in absence of phosphorus and lowering of pH. Continuous incorporation of FYM (T_4) enhanced organic carbon content which might be due to mineralization of organic matter added to soil (Prasad *et al.*, 1997). The increase in organic carbon content of the soil in the treatments receiving organics along with inorganic fertilizers can be attributed to the higher crop residues besides supply of organic matter. Total N in soil showed an increase in all the treatments except control. Addition of inorganic fertilizers alone had lower affect than manure or along with manure and may be attributed to higher loss of inorganic N through nitrification and denitrification. Besides the direct addition of N through FYM (0.5 per cent) the organic manures are also reported to contribute by chelating the soil nitrogen and thereby conserving it from losses (Bastia, 2002).

C/N ratio varied from 5.40 to 10.56 and 5.19 to 12.14 after maize and wheat harvest, respectively. Continuous application of inorganic fertilizers had narrowed the C/N ratio (5.41 to 6.93 and 5.19 to 6.92) whereas addition of full dose of FYM had widened it (8.52 to 9.50 and 8.90 to 10.47) after maize and wheat harvest, respectively which may be due to higher nitrogen build up as compared to organic carbon and *vice versa*. Liming reduced C/N ratio and this change may be interpreted as an index of increased biological activity (Moller, 1990). C/N ratio varies with treatments as it is the function of organic carbon and total nitrogen content in soil. Similar findings were obtained earlier by Lal, 1987.

Table 49.1: Influence of Long-term Organic and Inorganic Fertilization on Organic Carbon, Total N and C/N Ratio of Alfisol

Treatments	*Organic Carbon (g kg^{-1})*		*Total N (kg ha^{-1})*		*C/N Ratio*	
	After Maize	*After Wheat*	*After Maize*	*After Wheat*	*After Maize*	*After Wheat*
T_1	6.55	6.80	1391	1274	10.56	12.14
T_2	5.76	5.91	1863	1910	6.93	6.92
T_3	8.72	8.77	2959	3055	6.60	6.44
T_4	14.63	16.40	3459	3536	9.50	10.47
T_5	11.23	11.38	4007	4014	6.30	6.35
T_6	7.24	7.24	2506	2670	6.52	6.08
T_7	12.41	13.30	3554	3610	7.85	8.26
T_8	8.86	8.92	3673	3783	5.40	5.31
T_9	8.27	8.42	2876	3003	6.46	6.28
T_{10}	14.12	14.78	3362	3393	9.41	9.78
T_{11}	10.64	10.59	3991	4253	5.97	5.60
T_{12}	5.47	5.61	2266	2432	5.41	5.19
T_{13}	14.48	15.22	3805	3839	8.52	8.90
T_{14}	5.62	5.56	2042	2087	6.17	5.61
S.E.m±	0.1083	0.0796	31.00	32.76		
C.D. (P = 0.05)	0.3148	0.2314	90.12	95.25		

References

Basta, D.K. 2002. Effect of integrated nutrient supply on yield of rice (*Oryza sativa*) and soil fertility. *Madras Agric. J.*, 89(7–9): 383–385.

Jackson, M.L. 1967. *Soil Chemical Analysis.* Prentice Hall Inc., Englewood Cliffs, N.J., USA.

Lal, S. 1987. The effect of long-term manuring and fertilization on crop yield and fertility level of typic Haplustalf soil. *M.Sc (Ag.) Thesis*, BAU, Ranchi.

Moller, H. 1990. Effect of lime, fertilizers on C/N ratio, urease and saccharine activity of the humus fermentation layers in Norway spruce forests of the Deister region after one to two years. *Tuercenia,* 10: 487–496.

Piper, C.S. 1950. *Soil and Plant Analysis.* The University of Adelaide, Australia.

Prasad, J. and Mathur, B.S. 1997. Influence of long-term use of fertilizers, manure and lime on phosphate adsorption parameters in acid alfisols of Ranchi. *J. Indian Soc. Soil Sci.*, 45(1): 24–27.

Chapter 50

Efficacy of Insecticides and their Combination with NSKE for the Management of Insect Pests of Blackgram

Devendra Prasad, Dharmjeet Kumar, Rabindra Prasad, Binay Kumar, Rajesh Kumar and Niraj Kumar

Department of Entomology, Birsa Agricultural University, Ranchi – 834 006, Jharkhand

ABSTRACT

Findings of the field experiment conducted in the farm of Birsa Agricultural University, Kanke, Ranchi during *kharif* 2000 revealed that all the tested insecticides were proved to be effective in reducing the incidence of insect pests of blackgram (cv. BAU-1). Imidacloprid was found to be effective in suppressing the population of leaf hopper, leaf miner and thrips to the extent of 80.64, 86.67 and 83.50 per cent, respectively at 10 days after treatment. Higher reduction in the population of Bihar hairy caterpillar was obtained by the application of dichlorvos (81.83 per cent), chlorpyriphos (81.25 per cent) and lambda cyhalothrin (80.96 per cent). Chlorpyriphos, dichlorvos and lambda cyhalothrin were found to be equally effective in minimizing the pod damage as well as achieving higher grain yields resulted 193.74, 181.54 and 194.10 per cent more yields, respectively over untreated control. Considering the values of additional yields, cost of insecticides and their applications, lambda cyhalothrin, dichlorvos and chlorpyriphos were proved to be economical in achieving the higher benefit cost ratios of 9.33 : 1, 9.01 : 1 and 8.13 : 1, respectively. Due to higher cost of imidacloprid, the benefit cost ratio was reduced to the level of 5.88 : 1 which was nearer to that of NSKE (5.18 : 1).

Keywords: *Blackgram, Insect pests, Efficacy, Insecticides, NSKE, B : C ratio.*

Introduction

Blackgram [*Vigna mungo* (L.) Hepper] is one of the important pulse crop in India. The crop is heavily damaged by number of insect pests in the field. Dhuri and Singh (1983) recorded 30 insect pests infesting at different stages of the crop growth. Hameed and Durbey (1983) reported from Bihar that avoidable losses in yield due to pest complex in urdbean was found to be 23 per cent, while Rawat and Kaushik (1983) reported that *Spodoptera exigua* and *Diacrisia obliqua* could be responsible for reducing the yield upto 60 per cent in Madhya Pradesh. Considering the ill effects and utilities of chemical insecticides in reducing the yield loss, the present investigation was initiated to evaluate the efficacy of insecticide alone as well as reduced dose of insecticides combined with NSKE for the management of insect pests of blackgram.

Materials and Methods

A field experiment was conducted in a randomized block design having nine treatments and three replications in the plots of 4 m × 3 m with blackgram (cv BAU-1). Seeds was sown on 18th July, 2000. Recommended package of practices were followed to raise the crop. The efficacy of insecticides were evaluated alone as well as half dose of insecticides combined with NSKE 5 per cent. Ten plants were selected randomly and tagged in each plot for recording the incidence of insect pests. Absolute population per plant were counted in the case of thrips, pupae of leaf miner, leaf hopper (nymph), semi-looper and Bihar hairy caterpillar. Number of total pods as well as damaged pods were counted at the time of maturity on ten randomly selected plants per plot for calculating the percentage of pod damage. Grain yield were also recorded plot wise and converted into kg/ha. The number of insect pests were transformed into $\sqrt{X + 0.5}$, while percentage data were transformed into angular values. Avoidable loss in yield caused by insect pests were calculated by the following formula as suggested by Pawar *et al.* (1984).

$$\text{Avoidable loss in yield} = \frac{T - C}{T} \times 100$$

where,

T = Yield in the treated plot

C = Yield in the control plot

On the basis of market price of the insecticides and grain yield of blackgram, economics were worked out.

Results and Discussion

Incidence of insect pests were shown in Table 50.1 while the economics of different insecticides were presented in Table 50.2.

Blackgram was found to be attacked by seven major insect pests *viz.*, leaf hopper, *Empoasca kerri* Pruthi; leaf miner, *Acrocercops phaeospora* Meyr., Bihar hairy caterpillar, *Spilosoma obliqua* (Walker); thrips, *Megalurothrips distalis* (Karney) and pod boring insects *viz.*, blue butterfly, *Lampides boeticus* L., *Catochrysops cnejus* (Fab.) and gram pod borer, *Helicoverpa armigera* (Hubn.).

Leaf Hopper

The lowest jassid population (1.79/plant) was obtained with the application of NSKE 5 per cent + 1/2 dose of imidacloprid at 7 DAT which remained at par with NSKE 5 per cent + 1/2 dose of

Table 50.1: Efficacy of Insecticides and their Combination with NSKE Against Insects Pests of Blackgram

Treatment	Dose (kg or lit./ha)	No. of Leaf Hopper/Plant		No. of Leaf Miner Pupae/Plant		No. of Bihar Hairy Caterpillar/Plant		No. of Thrips/Plant		Incidence of Pod borer (%)
		Days After Treatment								
		7	10	7	10	7	10	7	10	
Chlorpyriphos	0.80	2.39 (1.70)	1.19 (1.30)	1.03 (1.24)	0.79 (1.14)	1.86 (1.54)	1.29 (1.34)	2.06 (1.60)	0.86 (1.17)	5.65 (13.50)
Dichlorvos	0.40	2.19 (1.64)	1.39 (1.37)	1.96 (1.57)	1.10 (1.26)	1.46 (1.40)	1.25 (1.32)	3.50 (2.00)	2.78 (1.81)	6.23 (14.48)
Lambda cyhalothrin	0.40	1.98 (1.57)	1.19 (1.30)	0.85 (1.16)	0.58 (1.04)	2.75 (1.80)	1.31 (1.35)	2.80 (1.82)	1.67 (1.47)	6.95 (15.27)
Imidacloprid	0.08	1.80 (1.52)	0.79 (1.14)	0.50 (1.00)	0.32 (0.91)	2.06 (1.60)	2.07 (1.60)	1.91 (1.55)	0.81 (1.14)	8.33 (16.66)
NSKE 5 % + ½ Chlorpyriphos	40+0.40	2.79 (1.81)	1.58 (1.44)	1.05 (1.24)	0.90 (1.18)	2.78 (1.81)	2.07 (1.60)	2.96 (1.86)	1.91 (1.55)	7.48 (15.86)
NSKE 5% + ½ Dichlorvos	40+0.20	2.09 (1.61)	1.39 (1.37)	1.96 (1.57)	1.33 (1.35)	3.78 (2.07)	1.92 (1.56)	3.40 (1.97)	2.72 (1.79)	7.63 (16.07)
NSKE 5% + ½ Lambda cyhalothrin	40+0.20	2.09 (1.61)	1.19 (1.30)	1.77 (1.51)	0.90 (1.18)	2.26 (1.66)	1.45 (1.40)	2.92 (1.85)	1.79 (1.51)	8.41 (16.86)
NSKE 5% + ½ Imidacloprid	40+0.04	1.79 (1.51)	0.99 (1.22)	0.80 (1.14)	0.59 (1.04)	2.14 (1.62)	2.05 (1.59)	2.75 (1.80)	1.06 (1.25)	8.85 (17.26)
NSKE 5%	40	2.98 (1.87)	2.19 (1.64)	1.97 (1.57)	1.78 (1.51)	3.83 (2.08)	3.54 (2.01)	3.80 (2.07)	3.32 (1.95)	10.16 (18.67)
Control	–	4.19 (2.17)	4.08 (2.14)	2.60 (1.76)	2.40 (1.70)	7.50 (2.83)	6.88 (2.72)	6.75 (2.69)	4.91 (2.33)	22.83 (28.47)
SEm (±)	–	0.06	0.07	0.06	0.03	0.07	0.04	0.08	0.02	0.78
CD (P = 0.05)	–	0.18	0.20	0.17	0.09	0.23	0.12	0.24	0.07	2.33
CV (%)	–	6.17	8.03	7.90	4.68	7.26	4.42	6.06	4.87	7.84

Number of insects were transformed into $\sqrt{x + 0.5}$.

Percentage of pod damage were transformed into angular values.

Figures in parentheses correspond to transformed values.

Table 50.2: Economics of Different Insecticidal Treatment Used Against Insect Pests Infesting Blackgram

Treatments	*Dose (kg or lit/ha)*	*Grain Yield (kg/ha)*	*Additional Yield Over Controt (kg/ha)*	*Avoidable Loss in Yield (%)*	*Value of Additional Yield (Rs. 9.00/ kg)*	*Cost of Pest Control Operation*			*Net Monetary Return*	*Benefit Cost Ratio (BCR)*
						Cost of Insecticide (Rs.)	*Cost of Labour (Rs.)*	*Total Cost (Rs.)*		
Chlorpyripho	0.80	1185.06	781.62	65.00	7034.58	520.00	250.00	770.00	6264.58	8.13
Dichlorvos	0.40	1135.67	732.23	64.47	6590.07	408.00	250.00	658.00	5932.07	9.01
Lambda cyhalothrin	0.40	1185.50	783.06	66.00	7047.54	432.00	250.00	682.00	6365.54	9.33
Imidacloprid	0.08	1130.70	727.26	64.32	6545.34	700.80	250.00	950.80	5594.54	5.88
NSKE 5% + ½ Chlorpyriphos	40+0.40	1021.80	618.36	60.52	5565.24	820.00	250.00	1070.00	4495.24	4.20
NSKE 5% + ½ Dichlorvos	40+0.20	916.78	513.34	55.99	4620.06	764.00	250.00	1014.00	3606.06	3.56
NSKE 5% + ½ Lambda cyhalothrin	40+0.20	1020.85	617.41	60.48	5556.69	776.00	250.00	1026.00	4530.69	4.41
NSKE 5% + ½ Imidacloprid	40+0.04	986.67	583.23	59.11	5249.07	910.00	250.00	1160.40	4088.67	3.52
NSKE 5%	40	959.46	556.02	57.95	5004.18	560.00	250.00	810.00	4194.18	5.18
Control	–	403.44	–	–	–	–	–	–	–	–
S.E.m (±)	–	31.72	–	–	–	–	–	–	–	–
CD (P = 0.05)	–	94.24	–	–	–	–	–	–	–	–
CV (%)	–	5.52	–	–	–	–	–	–	–	–

lambda cyhalothrin, NSKE 5 per cent + 1/2 dose of dichlorvos, imidacloprid, lambda cyhalothrin and dichlorvos. Imidacloprid was found to be the best treatment recording lowest incidence of leaf hopper (0.79/plant) at 10 DAT which was not different from the populations obtained by the application of chlorpyriphos, lambda cyhalothrin, NSKE 5 per cent + 1/2 dose of imidacloprid and NSKE 5 per cent + 1/2 dose of lambda cyhalothrin.

Leaf Miner

All the insecticides were found to be better than untreated control. The lowest population of the pest was recorded by the application of imidacloprid at 7 and 10 DAT. However, lambda cyhalothrin and NSKE 5 per cent + 1/2 dose of imidacloprid were found to be as good as imidacloprid alone at 7 DAT.

Bihar Hairy Caterpillar

The highest larval population of the pest was recorded in untreated control. The lowest larval population of *S. obliqua* (1.46 larvae/plant) was obtained with the application of dichlorvos which remained at par with chlorpyriphos (1.86/plant), imidacloprid (2.06/plant) and NSKE 5 per cent + 1/2 dose of imidacloprid (2.14/plant) at 7 DAT. Chlorpyriphos, dichlorvos, lambda cyhalothrin and NSKE 5 per cent + 1/2 dose of lambda cyhalothrin proved to be equally effective in controlling the larval population of *S. obliqua* at 10 DAT.

Thrips

Imidacloprid proved to be the best insecticide recording lowest thrips population at 7 and 10 DAT which remained at par with the population obtained by the application of chlorpyriphos. The performance of NSKE 5 per cent + 1/2 dose imidacloprid was also found to be better in reducing the incidence of the pest.

Pod Damage

The highest pod damage of 22.83 per cent caused by *Catochrysops cnejus, Lampides boeticus* was recorded in untreated control. The lowest incidence (5.65 per cent) of the pests was recorded with the application of chlorpyriphos which remained at par with the incidence obtained by the application of dichlorvos and lambda cyhalothrin.

Grain Yield

Lambda cyhalothrin, chlorpyriphos, dichlorvos and imidacloprid were found to be equally effective in producing the higher grain yields of 1186.50, 1185.06, 1135.67 and 1130.70 kg/ha, respectively (Table 50.2) as against lowest yield (403.44 kg/ha) was obtained in untreated control. The yield loss ranging between 55.99 to 66.00 per cent could be protected by the application of insecticides resulted 127.24 (NSKE 5 per cent + 1/2 dichlorvos) to 194.09 per cent (lambda cyhalothrin) more yields over untreated control.

Benefit Cost Ratio

On the basis of values of additional grain yield of blackgram, cost of IPM modules and their applications, lambda cyhalothrin was found to be economical in achieving highest benefit cost ratio (9.33 : 1) closely followed by dichlorvos (9.01 : 1) and chlorpyriphos (8.13 : 1). Due to the highest cost of imidacloprid, only 5.88 : 1 benefit cost ratio was obtained which was very nearer to NSKE 5 per cent (5.18 : 1).

As such, foregoing discussion amply demonstrated that all the treatments were proved to be better in reducing the incidence of insect pests as compared to untreated control. Imidacloprid was proved to be effective insecticide in reducing the population of leaf hopper, leaf miner and thrips to the extent of 80.64, 86.67 and 83.50 per cent, respectively at 10 DAT. Higher reductions in the population of Bihar hairy caterpillar was obtained with the application of dichlorvos (81.83 per cent), chlorpyriphos (81.25 per cent) and lambda cyhalothrin (80.96 per cent). Lambda cyhalothrin was also responsible for suppressing the population of leaf hopper, leaf miner and thrips to the extent of 70.83, 75.83 and 66.0 per cent, respectively. Chlorpyriphos was also found to be effective in reducing the thrips population to the extent of 82.48 per cent. Higher reductions in the population of leaf hopper (75.74 per cent), leaf miner (75.42 per cent), Bihar hairy caterpillar (70.20 per cent) and thrips (78.41 per cent) were obtained by the application of NSKE 5 per cent + 1/2 dose of imidacloprid followed by NSKE 5 per cent + 1/2 lambda cyhalothrin where 70.83, 62.50, 78.92 and 63.54 per cent reductions of respective insects were obtained. Chlorpyriphos, dichlorvos and lambda cyhalothrin were found to be equally effective insecticides in reducing the pod damage as well as achieving the higher grain yields resulting 193.74, 181.54 and 194.10 per cent more yields, respectively. Considering the value of additional yields, costs of insecticides and their application, lambda cyhalothrin, dichlorvos and chlorpyriphos were found to be economical in achieving higher benefit cost ratios of 9.33 : 1, 9.01 : 1 and 8.13 : 1, respectively. Due to high cost of imidacloprid, the benefit cost ratio was reduced to the level of 5.88 : 1 which was very nearer to B : C ratio obtained by NSKE (5.18 : 1). Benefit cost ratios of only 4.41 : 1, 4.20 : 1, 3.56 : 1 and 3.52 : 1 were obtained by the application of NSKE 5 per cent combined with lambda cyhalothrin, chlorpyriphos, dichlorvos and imidacloprid, respectively due to extra cost of synthetic insecticides.

Ujagir and Sehgal (1997) advocated that chemical control is the only method which is being used to manage the pests of *Vigna* group. Chander and Singh (1989) indicated that spraying with monocrotophos (0.04 per cent), dimethoate (0.03 per cent) and chlorpyriphos (0.05 per cent) at flowering reduced the level of infestation of pod borers and also effective against galerucid beetle for about 10 days. Further, Chander and Singh (1991) also reported that seed treatment and foliar spray with monocrotophos and dimethoate both proved effective against jassid for 12 and 9 days, respectively during *kharif* season. Nath and Vein (1996) obtained effective control of *M. testulalis* in blackgram by different insecticides including chlorpyriphos and dichlorvos. The findings of the present investigation is also supported by the results obtained by Khutwad *et al.* (2002) who obtained lower populations of aphids and jassids by foliar application of imidacloprid (200 SL) 0.02 per cent. Venkateswarlu *et al.* (1992) reported that Neem oil (1.0 per cent) in combination with endosulfan (0.07 per cent) was found to be the most effective in reducing the infestation due to pod borer complex and increasing the grain yield. More recently Lakshmi *et al.* (2002) indicated that neem (Neemazal) combined with chlorpyriphos at half of their original concentrations was not found effective against *M. vitrata* on 2nd day after spraying. However, its moderate efficacy was noticed by 8th day after spraying.

References

Chander, S. and Singh, Y. 1989. Control of thrips, *Megaleurothrips distalis* Karney and pod borer, *Euchrysops cnejus* Fab. and *Maruca testulalis* Geyer infesting green gram, *Vigna raidata* L. *Indian J. Ent.*, 51(2): 131–135.

Chander, S. and Singh, Y. 1991. Chemical control of stem fly, *Ophiomyia phaseoli* (Tryon.) and jassid, *Empoasca kerri* Pruthi, infesting green gram, *Vigna radiata* L. *Indian J. Ent.*, 53(2): 252–256.

Dhuri, A.V. and Singh, K.M. 1983. Pest complex and succession of insect pests in blackgram, *Vigna mungo* (L.) Hepper. *Indian J. Ent.*, 45(4): 396–401.

Hameed, S.F. and Durbey, S.L. 1983. Losses due to insect pests in North Bihar. In *Proc. Nat. Sem. Crop Losses Due to Insect Pests.* Special issue, *Indian J. Ent.*, 1: 136–146.

Khutwad, D.S., Nakat, R.V. and Chavan, B.P. 2002. Seed dressers and foliar sprays on sucking pests of green gram, *Vigna radiata* (L.) Willczek. *Pestology*, 26(7): 57–59.

Lakshmi, P.S.P., Rajasekhar, P. and Rao, R.V. 2002. Bioefficacy of certain insecticides against spotted pod borer on urdbean. *Indian J. Pulses Res.*, 15(2): 201–202.

Nath, H.K. and Yein, B.R. 1996. Efficacy of some insecticides against spotted pod borer, (*Maruca testulalis* Geyer) on blackgram. *J. Agric. Sci. Soc., North-East India*, 9(1): 107–109.

Pawar, V.M., Shrishikar, S.P. and Jadhav, G.D. 1984. Relative efficacy and economics of acephate (orthene 75 WP) for cotton bollworm. *Indian J. Pl. Prot.*, 12: 43–47.

Rawat, R.R. and Kaushik, U.K. 1983. Crop losses due to insect pests in the North-east Madhya Pradesh. In: *Proc. Nat. Sem. Crop Losses Due to Insect Pests.* Special issue, *Indian J. Ent.*, 1: 73–83.

Ujagir, R. and Sehgal, V.K. 1997. Insect pest management in pigeonpea and Vigna crops. In: *Recent Advances in Insect Pest Management in Major Crops*, (Eds.), Sehgal, V.K., Pathak, P.K., Sachan, G.C. and Kanaujia K.R., Department of Entomology, G.B. Pant Univ. of Agric. and Tech., Pantnagar, pp. 90.

Venkateshwarlu, P., Raghavaiah, G. and Nagalingum, B. 1992. Efficacy of neem oil alone and in combination with certain insecticides on pod borer complex of blackgram, *Vigna mungo* (L.) Hepper. *New Agriculturist*, 3(2): 147–152.

Chapter 51

Physiological Studies on New Plant Types Originating from Tropical Japonicas in Rice (*Oryza Sativa* L.)

P.R. Rao and B. Mishra***

**Scientist Plant Physiology, **Project Director,*
Directorate of Rice Research, Rajendranagar – 500 030, A.P., India

ABSTRACT

In this study five new plants types (NPTs) received from IRRI were compared with four high yielding varieties (HYVs) and one hybrid to analyze their growth and productivity during *kharif* 2003. The results revealed that among the cultivars studied hybrid (KRH-2) recorded maximum plant height (126 cm), maximum number of tillers (525 tillers/m^2) at 30 days after transplanting, maximum chlorophyll (40.3 as measure through Spad 502). When growth analysis done it was found that hybrid KRH-2 had CGR (23.34 gm/m^2/day), LAI (5.29), LAD (131 d/m^2) and NAR (0.2318 g/m^2/day). Hybrid KRH-2 produced maximum dry matter of 1568 g/m^2 as compared to NPTs and HYVs. The yield components *viz.*, grain no/m^2 and grain no/panicle was highest in KRH-2 as a result recorded highest yield of 816 g/m^2 compared to NPTs and HYVs, where in the experimental mean yield was 656 g/m^2. However, new plant types did show on par results with high yielding varieties for the above parameters even though hybrid showed superiority in terms of yield and other characters.

Keywords: *Rice, New Plant Types (NPTs), High yielding varieties (HYVs), Hybrid (KRH-2) and Growth analysis.*

Introduction

Rice (*Oryza Sativa* L.) is widely grown all over the world and staple food for more than half of the human race. It is grown in India under 44.62 m.ha with an annual production of 93.08 m.t. (http://agricoop.nic.in/statistica, 2003). However, stagnation in yield is the question of the time. Work is under progress at the rice research institutions all over the world to cross the yield barrier. Khush (1995) proposed a new plant type (NPT) having tropical Japonica back ground with high yield potential. These NPTs have low tillering capacity, no unproductive tillers, 200 to 250 grains per panicle, height of 90 to 100 cms, very sturdy stems, dark-green, thick and erect leaves, vigorous root system, growth duration of 100 to 130 days, multiple disease and insect resistance and acceptable grain quality. Most of these attributes were found in tropical Japonica germplasm (IRRI, 1992). Attempts are being made to cross them with indicas to improve the grain quality from long bold to long slender grains. Efforts are going on in this direction to improve these NPTs before making it a reality. In this study we have included 5 such developed lines and compared them with 4 popular high yielding varieties in India along with an Hybrid. In the present investigation, attempts have been made to compare the physiological differences among them and understand their growth and yield potential.

Materials and Methods

A field experiment was carried out during *Kharif* and *Rabi* 2003–2004 at Directorate of Rice Research, Rajendranagar. The soil was alkaline in reaction (pH 8.2), low in available nitrogen (217 kg/ha), medium in available Phosphorous (25.52 kg/ha) and Potassium (217.46 kg/ha). The experiment was laid out in RBD with four replications. The main treatment consisted of 5 NPTs, 4 HYVs and 1 Hybrid. Fertilizer dose was applied @ 120 : 60 : 40 (NPK kg/ha), Nitrogen was applied in 3 split doses initially after 10–15 days after planting thereafter at active tillering and PI stage. The spacing that was followed was 20 × 10 cms and recommended package of practices were followed.

The chlorophyll was measured using SPAD-502 Minolta company, the SPAD values less than 35 indicate need for topdressing with Nitrogen, it has been introduced in India for rice growers through CREMNET (Crop and Resource Management Network) is IRRI program. A close correlation exists between leaf chlorophyll concentration, leaf nitrogen and chlorophyll meter readings. The leaf area was measured by delta T image analysis. Growth analysis was done by using the formulae suggested by Redford (1967). Data was recorded for dry matter analysis at 30, 60, 90 days of transplanting and at maturity to record yield and yield components. The data was analyzed statistically.

Results and Discussion

Growth in terms of plant height increased rapidly up to 90 days of transplanting and later the increase was not much in all the cultivars (Table 51.1). The cultivars differed significantly in plant height at different stages of growth. KRH-2 recorded maximum plant height (126 cm) compared to NPTs and HYVs. This may be due to first generation hybrid vigour of the plant compared to other cultivars. Although it is not quiet proper to relate morphological characters directly to the yield, may early workers strongly feel that these may operate indirectly in improving the yield considering the plant height? It has been pointed out that this character was more in high yielding cultivars. Positive correlation was observed for this character with yield by Paramasivam and Rangaswamy (1988).

The tiller number increased up to 60 DAT and decreased marginally at 90 DAT. At 30 DAT KRH-2 had 525 tillers/m^2 and at 90 DAT it had 420 tillers/m^2. The difference in tiller production among cultivars may be attributed to varietal character. Hybrids have higher tillering ability than conventional varieties Chen *et al.*, 1989. Leaf area index gradually increased from 30 DAT to 60 DAT and later it decreased in all the cultivars at 90 DAT. NPT 21 had the high leaf area index 6.2 has compared to HYVs and Hybrid (5.3).

Table 51.1: Growth Characters of Rice NPTs, HYVs and Hybrid

Cultivars	Plant Height (cm) Days after transplanting			No. of Tillers/m^2 Days after transplanting		
	30	60	90	30	60	90
NPT18	47	74	102	377	408	335
NPT19	48	72	104	377	400	385
NPT20	45	70	109	485	583	440
NPT21	50	81	117	482	465	445
NPT22	48	82	111	467	493	412
IR-64	44	73	108	470	465	457
JAYA	43	73	108	500	438	390
NDR359	50	82	123	415	445	342
RASI	43	82	189	507	545	542
KRH-2	52	87	126	525	450	420
Ex. Mean	47.3	78	110.6	460	465	417
CD (0.5%)	4.9	5.4	4.36	88.5	102.6	89.2
CV (%)	7.3	4.8	2.7	13.3	15.2	14.8

NPT18: IR 7385-33-1-3-2; NPT19: IR 73896-51-2-1-3; NPT20: IR 73907-53-3-2-2; NPT21: IR 73930-31-3-2-2 NPT22: IR 73930-41-5-3-1.

The decrease of leaf area towards maturity may be due to lesser leaf number as a result of senescence in early-formed leaves. Lue *et al.* (1991) reported that the LAI of hybrid rice increased rapidly than conventional varieties, which was not true in our case. HYV NDR-359 had more (LAI) compared to hybrid and NPT 21 had the highest LAI among the ten cultivars taken up for the study (Table 51.2). Chlorophyll content of all the cultivars was above 35 at 30 DAT thereafter it decreased at 60 and 90 DAT. However, it was found that some of NPTs and hybrid maintained good chlorophyll content even at 90 DAT as compared to HYVs, which could have resulted in greater photosynthetic efficiency and production of dry matter (Tables 51.2 and 51.3). At 90 DAT it was found that NPTs had greater dry matter accumulation as compared to HYVs and Hybrids.

The dry matter accumulation in leaves increased up to 90 DAT and later decreased gradually towards maturity. The varietal difference in the leaf dry matter may be ascribed to difference in leaf size, leaf number and thickness. The gradual fall in leaf weight during subsequent growth stages may perhaps be due to the translocation of photosynthetic to the developing grain and respiration. In rice, all the leaves from the flag leaf down to third leaf export assimilate to panicle (Tanaka, 1958).

Total dry matter that can be produced and its further translocation to sink is the major factor that governs the economic yield of a crop variety. Even though NPT 21 produced greater dry matter than KRH-2 hybrid, KRH-2 had greater yield than NPT 21 primarily because of poor portioning to the grains. This suggests that KRH-2 had a good portioning ability to wards the reproductive structure. The present study confirms the views of Fisher and Turner (1998) that the genotype exerts major control over portioning.

Table 51.2: Growth Characters of Rice NPTs, HYVs and Hybrid

Cultivars	*Leaf Area Index Days after transplanting*			*Chlorophyll Days after transplanting*		
	30	*60*	*90*	*30*	*60*	*90*
NPT18	1.5	4.3	3.5	39.0	35.7	31.4
NPT19	1.5	4.4	3.6	39.4	38.0	32.2
NPT20	1.6	5.0	3.5	36.5	38.2	34.7
NPT21	1.6	6.2	4.5	34.7	36.4	34.6
NPT22	1.7	5.6	3.9	33.4	37.2	30.7
IR-64	1.4	4.6	3.0	39.5	37.9	31.6
JAYA	1.6	4.5	3.9	37.5	34.6	30.6
NDR359	1.3	5.5	4.8	38.5	34.9	28.2
RASI	1.4	3.3	2.3	38.3	39.3	23.2
KRH-2	1.6	5.3	3.4	40.3	39.1	33.3
Ex. Mean	1.6	4.9	3.6	37.7	37.1	31.0
CD (0.5%)	NS	0.7	0.8	3.5	2.4	3.5
CV (%)	11.8	11.1	16.4	6.4	4.6	7.9

NPT18: IR 7385-33-1-3-2; NPT19: IR 73896-51-2-1-3; NPT20: IR 73907-53-3-2-2; NPT21: IR 73930-31-3-2-2 NPT22: IR 73930-41-5-3-1.

Table 51.3: Dry Matter Accumulation in Various Plant Parts of Rice NPTs, HYVs and Hybrid

Cultivars	*30 Days After Transplanting (g/m²)*				*90 Days After Transplanting (g/m²)*				
	Root	*Leaf*	*Stem*	*Totat*	*Root*	*Leaf*	*Stem*	*Panicle*	*Total*
NPT18	31	58	80	138	55	208	627	262	1096
NPT19	33	62	97	160	52	220	683	520	1424
NPT20	35	63	116	179	57	181	630	427	1222
NPT21	16	59	80	139	24	230	665	340	1230
NPT22	31	58	74	133	31	204	621	385	1206
IR-64	28	57	89	146	43	161	429	383	975
JAYA	19	61	92	154	25	220	536	343	1100
NDR359	26	57	84	141	45	265	758	288	1312
RASI	32	63	81	144	33	121	422	564	1108
KRH-2	33	60	92	152	41	1.84	618	413	1216
Ex. Mean	28.5	60	89.1	149.1	40	199.8	597.6	392	1189.3
CD (0.5%)	10.6	NS	NS	NS	12.5	40.9	113.9	112	207.7
CV (%)	25.8	14.0	19.4	15.9	21.3	14.1	13.2	19.8	12.1

NPT18: IR 7385-33-1-3-2; NPT19: IR 73896-51-2-1-3; NPT20: IR 73907-53-3-2-2; NPT21: IR 73930-31-3-2-2 NPT22: IR 73930-41-5-3-1.

The crop growth rates gradually increased in all the cultivars up to 60 days and thereafter decreased when it came to 90 days (Table 51.4). At 30 DAT the CGR was maximum in case of KRH-2 (23.3) while it decreased to 12.1 at 90 DAT. At 90 DAT NPT 19 retained a high CGR of 23.7 as compared to all other varieties. Relative growth rates when measured it was found that at 30–60 DAT there was not much difference among the cultivars *i.e.*, NPTs, HYVs and Hybrid had similar RGR and RGR decreased in all the cultivars at 60–90 DAT and in some cases even negative.

However, RGR of hybrid KRH-2 was 0.0078 even at 90 DAT has compared to other cultivars. Net assimilation rate was highest in NDR-359 (HYV) at 30–60 DAT but KRH-2 maintained a constant higher NAR 0.2318 up to 90 DAT. RGR and NAR decreased with the age of crop, similar decrease with the age of the crop was reported by Sestak (1971) in other crops. Progressive increase in leaf area duration (LAD) was observed in all the cultivars up to 60–90 days and thereafter decreased. Maximum LAD was reported by NPT21 (161) while the hybrid KRH-2 reported 131 (Table 51.4). A significant positive correlation between LAD during post-flowering and yield was reported by Nighwan and Chandra (1980).

Table 51.4: Growth Analysis of NPTs, HYVs and Hybrid

Period (Days)	*Cultivars*										
	NPT18	*NPT19*	*NPT20*	*NPT21*	*NPT22*	*IR64*	*Jaya*	*NDR359*	*Rasi*	*KRH-2*	*CD at 0.5%*
	Crop Growth Rate (g/m²/day)										
30–60	17.20	18.42	19.37	20.84	19.47	19.98	16.91	91.18	17.32	23.34	1.57
60–90	14.73	23.73	15.39	15.53	16.31	7.64	14.62	19.86	14.79	12.13	1.50
	Relative Growth Rate (g/m²/day)										
30–60	0.0404	0.0407	0.0403	0.0491	0.0470	0.0421	0.0408	0.0483	0.0344	0.0453	0.013
60–90	0.0023	0.0013	0.0053	0.0038	0.0054	0.0075	0.0018	0.0030	0.0125	0.0078	0.025
	Net Assimilation Rate (g/m²/day)										
30–60	0.1598	0.1777	0.1839	0.2941	0.2386	0.1907	0.1667	0.3538	0.1012	0.2318	0.06
60–90	0.0033	0.0018	0.0120	0.0104	0.0142	0.0186	0.0018	0.0034	0.0221	0.0238	0.05
	Leaf Area Duration (d/m²)										
30–60	88	89	101	119	110	91	93	98	72	103	3.5
60–90	118	120	129	161	143	115	128	155	83	131	3.3

NPT18: IR 7385-33-1-3-2; NPT19: IR 73896-51-2-1-3; NPT20: IR 73907-53-3-2-2; NPT21: IR 73930-31-3-2-2 NPT22: IR 73930-41-5-3-1.

The maximum grain yield was obtained in KRH-2 816 (g/m^2) followed by NPT 20 (672 g/m^2) (Table 51.5). The rest of the cultivars where around the mean grain yield of all the cultivars (656 g/m^2). The yield differences in the ten cultivars were due to the significant difference in their yield components. At maturity KRH-2 hybrid had maximum dry matter and maximum yield as compared to NPTs and HYVs. The high yield of KRH-2 was attributed primarily because of high panicles per m^2 and number of grains per panicle. NDR-359 had high 1000-grain weight and high harvest index was found in Rasi. From the above study it can be concluded that hybrid KRH-2 performed better in terms of efficient portioning of assimilates to grains and grain number per panicle was maximum in KRH-2, which is a genetic character of the cultivar. At the same time observe that NPTs have some good physiological characters and they are not inferior to HYVs and they can be improved further to surpass the yield when compared to Hybrids.

Table 51.5: Yield and Yield Attributes of Rice NPTS, HYVs and Hybrid

Cultivars	*Grain Yield (g/m²)*	*Biological Yield (g/m²)*	*Panicles (m²)*	*Grain No./m² (10³)*	*Grain No./ Panicle*	*1000 gr. wt (g)*	*Harvest Index (%)*
NPTI8	635	1247	300	24.4	82	25.98	50.6
NPT19	668	1335	273	25.2	93	26.53	49.9
NPT20	672	1358	399	25.8	65	26.15	49.4
NPT21	623	1391	404	26.8	68	13.17	44.8
NPT22	661	1345	375	27.4	74	24.17	49.1
IR-64	656	1258	418	25.6	6.2	25.60	52.1
JAYA	546	1220	348	23.6	68	23.30	44.7
NDR359	625	1490	313	21.1	68	29.65	41.7
RASI	660	1264	439	30.2	69	21.85	52.3
KRH-2	816	1568	366	37.3	102	21.90	51.9
Ex. Mean	656	1347	363	26.7	75	24.83	48.6
CD (0.5%)	114.8	173.8	62.86	4.77	15.58	1.27	4.19
CV (%)	12.1	8.9	11.9	12.3	14.3	3.5	5.9

NPT18: IR 7385-33-1-3-2; NPT19: IR 73896-51-2-1-3; NPT20: IR 73907-53-3-2-2; NPT21: IR 73930-31-3-2-2 NPT22: IR 73930-41-5-3-1.

Acknowledgements

I am highly thankful to Dr. Parminder Virk, IRRI, Philippines for sending the seed material, Dr. B. Mishra, Project Director for allowing me to conduct these studies. I also thank Shri G.V.S. Siva Prasad and K. Ramulu, technical assistants, K. Yadaiah field worker for their technical and field work assistance.

References

Chen, O.H. and Liv, G.L. 1980. *The Characteristics of N Nutrition of Hybrid Rice*. Acta Agriculture University China, 8: 1–9.

Fisher, R.A. and Turner, C. 1978. Plant productivity in the arid and semi-arid zones. *Annual Review Plant Physiol.*, 29: 277–317.

http://agri.coop.nic.in/statistica. 2003.

IRRI. 1992. *Program Report for 1991*. Manila (Philippines). International Rice Research Institute.

Khush, G.S. 1995. Breaking the yield frontier of rice. *Geo J.*, 35(3): 329–332.

Paramasivam, K.S. and Rangaswamy, S.R.S. 1988. Genetic analysis of yield and its components in rice. *Oryza*, 25: 111–119.

Redford, P.J. 1967. Growth Analysis formulae their use and abuse. *Crop. Sci.*, 7: 171–175.

Tanaka. 1958. Studies on the characteristics of the physiological function of leaf of definite position on stem of the rice plant. Part II. Comparison of the photosynthetic activity of leaves at various positions on main stem and translocation of photosynthetic products within the plant. *J. Sci. Soil Manures, Japan*, 19: 327–333

Chapter 52

Effect of Planting Methods and Irrigation Levels on Water Use of Maize (*Zea mays*, L.)

Tarundeep Kaur and R.K. Mahey

Department of Agronomy and Agrometeorology, Punjab Agricultural University, Ludhiana – 141 004, India

ABSTRACT

The field study was conducted on maize at Punjab Agricultural University, Ludhiana to investigate the irrigation requirement of early (end May) sown maize. The trial was laid out in split plot design with seven planting methods *viz.*, flat, ridge, trench, trench-1R, trench-2R, bed-1R, bed-2R in main plots and three irrigation levels of I_{50}, I_{65} and I_{80} of CPE values in sub-plots. The results revealed that the flat planted crop resulted in highest water use and water expense followed by trench and bed methods. WUE and WEE by the crop was higher in bed-1R followed by trench, trench-1R and ridge planting methods. Flat method resulted in minimum WUE and WEE. Maximum water was used by I_{50} treatment followed by I_{65} and I_{80} while WUE was also highest under I_{50} and I_{80} treatments. Similarly water expense and WEE was higher under I_{50} treatment. Irrigation levels did not effect the grain yield significantly however the grain yield was significantly higher in bed-1R which was at par with trench, trench-1R and ridge but significantly higher than all other methods.

Keywords: *Maize, Trench/bed planting, Flat planting, Irrigation, Grain yield and Water use.*

Introduction

The growing season of *kharif* maize coincides with a period of high evaporative demand in the Punjab and its sowing in end May further puts stress on the already depleting under-groundwater. The planting of maize in *kharif* also experiences high rainfall during monsoon season which of them results in temporary flooding on flat land planted crop. Thus there is a need to try various planting methods in order to reduce its water requirement when sown in end of May and also to save it from temporary flooding damage caused due to monsoon rains. Most of the work on irrigation is based on either critical growth stages or soil moisture depletion without incorporating climatic parameters. IW : CPE ratio approach for scheduling irrigation takes these parameters into consideration.

Irrigation application is one of the most important factors influencing the yield of any crop and accounts for about 50 per cent of total energy invested in the agricultural sector (Singh *et al.*, 1996). Moreover water table is receding at an alarming rate of about 23 cm per annum in the state because of marked increase in the acreage under rice raised with tubewell irrigation (Gupta *et al.*, 1995). So the efficient use of irrigation water is a basic requirement to attain a higher level of production per unit of water. On the one hand frequent irrigations and on the other hand in efficient use of water can result in wastage of water and drying of soil profile before subsequent irrigation and thus both these situations can result in significant yield reduction. Trench sowing may also help in saving irrigation water and its application time in comparison to flood irrigation in flat sown crop. Keeping this in view, the present study was conducted.

Material and Methods

The experiment was conducted during *kharif* 2001 at Student's Research Farm, Department of Agronomy and Agrometeorology, Punjab Agricultural University, Ludhiana, Ludhiana. The trial was laid out in split plot design with seven planting methods *viz.*, flat planting 60 cm apart, ridge planting 60 cm apart (1 row), trench 60 cm apart (1 row), trench planting 67.5 cm apart (1 row and 2 rows), bed planting 67.5 cm apart (1 row and 2 rows) in main plots under three irrigation levels of 50 mm, 65 mm and 80 mm of CPE values in sub-plots. The treatments were replicated four times. The experimental field was normal with respect to pH and EC, low in organic carbon and available N and medium in available P and K.

The crop was applied one common establishment irrigation on 1st June, 2001. One life saving irrigation was also given to the whole field on 7th August, 2001. The depth of each irrigation in flat sown crop was 7.5 cm whereas in trenches and bed planted crop it was 5.0 cm. The stream of irrigation water applied to the crop was measured with Parshall flume. Single irrigation was applied to each three treatments of irrigation levels as well distributed and sufficient rainfall of 846.9 mm was received during the crop season.

The water use for the crop season under different treatments were calculated by the formula described by Singh *et al.* (1960). Water use efficiency (WUE), water expense, water expense efficiency and wetted area were calculated.

Results and Discussion

Water Use and Water Expense

The data on water use by the crop during the crop season showed that it varied with different planting methods and irrigation levels (Table 52.1). Maximum water use was obtained in flat method of planting which was higher than all the planting methods. The minimum water use was obtained in bed-1R planted treatment. The somewhat less water use values in bed and trench plantings as compared to flat might be due to less irrigation depth of 5 cm applied under these treatments as compared to flat planting where it was 7.5 cm and also more evaporation losses in flat treatment due to more exposed area. The crop that received irrigation at 50 mm CPE (I_{50}) used maximum water which was higher than 65 mm (I_{65}) and 80 mm (I_{80}) CPE irrigation treatments, because the differential irrigation in I_{50} treatment was given 12 days after sowing. However, the increase in water use was not so distinguishable as these three irrigation levels received only single differential irrigation during the crop season due to sufficient amount and well distribution of rainfall.

Table 52.1: Effect of Different Treatments on Grain Yield and Water Use of Maize

Treatment	*Grain Yield (q/ha)*	*Water Use (cm) (0–90 cm depth)*	*Water Use Efficiency (kg/ha/cm)*	*Total Water Expense (cm)*	*Water Expense Efficiency (kg/ha/cm)*
Planting method					
Flat	32.7	43.6	75.0	58.3	56.1
Ridge	38.9	41.3	94.1	51.0	76.3
Trench	43.0	42.1	102.2	51.4	83.7
Trench-1R	41.6	42.1	98.9	51.1	81.4
Trench-2R	33.3	42.1	79.1	51.5	64.7
Bed-1R	43.5	40.1	108.4	50.3	86.5
Bed-2R	35.8	41.4	86.6	51.2	69.9
CD (P = 0.05)	5.8	–	–	–	–
Irrigation level (CPE)					
50 mm	39.8	43.1	92.4	52.2	76.2
65 mm	37.6	42.9	87.6	51.9	72.4
80 mm	37.8	42.0	89.9	51.9	72.8
CD (P = 0.05)	NS	–	–	–	–

All interactions are non-significant.

Water expense was highest under flat planting method than under other planting methods. The lowest water expense was recorded under bed-1R planting method. It was considerably reduced under bed and trench methods. The water expense under flat planting method was 13.4, 14.1 and 15.9 per cent higher than trench, trench-1R and bed-1R planting methods. The higher water expense under flat planting by the crop was mainly due to more irrigation water applied, as the moisture depletion from the profile by the crop was more or less similar under various planting methods.

Water Use Efficiency

The maximum water use efficiency was obtained with bed-1R followed by trench, trench-1R and ridge methods (Table 52.1). Bed-1R, trench, trench-1R and ridge planting produced 44.5, 36.3, 32.9 and 25.5 q/ha more grains than flat with each cm of water used by the crop. Flat method recorded the lowest water use efficiency among all other planting methods. The highest water use efficiency was obtained with I_{50} irrigation schedule which was more than I_{65} and I_{80} treatments. The results are not so distinguishable as there was (846.9 mm) rainfall during the crop season.

Water Expense Efficiency

The data on water expense efficiency (WEE) by the crop showed that it varied with different planting methods. It ranged between 56.1 kg/ha/cm (flat) and 86.5 kg/ha/cm (bed-1R) of water. The irrigation water expense efficiencies were 54.2, 49.2, 45.1 and 36.0 per cent higher under bed-1R, trench, trench-1R and ridge respectively over flat method. This shows the greater efficiency of bed and trench planting methods. The water expense efficiency was higher under I_{50} irrigation level than I_{65} and I_{80} treatments.

Wetted Area

The wetted area was different with respect to different planting methods. In case of flat method, one hectare was assumed to be wetted, considering the whole plot area of 9.5 m × 4.05 m (38.475 m^2) as wetted. The wetted area reduced by 36.5 per cent under bed-1R and trench-1R planting as compared to flat planting (Table 52.2). Similarly 26.6 per cent less area was wetted in case of ridge and trench having 60 cm spacing as compared to flat.

Table 52.2: Effect of Planting Methods on Wetted Area

Treatment	*Wetted Area (m^2/ha)*
Planting method	
Flat	10000
Ridge	7338.8
Trench	7338.8
Trench–1R	6350.9
Trench-2R	6350.9
Bed-1R	6350.9
Bed-2R	6350.9

Irrigation water application varied from 15 to 22.5 cm under different planting methods. Irrigation through furrows (*viz.*, under trench, bed and ridge) resulted in saving of about 2.5 cm water compared to flat method. This saving in water applied through irrigation was due to 5 cm irrigation depth in furrows (ridge/bed/trench treatments) in comparison to 7.5 cm depth of flooded irrigation applied under flat method of planting. However, overall application of water through irrigation to the crop was less mainly due to incessant rains during the growing season. There was not any difference in irrigation water application under these irrigation treatments as they received only one differential irrigation. Also the total water expense under different irrigation level was not so discernible as there was heavy rainfall of 846.9 mm during the crop season.

Grain Yield

Bed-1R produced the maximum grain yield which was at par with trench, trench-1R and ridge but significantly superior than bed-2R, trench-2R and flat planting. Bed-1R, trench, trench-1R and ridge methods of planting produced 33.0, 31.5, 27.2 and 19.0 per cent higher grain yield respectively than flat method. The increase in grain yield was primarily due to higher growth and yield attributing characters. Bed planting and ridging might have created favourable conditions for the growth and development of the crop through the reduced loss of nutrients through leaching by avoiding direct contact of water with nutrients present near the plant base. Tisdall and Hodgson (1990) in Australia obtained higher yields of maize from crops grown on ridges compared with those on flat land. Joshi and Dastane (1966) and Thomas and Kaspar (1997) also observed that practice of ridging was beneficial for root development and grain yield. The grain yield of maize was not significantly influenced by different levels of irrigation due to well distributed 84.7 cm rainfall.

References

Gupta, R.D., Mahajan, G. and Goel, B.R. 1995. Availability and quality of groundwater in Punjab State. *Proc. Symposium on Water Management*, Punjab Agricultural University, Ludhiana, p. 18–42.

Joshi, M.S. and Dastane, N.G. 1966. Studies in excess water tolerance of crop plants. II. Effect of different durations of flooding at different stages of growth, under different layouts on growth, yield and quality of maize. *Indian J. Agron.*, 11: 70–79.

Singh, M., Gandhi, R.T. and Raheja, P.C. 1960. A critical review of the methods used to determine water requirements of crops and suggestions for planning future irrigation experiments in India. *Indian J. Agron.*, 4: 272–84.

Singh, S., Pannu, C.J.S., Singh, Sarjinder, Singh, I.P. and Kaur, S. 1996. Energy in Punjab Agriculture. *Annual Report*, Department of Farm Power and Machinery, Punjab Agricultural University, Ludhiana.

Thomas, A.L. and Kaspar, T.C. 1997. Maize nodal root response to time of soil ridging. *Agron. J.*, 89: 195–200.

Tisdall, J.M. and Hodgson, A.S. 1990. Ridge tillage in Australia: A review. *Soil and Tillage Res.*, 18: 127–44.

Chapter 53

The Impact of Organic Farming Practices on Fruit Quality

K. Boomiraj and A. Christopher Lourduraj

Department of Environmental Sciences, Tamil Nadu Agricultural University, Coimbatore – 614 003, T.N.

ABSTRACT

A field experiment was conducted at Tamil Nadu Agricultural University, Coimbatore during February 2003 to May 2003 with an objective to evaluate the impact of organic and inorganic sources of nutrients. Panchagavya spray and botanicals spray on the fruit quality or bhendi. The nutritional quality of bhendi fruit is decided by parameters like ascorbic acid, carbohydrate, protein and crude fibre. Significantly higher ascorbic acid content was noticed with the application of poultry manure + panchagavya (28.90 mg/100 g) (T_3), followed by neem cake + panchagavya (28.61 mg/100 g) (T_6). The carbohydrate content of fruit was not significantly influenced by the diffferent treatments, although poultry manure + panchagavya (T_3) treatment recorded numerically higher carbohydrate content. Application of poultry manure + panchagavya (T_3) (2.75 mg/100 g), (5.67 g/100 g) and neem cake + panchagavya, (2.72 mg/100 g), (5.65 g/100 g) (T_6) recorded the highest protein (T_3) (2.75 mg/100 g) and (T_6) (2.72 mg/100 g) and crude fibre content (T_3) (5.67 g/100 g) and (T_6) (5.65 g/100 g) of fruit respectively.

Keywords: *Bhendi, Ascrobic acid, Fibre, Protein, Carbohydrate, Panchagavya, Manure.*

Introduction

It is reported that the quality of the agricultural produce, particularly horticultural produce like flowers, vegetables and fruits improves when the nutrients are supplied through organic manures than in the form of fertilizer. This is because of supply of all the essential plant nutrient by the manures besides growth principles like enzyme, hormones, growth regulators etc. As a result, the metabolic function get regulated more effectively resulting in better synthesis of proximate constituents and consequent improvement in the quality of the produce (Kumaraswamy, 2002). Organic products provide lucrative business in the world market. Organically grown food is expected to fetch higher price because of its quality. Organically grown produce has got its own export values, for which a systemic research and development programme is respect of sustaining agricultural system through pure organic agriculture needs initiation (Rawat, 2002). Study on developing a package of technologies for 'organic' okra production including the use of organic manures, foliar nutrition through panchagavya spray and botanical pesticides had not been attempted earlier in Tamil Nadu. Hence, the present study was taken up.

Materials and Methods

A field experiment was conducted in a moderately deep, well drained sandy clay loam taxonomically classified under Typic ustropept, with a pH of 8.2. The soil was low (176 kg ha^{-1}), medium (19.5 kg ha^{-1}) and high (978 kg ha^{-1}) in available N, P and K, respectively. The following treatments were adopted T_1: Poultry manure alone; T_2: Neem cake alone; T_3: Poultry manure + Panchagavya spray; T_4: Poultry manure + Herbal leaf extract spray; T_5: Poultry manure + Neem formulation (TNAU NO 60 EC) spray; T_6: Neem cake + Panchagavya spray; T_7: Neem cake + Herbal leaf extract spray; T_8: Neem cake + Neem formulation (TNAU NO 60 EC) spray; T_9: Inorganic fertilizer (40 : 50 : 30 NPK kg ha^{-1}) + monocrotophos spray. The organic manures were analysed for their nutrient content and the quantity of organic manures to be applied was decided based-on N equivalent ratio. Thus, in the trial 2 tonnes ha^{-1} of poultry manure and 870 kg ha^{-1} of neem cake were applied in different treatments. Bhendi variety, Arka Anamika seeds (8 kg ha^{-1}) treated with *Azospirillum* (2 kg ha^{-1} of seeds) was sown at 60 cm × 30 cm spacing and seeds were placed at the height of 2/3rd of ridges height from top. Recommended dose of 40 : 50 : 30 kg N : P : K kg ha^{-1} was applied in the form of urea, single super phosphate and murate of potash for inorganic treatment (T_9) and in the form of neem cake and poultry manure for organic treatments (T_1–T_8) at nitrogen equivalent basis. The experimental field was irrigated with good quality water (pH 7.4 and EC 0.58 dSm^{-1}) from a bore-well situated near by the field. All other recommended packages of practices for bhendi crop (CPG, 1999) were followed.

Foliar spray of panchagavya 3 per cent, herbal leaf extract 10 per cent and TNAU NO 60 EC 3 per cent (neem formulation) 3 per cent were given on 30, 45, 60, 75 days after sowing (DAS) as per treatments for pest control. For inorganic treatment (T_9), monocrotophos was sprayed at the same interval. The sowing was taken on Feb 17th. The bhendi fruits at vegetable maturity were harvested at two days interval from 55 DAS and upto May 31st.

Quality Parameters

Ascorbic Acid Content

Ascorbic acid in the fruit was determined by 2,6 dichlorophenol indophenol dye titration method (Freed, 1966) and expressed in mg per 100 g.

Crude Protein Content

Protein content in the fruit was estimated as per the procedure suggested by Lowry *et al.* (1951).

Crude Fibre Content

Fruits were randomly collected at vegetable maturity and used for crude fibre estimation as per the procedure suggested by Goering and Vansoest (1970).

Carbohydrate Content

Carbohydrate content of fruit was estimated as per the procedure given by Somoigyi (1952).

Results and Discussion

Fruit Quality Parameters

Ascorbic Acid Content (Table 53.1 and Figure 53.1)

The highest and fruit ascorbic acid content of 28.90 mg/100 g was observed in bhendi applied with poultry manure + panchagavya (T_3) followed by neem cake + panchagavya (T_6) (28.61 mg/100 g), which did not significantly differ each other. The present investigation results go in support of Beaulah (2001), who reported that ascorbic acid content of moringa increased due to application of poultry manure + neem cake + panchagavya. The lowest ascorbic acid content was recorded in bhendi applied neem cake alone (T_2) (24.92 mg/100 g) which was on par with poultry manure alone (T_1) (25.52 mg/100 g) and neem cake + neem formulation (T_8) (25.56 mg/100 g).

Table 53.1: Effect of Organic and Inorganic Sources of Nutrition, Panchagavya and Botanicals Spray on Bhendi Fruit Quality

Treatments	*Ascorbic Acid Content (mg/100 g)*	*Carbohydrate Content (g/100 g)*	*Protein Content (mg/100 g)*	*Crude Fibre (g/100 g)*
T_1–Poulry manure alone	25.28	8.68	2.30	5.10
T_2–Neem cake alone	24.92	8.65	2.26	5.01
T_3–Poultry manure + Panchagavya	28.90	9.29	2.75	5.67
T_4–Poultry manure + Herbal leaf extract	27.39	9.14	2.54	5.48
T_5–Poultry manure + Neem formulation	26.58	8.92	2.49	5.27
T_6–Neem cake + Panchagavya	28.61	9.18	2.73	5.65
T_7–Neem cake + Herbal leaf extract	27.34	9.01	2.52	5.43
T_8–Neem cake + Neem formulation	25.56	8.90	2.47	5.26
T_9–Inorganic fertilizer + inorganic pesticides	27.36	9.09	2.54	5.54
$S.E._d$	0.78	0.26	0.18	0.32
CD (P = 0.05)	1.62	NS	0.08	0.14

In case of application of recommended NPK through inorganic source + monocrotophos, the ascorbic acid content was less compared to treatments wherein panchagavya was sprayed in addition to soil application of poultry manure and neem cake. Montagu and Goh (1990) also found that the rate of nitrogen through fertilizer significantly decreased vitamin C concentration and fruit visual quality in tomato. Prabhakaran and Pitchai (2002) reported that the application of recommended dose of nitrogen through poultry manure recorded higher ascorbic acid content and crude protein in tomato fruit compared with inorganic fertilizer.

Carbohydrate Content (Table 53.1 and Figure 53.1)

The carbohydrate content in fruit was not significantly influenced by different treatments. However, numerically higher carbohydrate content was observed in bhendi applied with poultry manure + panchagavya (T_3) (28.90 mg/100 g).

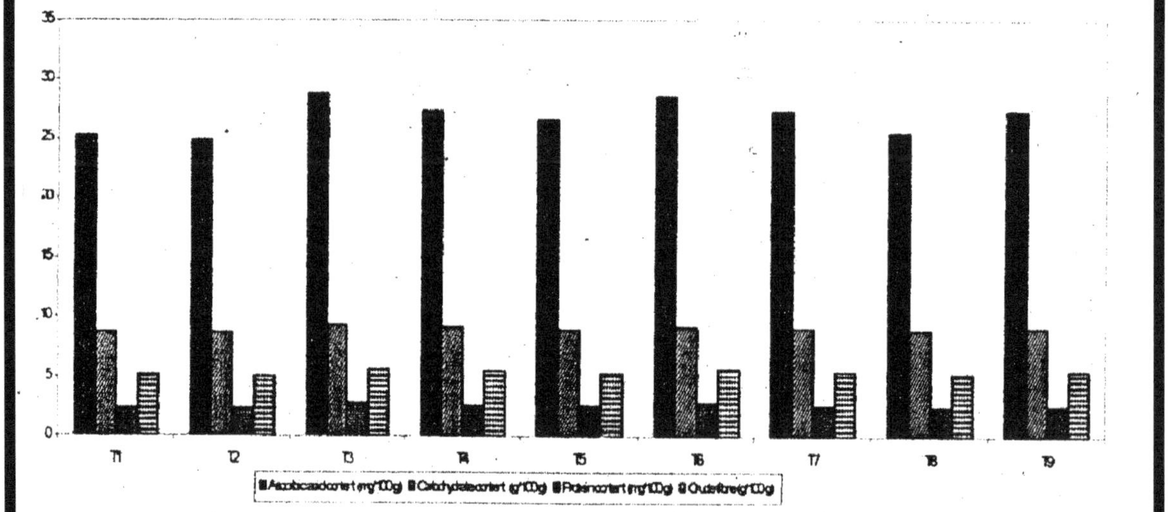

Figure 53.1: Effect of Organic and Inorganic Sources of Nutrition, Panchagavya and Botanicals Spray on Bhendi Fruit Quality

Protein Content (Table 53.1 and Figure 53.1)

Significantly higher protein content of 2.75 mg/100 g of fruit was recorded in bhendi applied poultry manure + panchagavya (T_3), followed by neem cake + panchagavya (T_6) (2.72 mg/100 g), which did not differ significantly with each other. Lowest protein content was recorded in neem cake alone (T_2) (2.53 mg/100 g), which was on par with poultry manure alone (T_1) (2.56 mg/100 g).

Crude Fibre (Table 53.1 and Figure 53.1)

The highest crude fibre content (5.67 g/100 g) was observed in bhendi applied poultry manure + panchagavya (T_3), followed by neem cake + panchagavya (T_6) (5.65 g/100 g), which did not differ significantly with each other. The lowest crude fibre was recorded in neem cake alone (T_2) (5.16 g/100 g) which was on par with poultry manure alone (T_1) (5.22 g/100 g), neem cake + neem formulation (T_8) (5.27 g/100 g), poultry manure + neem formulation (T_5) (5.29 g/100 g).

Application of poultry manure + panchagavya and neem cake + panchagavya recorded the highest protein and crude fibre content or fruit. Raj and Kumari (2001) and Beaulah (2001) have also reported increased protein content of okra fruit due to poultry manure application and increased crude fibre content due to poultry manure + neem to cake + panchagavya application, respectively.

The microorganisms present in the panchagavya (foliar spray) would have helped in the fixation of atmospheric nitrogen in leaves and its adsorption and utilization for protein synthesis. Increased nitrogen content would increase protein content. These results are in line with earlier findings of Roy and Seth (1971) in radish, Venter (1979) in carrot, Borne *et al.* (1987) in cabbage, Kohil *et al.* (1992) in pea, Beaulah (2001) in moringa and Somasundaram (2003) in maize, sunflower and green gram.

From this study the organic source of nutrients, panchagavya and herbal leaf extract spray improved the fruit quality of bhendi. The ascorbic acid content, protein content and crude fibre content of fruit was improved under poultry manure with panchagavya spray and was on par with neem cake with panchagavya spray. Similar trend was observed for carbohydrate content, though it failed to reach the level of significance. Finally the organic sources of nutrients *viz.*, poultry manure and neem cake with panchagavya and herbal leaf extract spray improved the quality aspects of bhendi fruit. Poultry manure with panchagavya spray was found to be the best with respect to fruit quality.

References

Beaulah. A. 2001. Growth and development of moringa (*Moringa oleifera* Lem.) under organic and inorganic system of culture. *Ph.D. Thesis*, Tamil Nadu Agril. Univ., Coimbatore.

Bome, V., K. Eid and A. Krarus. 1987. Nitrogen fertilization of white cabbage for saucer kraut production. *Soil and Fertilizers*, 50: 9492.

Kohli, V.K., I.K. Thakur and Y.R. Shukla. 1992. A note on response of pea (*Pisum sativum*, L.) to P and K application. *Hort. J.*, 5(1): 59–61.

Kumarasamy, K. 2002. Organic farming: Relevance and prospects. *Indian Soc. Soil Sci. News letter*, p. 12.

Montagu, K.D and K.M. Goh. 1990. Effects of forms and rates of inorganic nitrogen fertilizers on the yield and some quality indices of tomato. *J. Crop Hort. Sci.*, 18: 31–37.

Natarajan. K. 1999. Panchagavya. *Numvazhi Velanmai*, 11: 3–4.

Prabhakaran, C. and G.J. Pitchai. 2002. Effect of different organic nitrogen sources on pH, total soluble solids, titratable acidity, reducing and non-reducing sugars, crude protein and ascorbic acid content of tomato fruits. *J. Soils and Crops*, 12(2): 160–166.

Raj, A.K. and V.L.G. Kumar. 2001. Effect of organic manure and *Azospirillum* inoculation on yield and quality of okra [*Abelmoschus esculentus*) (L.) moench]. *Veg. Sci.*, 28(2): 179–181.

Roy, R.N.J. and Seth. 1971. Nutrient uptake and quality of radish (*Raphanus sativus* L.) as influenced by levels of N, P and K and methods of their application. *Indian J. Hort.*, 28(2): 144–149.

Rawat, A.K. 2002. Organic farming. In: *CAS Training of Organic Agriculture: A Paragon, for Sustainability*, held at JNRVV, Jabalpur, March 26–April 15th, 2002, p. 1–6.

Somasundaram, E. 2003. Evaluation of organic sources of nutrients and panchagavya spray on the growth and productivity of maize-sunflower-green gram system. *Ph.D. Thesis*, Tamil Nadu Agricultural University, Coimbatore.

Somogyi, M. 1952. Note on sugar determination. *J. Biol. Chemistry*, 200: 145–154.

Venter, F. 1979. Nitrate contents in carrots (*Daucus carota*) as influenced by fertilization. *Acta Hort.*, 93: 163–172.

Chapter 54

Resurgence of Red Spider Mite *Tetranychus cinnabarinus* Boisd. on Brinjal

B.M. Mhaske, A.P. Chavan, D.B. Kadam and B.N. Chaudhari

Department of Entomology, Mahatma Phule Krishi Vidyapeeth, Rahuri Distt. Ahmednagar – 413 722, Maharashtra, India

ABSTRACT

The results revealed that the resurgence of red vegetable mites observed with the treatments of some synthetic pyrethroieds and some botanicals including azadirachtin. Therefore, it is advisable to use chemical pesticides in sequences from different group, so as to avoid resurgence of mites on brinjal.

Keyword: *Synthetic pyrethroieds, Brinjal, Mites, Resurgence.*

Introduction

Red vegetable mite *Tetranychus cinnabarinus* Boisd. is a serious pest of many vegetables. Recently, due to the congenial condition and mono cropping in vegetable growing area of western Maharashtra; considerable damage by mites was noticed. The problem of resistance to various insecticides had not been overcome so far. The excessive reliance on chemical is leading to serious upset in natural balance (Singh and Mukherjee, 1991).

Materials and Methods

The experiment was laid out during summer 2002 with an object to study the efficacy of different insecticides against brinjal shoot and fruit borer on the Instructional Farm of Post Graduate Institute, Department of Entomology, Mahatma Phule Krishi Vidyapeeth, Rahuri, Dist. Ahmednagar, Maharashtra, India in the plots arranged in a randomized block design with twelve treatments each with three replication. Brinjal variety Mahyco-10 (Kalpataru) was transplanted in a 4.5 × 3.0 m^2 plots with 60 × 45 cm^2 spacing. Each insecticidal treatment was applied three times at an interval of 15 days. During this experiment resurgence of red spider mite was observed in the later stage of crop. The mite population per square inch per six leaves per plant was recorded with the help of 10 x lens from five plants in each treatment. The population of mites per square inch was worked and transformed in $\sqrt{n+1}$ and subjected to statistical analysis.

Results and Discussion

In the event from the Table 54.1, that the treatment with cypermethrin 60 g a.i. per ha recorded the highest number of red mites. However, lower doses of the azadirachtin 10 and 12.5 and deltamethrin 11 g a.i. per ha were also at par with it. Though, the resurgence of mites observed, no much adverse effect on the yield of the fruits was noticed. Singh and Singh (1992) have reported the effectiveness of dicofol and phosalone against the mite and found that they did not induce mite resurgence but rather became ineffective against the mite. Also in 1993, studies on seasonal incidence of *Tetranychus cinnabarinus* Boisd. on okra, the peak was noticed in the months of May and June, also found that temperature is the regulatory factor for population build up of mite. Kumar and Singh (1999) used neem-based formulation and conventional acaricides for the study of resurgence of red spider miles. The results indicated that phosalone, fenvalerate and dimethoate showed resurgence of spider mites. The neem based formulation (azadirachtin) showed encouraging performance but showed resurgence. Kumar *et al.* (2002) reported some resurgence with ethion 50 EC and sulper 80 WP. However, resurgence in mahua oil, pungum oil, neem oil, azadirachtin and NSKE did not show any effectiveness against this pest.

Table 54.1: Resurgence of *Tetranychus cinnabarinus* Boisd. Observed on Brinjal

Treatment	*Dose (g a.i. per ha)*	*Average Mites (Square per inch)*	*Yield (q/ha)*
Profenofos 50 EC	500	14.07 (3.88)*	198.58
Profenofos 50 EC	750	11.47 (3.53)	207.07
Profenofos 50 EC	1000	9.53 (3.24)	230.37
Azadirachtin 1% EC	10	227.73 (15.12)	201.05
Azadirachtin 1% EC	12.5	239.20 (15.50)	216.02
Azadirachtin 1% EC	15	48.33 (7.02)	221.27
Alpha endosulfan 35 EC	280	15.87 (4.11)	219.88
Monocrotophos 36 EC	145	18.53 (4.42)	186.55
Endosulfan 35 EC	280	16.27 (4.16)	230.21
Deltamethrin 2.8 EC	11	224.27 (15.66)	208.92
Cypermethrin 10 EC	60	245.60 (15.27)	216.48
Untreated control		46.87 (6.92)	142.57
S.E.±		0.32	11.01
C.D. at 5% level		0.93	32.38

*Figures in parantheses are $\sqrt{n+1}$ transformations.

It could be concluded from the results of present investigation and findings of above workers that resurgence of red vegetable mites observed with the treatment of some synthetic pyrethroieds and some botanicals including azadirachtin. Therefore, it is advisable to use chemical pesticides in sequences from different groups, so as to avoid resurgence of mites on brinjal.

References

Kumar, S. and R.N. Singh. 1999. Studied on resurgence of spider mites (*T. urticae*) on okra. *National Symposium on Biological Control of Insects in Agriculture, Forestry, Medicine and Veterinary Science.* Abstract and Souvenir–9.

Kumar, S., S. Prasad and R.N. Singh. 2002. Resurgence of two spotted mite due to acaricides and botanicals on okra. *Ann. Pl. Prot. Sci.,* 10(2): 239–242.

Singh, R.N. and J. Singh. 1992. Evaluation of some pesticides against carmine mite *T. cinnabarinus* Boisd. on lady's finger. *Pestology,* 16(2): 20–23.

Singh, R.N. and J. Singh. 1993. Incidence of *T. cinnabarinus* in relation to weather factor in Varanasi. *Pestology,* 17(8): 18–23.

Singh, J. and I.N. Mukherjee. 1991. Pest status of polyphagous mites in some Northern states of India. In: *Proc. 1st Asia Pacific Conference of Entomology,* Nov. 8–13, 1989. Thailand, 1: 192–203.

Chapter 55

Efficacy of Cashewnut Shell Liquid as Seed Protectant of Cowpea, *Vigna unguiculata* (Linn.) Against its Pest *Callosobrunchus maculatus* (Fab.)

Binu N. Nair and V.R. Prakasam

Department of Environmental Sciences, University of Kerala, Kariavattom, Thiruvananthapuram – 695 581

ABSTRACT

In this investigation on the effect of Cashewnut Shell Liquid (CNSL) on oviposition of the pest insect *Callosobrunchus maculatus* on cowpea seeds, it was found that there was reduction in egg laying by 66.02 per cent at a dose rate of 1.25 ml CNSL/kg; by 93.10 per cent at a dose rate of 2.5 ml CNSL/kg; and 100 per cent at a dose rate of 5 ml CNSL/kg. Similarly adult emergence from the eggs deposited on cowpea was reduced by 82.07 per cent at the lowest experimented dose of 1.25 ml CNSL/kg while it was nil at higher doses.

The percentage germination of CNSL treated cowpea seeds was reduced by 4.4 per cent at a dose rate of 1.25 ml CNSL/kg; 5.2 per cent at dose rate of 2.5 ml/kg and 8.9 per cent at a dose rate of 5 ml/kg. The study revealed the possibility of using CNSL as an ecofriendly substitute for harmful chemical pesticides in seed protection.

Keywords: *Callosobrunchus maculatus, Cowpea, Seed protectant, Germination.*

Introduction

Cowpea weevil, *Callosobrunchus maculatus* is a serious pest of stored pulses in tropical and sub-tropical regions. Both the larvae as well as the adults are infective (Raja and Ignacimuthu, 2001). Various chemical pesticides like malathion have been used as seed protectant against pulse beetle. Uncontrolled use of these chemicals can be detrimental to man and non-target species. Moreover phytotoxic effects are also produced by these chemical pesticides (Casida and Lykken, 1969; Chakravarthi, 1986). Therefore application of natural pesticides for seed protection is becoming more and more popular. Seed protection using Neem oil, Castor oil, Mustard oil, Sesame oil have been studied (Raghavani and Kapadia, 2003).

Surface application of CNSL, obtained through perforation of fresh cashewnuts, was found superior to ginger and neem against infestation and damage by *C. maculatus* in cowpeas without mentioning exact doses used (Echendu, 1991). The present study is an attempt to find out the efficacy of Cashewnut Shell Liquid (CNSL) extracted by roasting of cashewnuts (heat extracted CNSL) as a seed protectant of cowpea *Vigna unguiculata* against a pest weevil, *Callosobrunchus maculates.* The chemical composition of heat extracted CNSL is different from natural CNSL. During heat extraction the anacardic acid component of CNSL is decarboxylated to form cardanol and, therefore, the effect of such CNSL is likely to be different from natural CNSL.

Materials and Methods

Cowpea *Vigna unguiculata,* seeds of variety Sharika, purchased from a local market in Thiruvananthapuram, were used from the experiment. Initial stocking of *Callosobrunchus maculatus* was made from infested cowpea seeds obtained from the same market. Subsequent stocks were raised in the laboratory. Technical grade CNSL (heat extracted) obtained from Vijayalakshmi cashew industries, Kollam was used for the study.

Effect of CNSL on Oviposition and Adult Emergence

The cowpea seeds were surface sterilized with 0.02 per cent $HgCl_2$ for 5 minutes and washed thoroughly with distilled water. One kg of seeds was put in a glass jar of 1 litre capacity. Required amounts of CNSL (1.25, 2.5 and 5 ml) were added separately and shaken manually until a uniform smear of oil was produced on seed surfaces. These treated seeds were stored in the laboratory. A control consisting of untreated seeds was also maintained.

Three samples of 25 g seeds were then drawn from each treated stock into suitable plastic jars (6.00 × 4.00 cm) with perforated lid. Into them two pairs of adults *C. maculatus* were released. The number of eggs laid on treated and control seeds were recorded after 15th day. The experimental setup was kept undisturbed for 15 days in the laboratory allowing the development of the grub into the adults (Raja and Ignacimuthu, 2001). The number of F_1 adults emerged were then recorded on 30th day and till the day of last emergence (approximately 40 days).

The data obtained were analysed using 2-way ANOVA and significant differences were noted using least significance difference test (LSD).

Germination of CNSL Treated Cowpea Seeds

A sample of 70 seeds was randomly drawn from each set of treated and control seeds six months after storage. Ten seeds each were sown in a petridish containing two layered filter paper and kept in the laboratory at 30 ± 1°C. It was moistured daily. Percentage germination of seeds was calculated on 4th day.

Results and Discussion

The results of the effect of CNSL on oviposition and adult emergence of *C. maculatus* are given in Table 55.1. It showed that at doses of 1.25 and 2.5 ml CNSL/kg seeds, the mean number of eggs laid were 21.2 ± 1.94 and 4.3 ± 1.74 respectively, whereas the control value was 62.4 ± 3.73. It indicated that CNSL had affected the normal deposition of eggs. Further, at a higher dose (5.0 ml) the weevil failed to lay eggs on treated seeds.

Table 55.1: Showing Number of Eggs Laid by *Callosobrunchus maculatus* on Cowpea Seeds Treated with CNSL and Control

Dose of CNSL ml/kg	*Eggs Laid (Mean ± S.D on Seed Surfaces)*
1.25	21.2 ± 1.94
2.5	4.3 ± 1.74
5	Nil
Control	62.4 ± 3.73

($p < 0.05$) by LSD.

It can be noted from Table 55.2 that at a dose of 1.25 ml/kg, the number of adults emerged was 6.4 ± 2.37, whereas it was 37.6 ± 1.67 in control. But at higher doses of 2.5 and 5 ml/kg there was no emergence of adults. It is likely that the cardol fraction of CNSL might have produced toxicity leading to death of embryos. Cardol is a common constituent of both natural CNSL and heat extracted CNSL.

Table 55.2: Effect of CNSL on Adult Emergence of *Callosobrunchus maculatus*

Dose of CNSL ml/kg	*No. of Adults Emerged (Mean ± S.D)*
1.25	6.4 ± 2.37
2.5	nil
5	nil
Control	37.6 ± 1.67

($p < 0.05$) by LSD.

The result of germination test is given in Table 55.3. It is seen that at a dose of 1.25 ml/kg, the percentage germination was 85.3 ± 3.6; at a dose of 2.5 ml/kg percentage germination was 84.6 ± 41 and at the highest dose of 5 ml/kg the percentage germination was 81.2 ± 3.8. It is evident that CNSL has not affected the germination of treated seeds when compared to control. In this context it may be recollected that many of the synthetic seed protectants have interfered with germination of seeds through their phytotoxicity (Casida and Lykken, 1969; Patil and Shirashyad, 1989). From the reduced deposition of eggs, poor emergence of young ones from eggs and normal germination of CNSL treated seeds, it may be derived that technical CNSL could be a safe, economical and ecofriendly substitute for chemical pesticides to be used for seed protection from insect pests.

Table 55.3: Mean Percentage Germination of Seeds (CNSL Treated and Control)

Dose of CNSL ml/kg	*Mean ± S.D % Germination*
1.25	85.3 ± 3.6
2.5	84.6 ± 4.1
5	81.2 ± 3.8
Control	89.3 ± 4.6

($p < 0.05$) by LSD.

Acknowledgement

Sincere thanks are due to Mr. Thomas George and Mr. Abilash R., Research Scholars, Department of Environmental Sciences for their co-operation and help in carrying out this work.

References

Casida, J.E and L. Lykken. 1969. Mechanism of organic pesticide chemicals in higher plants. *Ann. Rev. Pl. Physiol.*, 20: 607–636.

Chakravarthi, S.K. 1986. Effect of Bavistin 25 DS on germination and seedling vigour of wheat. *Pesticides*, 3: 23–26.

Echendu, T.N.C.. 1991. Ginger, cashew and neem as surface protestants of cowpea against infestation and damage by *C. maculates. Trop. Sci.*, 31: 209–211.

Patil, T.M. and V.S. Shirashyad. 1989. Effect of Methyl parathion and phosphamidon on seed germination and on the activity of peroxidase and α-amylase in some vegetable seeds. *Geobios*, 16(2–3): 57–60.

Raghavani, B.R. and M.N. Kapadia. 2003. Efficacy of different vegetable oils as seed protectants of pigeon pea against *C. maculatus. Ind. J. Plant Prot.*, 31: 115–118.

Raja, N. and S. Ignacimuthu. 2001. Use of *madhuca longifolia* seed oil in controlling pulse beetle *Callosobrunchus maculatus*, L. *Entomon.*, 26(384): 279–283.

Index

H

I

S

www.ingramcontent.com/pod-product-compliance
Ingram Content Group UK Ltd.
Pitfield, Milton Keynes, MK11 3LW, UK
UKHW060120010826
14089UKWH00005B/81